Potential Theory

Potential Theory

Edited by
Josef Král
Czechoslovak Academy of Sciences
Prague, Czechoslovakia

and

Jaroslav Lukeš
Ivan Netuka and
Jiří Veselý
Charles University
Prague, Czechoslovakia

Plenum Press • New York and London

Library of Congress Cataloging in Publication Data

Potential theory.

"Proceedings of a conference on potential theory, held July 19-24, 1987, in Prague, Czechoslovakia."
"Organized by the Faculty of Mathematics and Physics, Charles University."
Includes bibliographical references and index.
1. Potential, Theory of—Congresses. I. Král, Josef. II. Universita Karlova. Matematicko—fyzikální Fakulta.
QA404.7.P66 1988 515.7 88-5836
ISBN 0-306-42838-5

Proceedings of a conference on Potential Theory, held July 19-24, 1987, in Prague, Czechoslovakia

A Division of Plenum Publishing Corporation
233 Spring Street, New York, N.Y. 10013

Printed in the United States of America

PREFACE

Within the tradition of meetings devoted to potential theory, a conference on potential theory took place in Prague on 19-24, July 1987. The Conference was organized by the Faculty of Mathematics and Physics, Charles University, with the collaboration of the Institute of Mathematics, Czechoslovak Academy of Sciences, the Department of Mathematics, Czech University of Technology, the Union of Czechoslovak Mathematicians and Physicists, the Czechoslovak Scientific and Technical Society, and supported by IMU.

During the Conference, 69 scientific communications from different branches of potential theory were presented; the majority of them are included in the present volume. (Papers based on survey lectures delivered at the Conference, its program as well as a collection of problems from potential theory will appear in a special volume of the Lecture Notes Series published by Springer-Verlag). Topics of these communications truly reflect the vast scope of contemporary potential theory. Some contributions deal with applications in physics and engineering, other concern potential-theoretic aspects of function theory and complex analysis. Numerous papers are devoted to the theory of partial differential equations. Included are also many articles on axiomatic and abstract potential theory with its relations to probability theory.

The present volume may thus be of intrest to mathematicians specializing in the above-mentioned fields and also to everybody interested in the present state of potential theory as a whole.

Josef Král
Ivan Netuka
Jaroslav Lukeš
Jiří Veselý

Prague, October 1987

CONTENTS

ON WEIGHTED BEPPO LEVI FUNCTIONS

Hiroaki Aikawa

Department of Mathematics, Gakushuin University
Mejiro 1-5-1, Toshima-ku, Tokyo 171, Japan

1. Introduction and statements of results

Let $1 < p < \infty$ and let w be a weight satisfying the Muckenhoupt A_p condition. We write

$$\|f\|_{L^p,w} = \left(\int_{R^n} |f(x)|^p w(x)dx\right)^{1/p}, \quad L^p(R^n, w) = \{f;\ \|f\|_{L^p,w} < \infty\}.$$

According to Deny and Lions [3] we denote by $BL_m(L^p(R^n, w))$ the space of distributions whose partial derivatives of m-th order all belong to $L^p(R^n, w)$, i.e.

$$BL_m(L^p(R^n, w)) = \{f;\ \sum_{|\alpha|=m} \|D^\alpha f\|_{L^p,w} < \infty\},$$

where $D^\alpha = \partial^{|\alpha|}/\partial x_1^{\alpha_1}\cdots\partial x_n^{\alpha_n}$ and $|\alpha| = m$. Since $w^{1/(1-p)}$ is locally integrable, it follows from Hölder's inequality and Kryloff's theorem [9; Chapitre VI, Théorème 15] that a distribution in $BL_m(L^p(R^n, w))$ is a locally integrable function. Therefore we call a locally integrable function in $BL_m(L^p(R^n, w))$ a Beppo Levi function of order m with weight w.

The weighted Sobolev space

$$V_p^m(R^n, w) = \{f;\ \sum_{|\alpha|\leq m} \|D^\alpha f\|_{L^p,w} < \infty\}$$

may be more familiar. Let $\mathcal{E}$ be the space of all infinitely differentiable functions and $\mathcal{D}$ the space of all functions in $\mathcal{E}$ of compact support. In the same way as in the case $w \equiv 1$, we can prove

Theorem A (cf. [11; Chapter V]). *Let G_m be the Bessel kernel of order m and let*

$$J_m(L^p(R^n, w)) = \{G_m * g;\ g \in L^p(R^n, w)\}.$$

Then

(i) $V_p^m(R^n, w) = J_m(L^p(R^n, w))$.

(ii) $\mathcal{D}$ is dense in $V_p^m(R^n, w)$.

In this article we shall deal with a similar problem for $BL_m(L^p(R^n, w))$. The counterpart of $J_m(L^p(R^n, w))$ may be the space of Riesz potentials. Let us give some definitions. Hereafter we restrict ourselves to the case $0 < m < n$. The Riesz kernel of order m is defined by $h_m(x) = \gamma(m)^{-1}|x|^{m-n}$ with $\gamma(m) = 2^m \pi^{n/2}\Gamma(\frac{m}{2})/\Gamma(\frac{m-n}{2})$. We observe that the Fourier transform of h_m is equal to $(2\pi|x|)^{-m}$ ([11; p.117]). If g is a locally integrable function such that

$$(1) \qquad \int_{R^n} (1 + |x|)^{m-n}|g(x)|dx < \infty,$$

then we write

$$I_m(g) = \int_{R^n} h_m(x-y)g(y)dy.$$

It is easy to see that (1) is a necessary and sufficient condition for $I_m(g)$ to be defined. Let

$$I_m(L^p(R^n, w)) = \{I_m(g);\ g \in L^p(R^n, w),\ g \text{ satisfies } (1)\}.$$

Since $D^\alpha h_m$ is the sum of a principal value distribution and a multiple of the Dirac measure at the origin for $|\alpha| = m$, it follows from the theory of singular integrals (cf. [2], [11]) that

$$(2) \qquad I_m(L^p(R^n, w)) \oplus P_{m-1} \subset BL_m(L^p(R^n, w)),$$

where P_{m-1} is the space of all polynomials of degree not greater than m - 1 and $\oplus$ means the direct sum. Now we are lead to the following problem: Does $BL_m(L^p(R^n, w))$ coincide with $I_m(L^p(R^n, w)) \oplus P_{m-1}$? If this is true, then it may be considered to be the counterpart of Theorem A (i).

This problem is closely related to the counterpart of Theorem A (ii), i.e. the approximation property of $BL_m(L^p(R^n, w))$. Let us introduce a norm on $BL_m(L^p(R^n, w))$. As was observed by Deny and Lions [3], the quotient space $BL_m(L^p(R^n, w))/P_{m-1}$ is a Banach space with norm $\sum_{|\alpha|=m}\|D^\alpha f\|_{L^p,w}$. However we shall soon see that this quotient space is not relevant for our purposes. We define a norm $|||f|||$ for $f \in BL_m(L^p(R^n, w))$ by

$$|||f||| = \int_{|x|<1} |f(x)|dx + \sum_{|\alpha|=m} \|D^\alpha f\|_{L^p,w}$$

and regard $BL_m(L^p(R^n, w))$ itself as a Banach space with this norm. By $\bar{\mathcal{D}}$ we denote the closure of $\mathcal{D}$ with respect to $|||\cdot|||$. We show

Theorem 1. *The weighted Beppo Levi space $BL_m(L^p(R^n, w))$ is decomposed as*

$$BL_m(L^p(R^n, w)) = \overline{\mathcal{D}} + \mathcal{P}_{m-1},$$

i.e. for any $f \in BL_m(L^p(R^n, w))$ there are a sequence φ_j in $\mathcal{D}$ and $P \in \mathcal{P}_{m-1}$ such that $|||f - \varphi_j - P||| \to 0$.

We note that the decomposition in Theorem 1 is not necessarily unique. It is not so difficult to see that $\mathcal{D}$ is included in $I_m(L^p(R^n, w))$ and is dense there. However the closure of $\mathcal{D}$ may not be included in it, i.e. a polynomial may be approximated by $\mathcal{D}$ (cf. [6; Theorem 11.2.2]). If $\overline{\mathcal{D}} = I_m(L^p(R^n, w))$, then (2) and Theorem 1 lead to

$$BL_m(L^p(R^n, w)) = I_m(L^p(R^n, w)) \oplus \mathcal{P}_{m-1} = \overline{\mathcal{D}} \oplus \mathcal{P}_{m-1}, \tag{3}$$

so that the problem raised after (2) is affirmatively solved. Let us give some criteria for (3) to hold. According to Adams [1], we define a weighted L^p-capacity by

$$R^w_{m,p}(E) = \inf\{\|g\|^p_{L^p,w} \; ; \; g \geq 0 \text{ and } I_m(g) \geq 1 \text{ on } E\}.$$

Let $p' = p/(p-1)$. We shall show

Theorem 2. *The following statements are equivalent:*

(a) $\overline{\mathcal{D}} \cap \mathcal{P}_{m-1} = \{0\}$, *and hence* $BL_m(L^p(R^n, w)) = \overline{\mathcal{D}} \oplus \mathcal{P}_{m-1}$.

(b) $\overline{\mathcal{D}} = I_m(L^p(R^n, w))$.

(c) $BL_m(L^p(R^n, w)) = I_m(L^p(R^n, w)) \oplus \mathcal{P}_{m-1}$.

(d) w *satisfies* $\int_{R^n} (1 + |x|)^{(m-n)p'} w(x)^{1/(1-p)} dx < \infty$.

(e) *Every* $g \in L^p(R^n, w)$ *satisfies* (1).

(f) *There is a set* E *such that* $R^w_{m,p}(E) > 0$.

Remark. Let $w \equiv 1$. Sobolev [10] proved Theorem 1 by 'reduction method' (see §2). It is not so difficult to see that each statement of Theorem 2 holds if and only if $mp < n$.

2. Proof of Theorems

By making use of a usual mollifier we see that $\mathcal{E} \cap BL_m(L^p(R^n, w))$ is dense in $BL_m(L^p(R^n, w))$. Then one may think that for $f \in \mathcal{E} \cap BL_m(L^p(R^n, w))$ the approximation by $\mathcal{D}$ can be achieved by taking the 'reduction of f', i.e. multiplying f by a suitable function φ_j in $\mathcal{D}$ tending to 1 on R^n. In view of Leibniz's formula, this approach, however, requires further information for the integrability of lower derivatives of f. This was actually done by Sobolev [10] in case $w \equiv 1$ (see also Kurokawa [5]). But it seems difficult to extend his method to the general case. Instead of the reduction method we shall employ an integral representation for $BL_m(L^p(R^n, w))$ as a main tool.

Let us begin with modifying the Riesz kernel. The following technique is found in [4; Chapter IV], [5] and [7]. Observe that if $y \neq 0$, then $h_m(x - y)$ has a multiple power series expansion in $x_1, x_2, \dots, x_n$, convergent in a neighborhood of the origin. We write

$$h_m(x - y) = \sum_{\nu=0}^{\infty} a_\nu(x, y),$$

where, for fixed ν and $y \neq 0$, $a_\nu(x, y)$ is a homogeneous polynomial in x_1 to x_n of degree ν and continuous in x, y jointly for $y \neq 0$ (cf. [4; Lemma 4.1]). For $q \geqq 1$ we set

$$h_{m,q}(x, y) = \begin{cases} h_m(x - y) & \text{if } |y| \leqq 1 \\ h_m(x - y) - \sum_{\nu=0}^{q-1} a_\nu(x, y) & \text{if } |y| > 1. \end{cases}$$

If a locally integrable function g satisfies

$$\int_{R^n} (1 + |x|)^{m-n-q}|g(x)|dx < \infty, \tag{4}$$

then we write

$$I_{m,q}(g) = \int_{R^n} h_{m,q}(x, y)g(y)dy.$$

It is easy to see that (4) is a necessary and sufficient condition for the above integral to converge, and that every $g \in L^p(R^n, w)$ satisfies (4) with $q = m$. For notational convenience we let $I_{m,0}(g) = I_m(g)$ and $I_{m,q}(L^p(R^n, w)) = \{I_{m,q}(g);\ g \in L^p(R^n, w),\ g \text{ satisfies } (4)\}$. Let

$$g_f = \sum_{|\alpha|=m} c_\alpha (D^\alpha h_m)*D^\alpha f \quad \text{for } f \in BL_m(L^p(R^n, w)),$$

where $c_\alpha = (-1)^m m!/\alpha!$. By virtue of [2] and [11; Chapter III, Theorem 6] we see that $g_f \in L^p(R^n, w)$. We have

Theorem 3. *Let* $0 \leq q \leq m$ *and let* $f \in BL_m(L^p(R^n, w))$. *If* g_f *satisfies* (4), *then* $f = I_{m,q}(g_f) + P$ *with some* $P \in \mathcal{P}_{m-1}$. *This representation is unique. Moreover every* $f \in BL_m(L^p(R^n, w))$ *satisfies* (4) *with* $q = m$, *and hence* $BL_m(L^p(R^n, w))$ *is decomposed as*

$$BL_m(L^p(R^n, w)) = I_{m,m}(L^p(R^n, w)) \oplus \mathcal{P}_{m-1}.$$

Now a preliminary version of Theorem 1 readily follows from Theorem 3 and the fact that $\mathcal{D}$ is dense in $I_m(L^p(R^n, w))$. In fact, take $f \in BL_m(L^p(R^n, w))$ and represent it as

$$f = I_{m,m}(g_f) + P \quad \text{with } P \in \mathcal{P}_{m-1}.$$

It is easy to find ψ_j in $\mathcal{D}$ such that $\psi_j \to g_f$ in $L^p(R^n, w)$. Since ψ_j satisfies (1), it follows that $I_m(\psi_j)$ exists and $I_{m,m}(\psi_j)$ differs from $I_m(\psi_j)$ only by a polynomial P_j in $\mathcal{P}_{m-1}$. Taking $\varphi_j \in \mathcal{D}$ such that $|||\varphi_j - I_m(\psi_j)||| < 1/j$, we obtain

$$|||f - \varphi_j - (P_j + P)||| \to 0.$$

Therefore, what remains is to consider the approximation property of polynomials. In case $w \equiv 1$, this was treated by Maz'ja [6; Theorem 11.2.2] in a slightly different context. We consider the problem in connection with the integral representation given in Theorem 3. For convenience we let $\mathcal{P}_{-1} = \{0\}$ and

$$A_{p,k} = \{w \in A_p;\ \int_{R^n} (1 + |y|)^{(k-n)p'} w(y)^{1/(1-p)} dy < \infty\},$$

where A_p stands for the class of weights satisfying the Muckenhoupt A_p condition. We see that $A_{p,m} \subset A_{p,m-1} \subset \cdots \subset A_{p,0} = A_p$, where the last equality is proved by [8] and the reverse Hölder inequality. We have

Theorem 4. *For* $0 \leq q \leq m - 1$ *the following statements are equivalent:*

(a) $\overline{\mathcal{D}} \cap \mathcal{P}_{m-1} \subset \mathcal{P}_{q-1}$, *i.e. polynomials of degree greater than* $q - 1$ *are not approximated by* $\mathcal{D}$.

(b) $BL_m(L^p(R^n, w)) = I_{m,q}(L^p(R^n, w)) \oplus \mathcal{P}_{m-1}$.

(c) $\overline{\mathcal{D}} \subset I_{m,q}(L^p(R^n, w)) \oplus \mathcal{P}_{q-1}$.

(d) $w \in A_{p,m-q}$.

(e) *Every* $g \in L^p(R^n, w)$ *satisfies* (4).

Corollary. The closure $\overline{\mathcal{D}}$ *coincides with one of* $I_{m,q}(L^p(R^n, w)) \oplus \mathcal{P}_{q-1}$, $0 \leqq q \leqq m$. *More precisely, for* $1 \leqq q \leqq m$ *the coincidence* $\overline{\mathcal{D}} = I_{m,q}(L^p(R^n, w)) \oplus \mathcal{P}_{q-1}$ *holds if and only if* $w \in A_{p,m-q} \setminus A_{p,m-q+1}$; *the coincidence* $\overline{\mathcal{D}} = I_{m,0}(L^p(R^n, w)) \oplus \mathcal{P}_{-1} = I_m(L^p(R^n, w))$ *holds if and only if* $w \in A_{p,m}$.

Now Theorems 1 and 2 are easy consequences of Theorem 4 and its corollary. In fact $\overline{\mathcal{D}}$ coincides with some $I_{m,q}(L^p(R^n, w)) \oplus \mathcal{P}_{q-1}$ by Corollary, and in this case $BL_m(L^p(R^n, w)) = I_{m,q}(L^p(R^n, w)) \oplus \mathcal{P}_{m-1} = \overline{\mathcal{D}} + \mathcal{P}_{m-1}$ by Theorem 4, so that Theorem 1 follows. Theorem 2 also follows from Theorem 4 except the characterization by the weighted Riesz capacity. However the equivalence between (e) and (f) of Theorem 2 is easily proved by definition. Detailed proofs of Theorems 3 and 4 will appear elsewhere.

References

1. D. R. Adams, Weighted nonlinear potential theory, Trans. Amer. Math. Soc. 297 (1986), 73-94.
2. R. R. Coifman and C. Fefferman, Weighted norm inequalities for maximal functions and singular integrals, Studia Math. 51 (1974), 241-250.
3. J. Deny and J. L. Lions, Les espaces du type de Beppo Levi, Ann. Inst. Fourier 5 (1955), 305-370.
4. W. K. Hayman and P. B. Kennedy, Subharmonic functions, Vol. 1, Academic Press, London, (1976).
5. T. Kurokawa, Riesz potentials, higher Riesz transforms and Beppo Levi spaces, to appear.
6. V. Maz'ja, Sobolev spaces, Springer, Berlin, (1985).
7. Y. Mizuta, Integral representations of Beppo Levi functions of higher order, Hiroshima Math. J. 4 (1974), 375-396.
8. B. Muckenhoupt and R. L. Wheeden, Weighted norm inequalities for fractional integrals, Trans. Amer. Math. Soc. 192 (1974), 261-274.
9. L. Schwartz, Théorie des distributions, Hermann, Paris, (1966).
10. S. L. Sobolev, The density of finite functions in the $L_p^{(m)}$ space, Soviet Math. Dokl. 4 (1963), 313-316.
11. E. M. Stein, Singular integrals and differentiability properties of functions, Princeton Univ. Press, Princeton, (1970).

GENERALIZED CAUCHY-RIEMANN EQUATIONS AND THE POSITIVE $\underset{\sim}{k}$-SUBHARMONIC FUNCTIONS

Ömer Akın
Department of Mathematics
Fırat University
Elazığ-Turkey

A. Okay Çelebi
Department of Mathematics
METU
Ankara-Turkey

SUMMARY

In this paper we will give a statement about the positive solutions of the generalized Cauchy-Riemann system in $\mathbb{R}^{2n}$. It will also be shown that the positive solutions are related with the "$\underset{\sim}{k}$-subharmonicity".

INTRODUCTION

In this article we will discuss the positive solutions of

$$L_{\underset{\sim}{k}}[u] = \sum_{i=1}^{2n} \frac{\partial^2 u}{\partial x_i^2} + \sum_{i=1}^{n} \frac{2k_i}{x_{n+i}} \frac{\partial u}{\partial x_{n+i}} \geqq 0 \tag{1}$$

by use of solutions of the generalized Cauchy-Riemann system in a hyper-octant in $\mathbb{R}^{2n}$, where $\underset{\sim}{k} := (k_1, \ldots, k_n)$.

We already know that the properties of the solutions of $L_{\underset{\sim}{k}}[u]=0$ have been investigated by many authors[1-4]. In the recent works of one of the authors[5,6], he has considered the properties of the solutions of

$$L_{p,q}[u] = \sum_{i=1}^{n} \frac{\partial^2 u}{\partial x_i^2} + \frac{p}{x_{n-1}} \frac{\partial u}{\partial x_{n-1}} + \frac{q}{x_n} \frac{\partial u}{\partial x_n} = 0$$

using the ideas given in Brelot[7] and the Martin's technique of "fine" and "non-tangential" limits.

In two dimensional case, we know that the functions

$$u(r, \Theta) \sim \sum_{n=0}^{\infty} a_n r^n \cos n\Theta \quad \text{and} \quad v(r, \Theta) \sim \sum_{n=0}^{\infty} a_n r^n \sin n\Theta$$

are harmonic in the unit disc; and v is the conjugate of u in the same domain. A generalized form of the Cauchy-Riemann system is defined as

$$\sum_{i=1}^{n} \frac{\partial u_i}{\partial x_i} = 0, \quad \frac{\partial u_i}{\partial x_j} = \frac{\partial u_j}{\partial x_i}; \quad i, j=1,\ldots,n \tag{2}$$

where $u_1, \ldots, u_n$ are called the conjugate harmonic functions[8]. In this paper we will define a general form of the system (2) by

$$U_{x_i} = V_{x_{n+i}} + \frac{2k_i}{x_{n+i}} V,$$
$$U_{x_{n+j}} = -V_{x_j}, \quad i, j=1,2,\ldots,n \tag{3}$$

and utilize the solutions of (3) to find the positive $\underset{\sim}{k}$-subharmonic functions.

THE MAIN RESULT

The main result in this article is summarized in the following:

THEOREM: Let $U, V \varepsilon C^2$ be a non-trivial solution of the system (3) in the hyper-octant $x_{n+i} > 0$, $i=1,2,\ldots,n$ where $k_i \geqq 0$ are constants. Let us form

$$F(x_1, \ldots, x_{2n}) = U(x_1, \ldots, x_{2n}) + iV(x_1, \ldots, x_{2n}).$$

If $\gamma \geqq \max_i \frac{2k_i}{2k_i+1}$ holds, then

$$L_{\underset{\sim}{k}}[|F|^\gamma] \geqq 0 \tag{4}$$

is satisfied.

(Such a function $|F|^\gamma$ will be called as "$\underset{\sim}{k}$-subharmonic" in class C^2 related to the operator $L_{\underset{\sim}{k}}$.)

PROOF - We can use the vector notations

$$F = (U, V), \quad F_{x_i} = (U_{x_i}, V_{x_i}), \quad i=1,2,\ldots,2n .$$

Thus the inner product will give us

$$F \bullet F_{x_i} = U \bullet U_{x_i} + V \bullet V_{x_i} .$$

So

$$\frac{\partial}{\partial x_i} |F|^\gamma = \gamma |F|^{\gamma-2}(F \bullet F_{x_i}),$$

$$\frac{\partial^2}{\partial x_i^2} |F|^\gamma = \gamma[(\gamma-2)|F|^{\gamma-4}(F \bullet F_{x_i})^2 + |F|^{\gamma-2}(|F_{x_i}|^2 + F \bullet F_{x_i x_i})]$$

Hence we get

$$L_{\underset{\sim}{k}}[|F|^\gamma] = \gamma|F|^{\gamma-4}[(\gamma-2)\sum_{i=1}^{2n} (F \bullet F_{x_i})^2 + |F|^2(\sum_{i=1}^{2n}|F_{x_i}|^2 + F \bullet L_{\underset{\sim}{k}}[F])] . \tag{5}$$

Since it is easy to find

$$L_{\underset{\sim}{k}}[U] = 0 \quad \text{and} \quad L_{\underset{\sim}{k}}[V] = 2V \sum_{i=1}^{n} \frac{k_i}{x_{n+i}^2}$$

from the system (3),

$$F \bullet L_{\underset{\sim}{k}}[F] = 2V^2 \sum_{i=1}^{n} \frac{k_i}{x_{n+i}^2} .$$

Substituting this value in (5) we get

$$L_{\underset{\sim}{k}}[|F|^{\gamma}] = \gamma|F|^{\gamma-4}\left[(\gamma-2)\sum_{i=1}^{2n}(F\cdot F_{x_i})^2+|F|^2\left(\sum_{i=1}^{2n}|F_{x_i}|^2+2V\sum_{i=1}^{n}\frac{k_i}{x_{n+i}^2}\right)\right] . \quad (6)$$

It is evident from (6) that $|F|^{\gamma}$ is a $\underset{\sim}{k}$-subharmonic function if $\gamma \geqq 2$. So it remains to prove that the function $|F|^{\gamma}$ is $\underset{\sim}{k}$-subharmonic, i.e. (4) holds, in the case of $\gamma < 2$. Thus to complete the proof we have to show that the inequality

$$\sum_{i=1}^{2n}(F\cdot F_{x_i})^2 \leqq \frac{|F|^2}{2-\gamma}\left[\sum_{i=1}^{2n}|F_{x_i}|^2 + 2V^2\sum_{i=1}^{n}\frac{k_i}{x_{n+i}^2}\right] , \quad (7)$$

holds.

Now let A_i denote the following matrices:

$$A_i = \begin{bmatrix} U_{x_i} & V_{x_i} \\ -U_{x_{n+i}} & -V_{x_{n+i}} \end{bmatrix} , \quad i=1,2,\dots,n . \quad (8)$$

These matrices are symmetric and

$$t_r(A_i) = \frac{2k_i}{x_{n+i}} V . \quad (9)$$

Also, we will recall the norms

$$||B|| = \max_n \left[\sum_{i=1}^{n}|b_{in}|^2\right]$$

and

$$|||B||| = \left[\sum_{j,k=1}^{n}|b_{jk}|^2\right]$$

for the matrix $B=(b_{jk})$ where the later one is known as the "Hilbert-Schmidt norm". Utilizing these notations in (7) we get

$$\sum_{i=1}^{n}|A_iF|^2 \leqq \frac{1}{2-\gamma}|F|^2\left[\sum_{i=1}^{n}\left(|||A_i|||^2 + \frac{[tr(A_i)]^2}{2k_i}\right)\right] \quad (10)$$

or

$$\sum_{i=1}^{n}||A_i||^2 \leqq \frac{1}{2-\gamma}\left[\sum_{i=1}^{n}\left(|||A_i||| + \frac{[tr(A_i)]^2}{2k_i}\right)\right] . \quad (11)$$

Since $|||A_i|||$, $|||A_i|||$ and $tr(A_i)$ are invariant under orthogonal transformations of coordinates, we may assume that A_i can be reduced into the diagonal form

$$A_i = \begin{bmatrix} \mu_{2i-1} & 0 \\ 0 & \mu_{2i} \end{bmatrix} , \quad i=1,2,\dots,n . \quad (12)$$

Thus (11) can be written as

$$\sum_{j=1}^{n} \max\{\mu_{2j-1}^2, \mu_{2j}^2\} \leqq \frac{1}{2-\gamma}\left[\sum_{j=1}^{2n} \mu_j^2 + \sum_{j=1}^{n} \frac{(\mu_{2j-1} + \mu_{2j})^2}{2k_j}\right] . \tag{13}$$

Since

$$\mu_i = -\mu_{i+1}+(\mu_i+\mu_{i+1}), \quad i=1,2,\ldots,2n-1,$$

the inequality

$$\mu_{2i-1}^2 \leqq \alpha_{i-1}\mu_{2i}^2 + \beta_{i-1}(\mu_{2i-1} + \mu_{2i})^2 \tag{14}$$

holds whenever

$$\alpha_{i-1} > 0, \quad \beta_{i-1} > 0 \quad \text{and} \quad \frac{1}{\alpha_{i-1}} + \frac{1}{\beta_{i-1}} = 1, \quad i=1,2,\ldots,n .$$

Adding $\alpha_{i-1}\mu_{2i-1}^2$ to the both sides of (14) we find

$$\mu_{2i-1}^2 \leqq \frac{\alpha_{i-1}}{1+\alpha_{i-1}}\left[\mu_{2i-1}^2 + \mu_{2i}^2 + \frac{\beta_{i-1}}{\alpha_{i-1}}(\mu_{2i-1}+\mu_{2i})^2\right] . \tag{15}$$

We know from the hypothesis that

$$2 > \gamma \geqq \frac{2k_i}{1 + 2k_i}, \quad i=1,2,\ldots,n,$$

so

$$\frac{1}{2-\gamma} \geqq \frac{2k_i + 1}{2k_i + 2} .$$

But if we choose $\alpha_{i-1} = 2k_i+1$ we can show that

$$\frac{1}{2-\gamma} \geqq \frac{\alpha_{i-1}}{1 + \alpha_{i-1}}, \quad \frac{\beta_{i-1}}{\alpha_{i-1}} = \frac{1}{2k_i} . \tag{16}$$

From (15) and (16) we can obtain the result (13) easily which completes the proof.

EXAMPLE - Let us define the function $F = U + iV$ by

$$U = \sum_{i=1}^{n} (\alpha_i x_i + \frac{\gamma_i}{1-2k_i} x_{n+1}^{1-2k_i}),$$

$$V = \sum_{i=1}^{n} (\frac{\alpha_i x_{n+1}}{1 + 2k_i} - \gamma_i x_i x_{n+i}^{-2k_i})$$

where α_i and γ_i are arbitrary constants. It can easily be shown that the system (3) holds. So, if $\gamma \geqq \max_i 2k_i/1+2k_i$ then

$$|F|^{\gamma} = \left[\left\{\sum_{i=1}^{n} \left(\alpha_i x_i + \frac{\gamma_i}{1-2k_i} x_{n+i}^{1-2k_i}\right)\right\}^2 + \left\{\sum_{i=1}^{n} \left(\frac{\alpha_i x_{n+i}}{1+2k_i} - \gamma_i x_i x_{n+1}^{-2k_i}\right)\right\}^2\right]^{\gamma/2}$$

is a $\underset{\sim}{k}$-subharmonic function.

REFERENCES

1. R.J. Weinacht, Fundamental solutions for a class of equations with several singular coefficients, J. Austr. Math. Soc. 8 (1968), pp.575-583.

2. M.S. Hall, D.W. Quinn, and R.J. Weinacht, Poisson integral formulas in generalized bi-axially symmetric potential theory, SIAM J. Math. Anal. 5 (1974), pp.111-118.

3. S.V. Parter, On the existence and uniqueness of axially symmetric potentials, Arc. Rat. Mech. Anal. 20 (1965), pp.279-286.

4. A.O. Çelebi, Some properties for solutions of $\sum_{i=1}^{n}\left(\frac{\partial^2}{\partial x_i^2} + \frac{k_i}{x_i}\frac{\partial}{\partial x_i}\right)u=0$, Comm. Fac. Sci. Univ. Ankara Sér. A, 19 (1970), pp.31-39.

5. Ö. Akın, The integral representation of the positive solutions of the generalized Weinstein equation on a quarter-space, to appear.

6. Ö. Akın, Boundary behaviour of solutions to the generalized Weinstein equation, preprint, March 1987.

7. B. Brelot-Collin and M. Brelot, Representation integrale de solutions positives de l'equation

$$L_k[u] = \sum_{i=1}^{n} \frac{\partial^2 u}{\partial x_i^2} + \frac{k}{x_n}\frac{\partial u}{\partial x_n} = 0,$$

Bull. Ac. Royale des. Sci. de Bruxelles, 58 (1972), pp.317-326.

8. B. Muchanhogt and E.M. Stein, Classical expansions and their relations to conjugate harmonic functions, Trans. Amer. Math. Soc., 118 (1965), pp.17-91.

ITERATION METHODS FOR POTENTIAL PROBLEMS

Thomas S. Angell* and Ralph E. Kleinman*

Department of Mathematical Sciences
University of Delaware
Newark, Delaware 19716 U.S.A.

Gary F. Roach

Department of Mathematics
University of Strathclyde
Livingstone Tower, 26 Richmond Street
Glasgow G1 1XH, Scotland, U.K.

Two iteration methods are considered as a means of constructing solutions of various integral equation formulations of the exterior Dirichlet and Neumann potential problems. These integral equations may be of the second kind, in which case the integral operators will in general be non self-adjoint, or of the first kind, which in the Dirichlet case leads to self-adjoint operators.

One iteration method, the so-called eigenvalue shifting technique, is applicable directly regardless of whether the operator is self-adjoint or not. The second approach which we designate Bialy's method, although others could be credited, requires symmetrizing the operators.

When the boundary integral equations are uniquely solvable both iteration methods lead to convergent results. Even when the equations are not uniquely solvable the iteration will converge if the inhomogeneous term is in the range of the operator. However when unique solvability is restored by means of a modified Green's function technique convergence is considerably enhanced. The methods are illustrated by worked examples for the sphere. These examples indicate that the eigenvalue shifting technique leads to a more rapidly convergent iteration than Bialy's method for second kind equations. The iterative solution of first kind equation exhibited the poorest rate of convergence.

1. FORMULATION OF PROBLEMS

Let $\Omega \subset \mathbf{R}^3$ be the unbounded region exterior to a closed, bounded, Lyapunov surface $\partial\Omega$. Let $R = R(p,q)$ denote the distance between two typical points $p,q \in \mathbf{R}^3$ and $\hat{n}_p$ the unit normal, erected at $p \in \partial\Omega$, directed into Ω.

We consider the following typical potential problems: determine a quantity

$$\Delta u(p) = 0\ , \quad p \in \Omega \tag{1.1}$$

*Work supported under Air Force Office of Scientific Research Grant AFOSR-86-0269.

$$u(p) = 0(1/r) \quad , \quad \frac{\partial u}{\partial n_p}(p) = 0(1/r^2) \quad , \quad r = |OP| \tag{1.2}$$

and either

$$u(p) = f(p) \, , \qquad p \in \partial\Omega \qquad \text{(Dirichlet Problem)} \tag{1.3}$$

or

$$\frac{\partial u(p)}{\partial n_p} = f(p) \, , \qquad p \in \partial\Omega \qquad \text{(Neumann Problem)} \; . \tag{1.4}$$

Here we have assumed that $p \in \partial\Omega$ has spherical polar coordinates $(r_p , \theta_p , \phi_p)$ relative to a Cartesian coordinate system erected with origin in the complement in $\mathbf{R}^3$ of $\overline{\Omega} := \Omega \cup \partial\Omega$. Furthermore, $\partial/\partial n_p$ denotes differentiation in the direction of the unit normal $\hat{n}_p$, $p \in \partial\Omega$.

A fundamental solution of the Laplace equation (1.1) is given by

$$\gamma(p,q) = -\frac{1}{2\pi R} \quad , \quad R = |p-q| \; . \tag{1.5}$$

If $w \in C(\partial\Omega)$ then we can define single and double layer distributions of density w as follows:

$$(Sw)(p) := \int_{\partial\Omega} \gamma(p,q) w(q) dS_q \quad , \quad p \in \mathbf{R}^3 \setminus \partial\Omega \tag{1.6}$$

$$(Dw)(p) := \int_{\partial\Omega} \frac{\partial \gamma}{\partial n_q}(p,q) w(q) dS_q \quad , \quad p \in \mathbf{R}^3 \setminus \partial\Omega \; . \tag{1.7}$$

It is convenient to introduce the integral K defined by

$$(Kw)(p) := \int_{\partial\Omega} \frac{\partial \gamma}{\partial n_p}(p,q) w(q) dS_q \quad , \quad p \in \partial\Omega \tag{1.8}$$

and its $L_2(\partial\Omega)$ adjoint

$$(K^* w)(p) := \int_{\partial\Omega} \frac{\partial \gamma}{\partial n_q}(p,q) w(q) dS_q \quad , \quad p \in \partial\Omega \; . \tag{1.9}$$

The jump conditions satisfied by the single and double layer distributions at the boundary $\partial\Omega$ can now be conveniently written in the form

$$\frac{\partial}{\partial n_p^{\pm}} \{Sw\}(p) = (\pm I + K) w(p) \, , \; p \in \partial\Omega \tag{1.10}$$

$$\lim_{p \to p^{\pm}} \{Dw(p)\} = (\mp I + K^*) w(p) \, , \; p \in \partial\Omega \tag{1.11}$$

where $+(-)$ denotes that p tends onto $\partial\Omega$ from the exterior (interior) of $\partial\Omega$.

With this notation an appropriate form of Green's Theorem for the exterior problems (1.1) to (1.4) is

$$(S\partial u)(p) - (Du)(p) = 2u(p) \, , \; p \in \Omega \tag{1.12}$$

$$(S\partial u)(p) - (K^* u)(p) = u(p)\ , \quad p \in \partial\Omega \tag{1.13}$$

where

$$\partial u \ := \partial u / \partial n \quad .$$

If now either a Green's theorem method or a layer-theoretic method is used to solve the boundary value problems (1.1) to (1.4) then we obtain the following representations of solutions and their associated boundary integral equations [1].

Table 1

Boundary Conditions	Representation of Solutions	Boundary Integral Equation
Dirichlet		
	$u = \frac{1}{2}(Sw - Df)$	$(I-K)w = -\frac{\partial}{\partial n}Df$
		or $Sw = (I+K)^*)f$
$u = f$		
	$u = -Dw$	$(I-K^*)w = f$
	$u = Sw$	$Sw = f$
Neumann		
	$u = \frac{1}{2}(Sf - Dw)$	$(I+K^*)w = Sf$
$\frac{\partial u}{\partial n} = f$		
	$u = Sw$	$(I+K)w = f$

The main aim of this note is to compare various iteration schemes for solving the boundary integral equations given in this table and to comment on their relative efficiency. However there are other equally interesting boundary integral equations which can be associated with the exterior problems considered here. Iterative methods for solving these equations will also be discussed. We include a discussion of first kind equations for the Dirichlet problem but omit them for the Neumann problem to avoid a lengthy treatment of the non-compact operators involved.

2. AN EIGENVALUE SHIFTING TECHNIQUE

Typical of the equation we wish to solve by iteration is

$$(I - K)w = f \quad . \tag{2.1}$$

Straightforward iteration yields

$$w_{n+1} = Kw_n + f \quad . \tag{2.2}$$

If we choose the initial point $w_o = 0$, this procedure yields, simply,

$$w_{n+1} = \sum_{j=0}^{n} K^j f \quad . \tag{2.3}$$

A sufficient condition for the convergence of this iteration scheme is that

$$\|K\| < 1 \tag{2.4}$$

and the following result is well known [2].

<u>*Theorem 2.1*</u>: When the iteration scheme (2.2) is convergent, then for any initial approximation w_o, the successive approximations converge to a unique solution w^* of (2.1). We then have the following bound

$$\|w^* - w_n\| \le \|(I-K)^{-1}\| \, \|K\|^n \, \|w_1 - w_o\| \quad n = 1, 2, \cdots \quad . \tag{2.5}$$

If in particular

$$\|K\| \le q < 1 \tag{2.6}$$

then

$$\|w^* - w_n\| \le \frac{q^n}{1-q} \|w_1 - w_o\| \quad . \tag{2.7}$$

Difficulties occur when a condition such as (2.4) is not satisfied as well as near eigenvalues of K. These difficulties can be reduced if not entirely removed by the following "shifting" technique. Rewrite equation (2.1) to obtain, for c real,

$$w - w + (1-c)(I-K))w = (1-c)f \quad .$$

Successive manipulations yield

$$w - cw - (1-c)Kw \quad = (1-c)f \quad ,$$

$$[I-(cI+(1-c)K)w = (1-c)f \quad .$$

and finally

$$(I-U)w = (1-c)f \tag{2.8}$$

where

$$U = cI + (1-c)K \quad . \tag{2.9}$$

If we consider the eigenvalue problem associated with K in the form

$$Ku = \lambda u$$

then any eigenvector for K is likewise an eigenvector for u. Indeed if $u \ne 0$ is an eigenfunction for K then

$$cu + (1-c)Ku = \lambda(1-c)u + cu$$

which reduces to

$$Uu = [\lambda(1-c) + c]u \quad . \tag{2.10}$$

We can deduce that

$$\sigma_p(U) = \{c+(1-c)\lambda : \lambda \in \sigma_p(K)] \tag{2.11}$$

where

$$\sigma_p(A) := \text{the point (eigenvalue) spectrum of an operator } A \ .$$

Now $r_\sigma(K)$, the spectral radius of the compact operator K, is defined by (see [3])

$$r_\sigma(K) = sup\{ |\lambda| : \lambda \in \sigma_p(K)\} = \lim_{n\to\infty} \|K^n\|^{1/n} \ . \tag{2.12}$$

The iteration scheme (2.2) converges when $r_\sigma(K) < 1$ and diverges when $r_\sigma(K) > 1$, [2]. Thus when $r_\sigma(K) \geq 1$ it would seem reasonable that c in (2.8) could be chosen to ensure $r_\sigma(U) < 1$, in which case an iteration scheme such as (2.2) but with U replacing K would converge and so provide a solution to (2.8) as indicated by Theorem 2.1. Conditions under which c can indeed be chosen in this way, thus indicating that a troublesome eigenvalue has been "shifted", are discussed in [4] and more generally in [5].

In a later section we illustrate this procedure for the particular case of spherical geometry. First, we introduce, in the next section, an iteration scheme which is particularly appropriate for use with nonselfadjoint operators.

3. BIALY'S METHOD

A method for handling nonselfadjoint operators was indicated in [6] but had perhaps its first systematic application in the work of Bialy [7]. In fact the technique goes back at least to von Mises and Pollaczek-Geirunger [8] who discuss the matrix case. Essentially the method is developed as follows. Given a Hilbert space H and a linear, not necessarily self-adjoint operator, $K : H \to H$ it is required to construct a solution of

$$Lw := (I + K)w = f \ . \tag{3.1}$$

We rewrite (3.1) in the form

$$w = w - \alpha L^* Lw + \alpha L^* f \tag{3.2}$$

$$\text{or} \quad [I - (I - \alpha L^* L)]w = \alpha L^* f \ .$$

The operator $L^* L$ is clearly self-adjoint and the intention now is to choose the parameter α so that the operator

$$B_\alpha := (I - \alpha L^* L) \tag{3.3}$$

has a minimum norm. The operator $L^* L$ has upper and lower bounds M and m respectively defined by

$$M := sup\{(L^* Lu, u) : \|u\| = 1\} \tag{3.4}$$

and

$$m := inf\{(L^* Lu, u) : \|u\| = 1\} \ . \tag{3.5}$$

Consequently B_α has an upper bound $(1-\alpha m)$ and a lower bound $(1-\alpha M)$. Therefore B_α will have a minimum norm when

$$\alpha = \frac{2}{M+m} \tag{3.6}$$

which will be

$$\|B_\alpha\| = \frac{M-m}{M+m} \quad . \tag{3.7}$$

Hence the Neumann series associated with (3.2) (written in terms of B_α) will converge whenever we can ensure that the right hand side of (3.7) is less than unity. Inherent in this approach is the unique solvability of (3.1) for in this case w is a solution to equation (3.1) if and only if it is a solution of equation (3.2). In case unique solvability does not obtain, one must introduce least squares solutions as explained below in Section 5. This was in fact already done by Bialy.

4. POTENTIAL PROBLEMS FOR A SPHERE

We consider first an exterior Neumann problem (ENP). The associated boundary integral equation (cf. Table 1) is

$$(I + K)w = f \quad . \tag{4.1}$$

From (1.5) and (1.9) we have

$$(Kw)(p) = -\frac{1}{2\pi}\int_{\partial\Omega}\frac{\partial}{\partial n_p}\frac{1}{|p-q|}w(q)dS_q \quad . \tag{4.2}$$

Standard results [11] yield

$$\frac{1}{|p-q|} = \sum_{n=0}^{\infty}\left\{\frac{r_<^n}{r_>^{n+1}}\right\}P_n(\cos\theta)\ , \quad \cos\theta = \hat{r}_p \cdot \hat{r}_q \quad .$$

Moreover on a sphere of radius a,

$$\left.\frac{\partial}{\partial r_p}\frac{1}{|p-q|}\right|_{r_p = r_q = a} = \frac{1}{2}\sum_{n=0}^{\infty}\frac{\partial}{\partial r_q}\left\{\frac{r_q^n}{r_p^{n+1}} + \frac{r_p^n}{r_q^{n+1}}\right\}_{r_p = r_q = q} P_n(\cos\theta)$$

$$= \frac{1}{2}\sum_{n=0}^{\infty}\frac{(n-[n+1])}{a^2}P_n(\cos\theta)$$

$$= -\frac{1}{2a^2}\sum_{n=0}^{\infty}P_n(\cos\theta) \quad . \tag{4.4}$$

Consequently, if $Y_n(\theta_p,\phi_p)$ is a spherical harmonic,

$$KY_n(\theta_p\ ,\ \phi_p) = \frac{1}{4\pi a^2} \int_0^{\pi} d\theta \int_0^{2\pi} d\phi \sin\theta\, a^2 \sum_{m=0}^{\infty} P_m(\cos\theta) Y_n(\theta_p\ ,\ \phi_p)$$

$$= \frac{1}{4\pi} \frac{4\pi}{2n+1} Y_n(\theta_p\ ,\ \phi_p) = \frac{1}{2n+1} Y_n(\theta_p\ ,\ \phi_p) \tag{4.5}$$

and we conclude that K^* has the eigenpairings

$$\left\{Y_n\ ,\ \frac{1}{2n+1}\right\}_{n=0}^{\infty} \tag{4.6}$$

the eigenvalues accumulating as expected at the origin. The fact that K has an eigenvalue $\lambda = 1$ is troublesome when trying to solve (4.1). This is not because the unique solvability of (4.1) is affected but rather that $r_\sigma(K) = 1$, which is an indeterminate situation. To try to improve matters we use the method introduced in §2. This leads to the equation

$$(I-U)w = (1-c)f \tag{4.7}$$

where

$$U = cI - (1-c)K\ . \tag{4.8}$$

Bearing in mind (4.5) we find that

$$UY_n(\theta_p\ ,\ \phi_p) = (cI-(1-c)K)Y_n(\theta_p \phi_p) = (c-\frac{(1-c)}{2n+1})Y_n(\theta_p\ ,\ \phi_p)$$

$$= \left[\frac{(2n+2)c-1}{2n+1}\right] Y_n(\theta_p\ ,\ \phi_p) \tag{4.9}$$

and conclude that U has the eigenpairings

$$\left\{Y_n\ ,\ \frac{(2n+2)c-2}{2n+1}\right\}_{n=0}^{\infty}\ . \tag{4.10}$$

Two cases are of particular interest [3], [4], namely $c = 1/2$, where the eigenpairs are

$$\left\{Y_n\ ,\ \frac{n}{2n+1}\right\}_{n=0}^{\infty} \tag{4.11}$$

and in this case

$$r_\sigma(U) = \frac{1}{2}\ , \tag{4.12}$$

and $c = 1/3$ where the eigenpairs are

$$\left\{Y_n\ ,\ \frac{2n-1}{3(2n+1)}\right\}_{n=0}^{\infty}$$

and in this case

$$r_\sigma(U) = \frac{1}{3} \quad . \tag{4.13}$$

In both these cases $r_\sigma(U) < 1$ and the Neumann series associated with (4.1) converges. However since the rate of convergence is governed by the spectral radius the case $c = \frac{1}{3}$ offers the better prospects.

It is interesting to notice that

$$r_\sigma(U) = sup\{\, |\nu| : \nu \in \sigma(U)\} \tag{4.14}$$

$$= sup\{\, |c - \lambda_n(1-c)| : \lambda_n \in \sigma(K)\} \tag{4.15}$$

since

$$Ku = \lambda u$$

implies

$$[cI - (1-c)K]u = Uu = [c - \lambda(1-c)]u \quad .$$

Consequently, bearing in mind (4.6) we see that for this class of problem any value of $|c| < 1$ will ensure a convergent Neumann series since this inequality implies

$$r_\sigma(U) = sup\{\, |c| \} \quad . \tag{4.16}$$

We now apply Bialy's method, §3, to (4.1). In this case we have

$$L := (I+K)$$

with $K : L_2(\partial\Omega) \to L_2(\partial\Omega)$ a compact, nonselfadjoint operator. However L^*L as an operator in $L_2(\partial\Omega)$ is self-adjoint. Furthermore, we can write

$$L^*Lw = (I+K^*)(I+K)w = (I+K+K^*+K^*K)w =: (I+A)w \tag{4.17}$$

where A is compact self-adjoint and as such has upper and lower bounds given respectively by [2]:

$$M(A) = sup\{(Au,u) : \|u\| = 1\} = sup\{\lambda : \lambda \in \sigma_p(A)\} =: \lambda_{max}(A) \tag{4.18}$$

$$m(A) = inf\{(Au,u) : \|u\| = 1\} = inf\{\lambda : \lambda \in \sigma_p(A)\} =: \lambda_{min}(A) \quad . \tag{4.19}$$

Consequently, if we consider the two eigenvalue problems

$$Aw = \lambda(A)w \tag{4.20}$$

and

$$Tu := L^*Lu = \lambda(T)u \quad . \tag{4.21}$$

We find that

$$\lambda(A) = \lambda(T) - 1 \tag{4.22}$$

and

$$M(T) = 1 + M(A) = 1+\lambda_{\max}(A) = \lambda_{\max}(T)$$

$$m(T) = 1+m(A) = 1+\lambda_{\min}(A) = \lambda_{\min}(T) \quad . \tag{4.23}$$

We now calculate the eigenvalues of T, recognizing (4.5) and the fact that on a sphere

$$K = K^* \tag{4.24}$$

$$TY_n(\theta_p\ ,\ \phi_p) = (I+K^*)(I+K)Y_n(\theta_p\ ,\ \phi_p) = (I+K^*)(1+\frac{1}{2n+1})Y_n(\theta_p\ ,\ \phi_p)$$

$$= \left\{\frac{2n+2}{2n+1}\right\}^2 Y_n(\theta_p\ ,\ \phi_p) \quad . \tag{4.25}$$

Consequently $T = L^*L$ has the eigenpairings

$$\left\{Y_n\ ,\ \left[\frac{2n+2}{2n+1}\right]^2\right\}_{n=0}^{\infty} \tag{4.26}$$

hence

$$M(T) = \lambda_{\max}(T) = 4 \quad \text{and} \quad m(T) = \lambda_{\min}(T) = 1 \quad . \tag{4.27}$$

Thus rewriting (4.1) in the form

$$(I-B_\alpha)w = \alpha L^*f \quad , \quad B_\alpha = I-\alpha L^*L \tag{4.28}$$

we choose α according to (3.6) to obtain

$$\alpha = \frac{2}{M(T)+m(T)} = \frac{2}{5} \quad . \tag{4.29}$$

This will imply that B_α has a minimum norm given by

$$\|B_\alpha\| = \frac{M(T)-m(T)}{M(T)+m(T)} = \frac{3}{5} \quad . \tag{4.30}$$

Since L^*L is self-adjoint so is B_α and we have

$$r_\sigma(B_\alpha) = \|B_\alpha\| = \frac{3}{5} \quad . \tag{4.31}$$

Since $r_\sigma(B_\alpha) < 1$ the Neumann series associated with (4.28) will converge but not as rapidly as the series governed by (4.12) and (4.13). Thus the eigenvalue shifting technique is more efficient than Bialy's method for this problem.

We now turn attention to the exterior Dirichlet problem (EDP). The associated boundary integral equation is (c.f. Table 1)

$$(I-K^*)w = f \quad . \tag{4.32}$$

Since $K^* = K$ on a sphere the previous calculations indicate that

$$\sigma_p(K^*) = \sigma_p(K) = [0,1] \quad . \tag{4.33}$$

Thus (4.32) is not uniquely solvable. Consequently Bialy's method cannot be used to construct solutions to (4.32) (but see the discussion in Section 5). Furthermore, the shifting technique of §2 is also unavailable because if we define

$$\lambda^+ = inf\{\lambda > 0:\ \lambda \in \lambda(K)\}$$

$$\lambda^- = sup\{\lambda \le 0:\ \lambda \in \sigma(K)\} \tag{4.34}$$

then the following result can be established [5].

Theorem 4.1: If $0 < \lambda^+ \le 1$ then there is no choice of c for which $r_\sigma(U) < 1$.

The lack of unique solvability of (4.32) contradicts the unique solvability of the exterior Dirichlet problem and suggests the use of a modified Green's function technique [9], [10] in order to make the methods of §2 and §3 available. To this end change notation slightly and write (1.5) in form

$$\gamma_o(p,q) = -\frac{1}{2\pi R} = -\frac{1}{2\pi|p-q|}$$

where the zero subscript denotes an unmodified (free space) Green's function. The associated integral operators defined by the right hand side of (1.6) and (1.7) will be similarly subscripted. Thus we obtain in place of (4.32)

$$(I-K_o^*)w(p) = f(p)\,, \quad p \in \partial\Omega$$

where

$$(K_o^* w)(p) = \int_{\partial\Omega} \frac{\partial\gamma_o}{\partial n_p}(p,q)w(q)dS_q \ .$$

To restore unique solvability we introduce the modified Green's function

$$\gamma_1(p,q) = \gamma_o(p,q) + \frac{c_o}{r_p r_q} \tag{4.35}$$

where c_o has yet to be chosen. The integral operators corresponding to (1.6) and (1.7) but using γ_1 rather than γ_o will be similarly subscripted. If now we assume that the EDP has a solution of the form

$$u(p) = -(D_1 w)(p)\ , \quad p \in \mathbf{R}^3 \setminus \partial\Omega \tag{4.36}$$

then this leads to a consideration of the boundary integral equation

$$(I-K_1^*)w(p) = f(p)\ , \quad p \in \partial\Omega \tag{4.37}$$

where

$$(K_1^* w)(p) = \int_{\partial\Omega} \frac{\partial\gamma_1}{\partial n_q}(p,q)w(q)dS_q \tag{4.38}$$

and for $\partial\Omega$ a sphere of radius a

$$\frac{\partial\gamma_1}{\partial n_p}(p,q)\Big|_{r_p=r_q=a} = \frac{\partial\gamma_o}{\partial r}(p,q) - \frac{c_o}{a^3} \ . \tag{4.39}$$

To determine the eigenvalues of K_1^* we recall the above calculation and standard results for integrals of spherical harmonics [11] to obtain

$$K_1^* Y_n(\theta_p \phi_p) = K_o^* Y_n(\theta_q, \phi_p) - \frac{c_o}{a^3} \int_{\partial\Omega} Y_n(\theta_q, \phi_q) dS_q$$

$$= \begin{cases} K_o^* Y_n(\theta_p, \phi_p) \ , \ n > 0 \\ \\ K_o^* Y_o(\theta_p, \phi_p) - \frac{c_o}{a^3}(4\pi a^2) Y_o(\theta_p, \phi_p) \ , \ n = 0 \end{cases}$$

$$= \mu_n Y_n(\theta_p, \phi_p) \tag{4.40}$$

where

$$\mu_n = \begin{cases} \frac{1}{2n+1} \ , \ n > 0 \\ \\ 1 - 4\pi co/a \ , \ n = 0 \ . \end{cases} \tag{4.41}$$

We now choose c_o to ensure that (4.37) is uniquely solvable. In this case since

$$\{\mu_n\}_{n=1}^{\infty} = \{\frac{1}{3}, \frac{1}{5}, \cdots 0\}$$

the most suitable choice of c_o would be that which satisfies

$$1 - \frac{4\pi c_o}{a} = \frac{1}{3}$$

that is

$$c_o = \frac{a}{6\pi} \ . \tag{4.42}$$

This choice of c_o implies that

$$r_\sigma(K_1^*) = \frac{1}{3} \tag{4.43}$$

and we see that (4.37) is uniquely solvable with an associated Neumann series which is as rapidly convergent as Neumann series obtained by an eigenvalue shifting technique for more straightforward cases. Furthermore, having chosen c_o we can now apply the eigenvalue shifting technique with a view to improving matters still further. Specifically we consider (4.37) when (4.43) holds. Proceeding as in §2 we are lead to the equation

$$(I-U)w = (1-c)f \ , \quad U = cI + (1-c)K_1^* \tag{4.44}$$

where

$$r_\sigma(U) = sup\{|\lambda|:\lambda \in \sigma(U)\} = sup\{|c + \mu(1-c)|:\mu \in \sigma(K_1^*)\} \quad . \tag{4.45}$$

We now want to choose c so that

$$|c+\mu(1-c)| < \frac{1}{3} = r_\sigma(K_1^*) \quad . \tag{4.46}$$

Bearing in mind (4.41) and (4.42) we see that

$$\mu_{\min} = 0 \quad \text{and} \quad \mu_{\max} = \frac{1}{3}$$

which implies that (4.46) can be satisfied however

$$-\frac{1}{3} < c < 0 \quad . \tag{4.47}$$

For example with $c = -\dfrac{1}{4}$ we obtain $r_\sigma(U) = \dfrac{1}{6}$ and as a consequence an improved convergence rate can be obtained by using (4.44) rather than (4.37).

However if we use the method of §3 we are lead to the equation

$$(I-B_\alpha)w = \alpha L^* w \quad , \quad B_\alpha = (I-\alpha L^* L) \tag{4.48}$$

here

$$r_\sigma(B_\alpha) = sup\{|\nu|:\nu \in \sigma(B_\alpha)\} = sup\{|(1-\alpha)\lambda|:\lambda \in \sigma(L^* L)\} \quad .$$

Bearing in mind the above calculations and (4.42) we obtain

$$L^* L Y_n(\theta_n,\phi_p) = (I-K_1)(I-K_1^*)Y_n(\theta_p,\phi_p) = \left\{\frac{2n}{2n+1}\right\}^2 Y_n(\theta_p,\phi_p) \quad .$$

Thus

$$\lambda_n = \begin{cases} \left\{\dfrac{2n}{2n+1}\right\}^2 & , \quad n > 0 \\[2ex] (\dfrac{4}{9}) & , \quad n = 0 \quad . \end{cases}$$

Hence

$$M = \lambda_{\max} = 1$$

$$m = \lambda_{\min} = \frac{4}{9}$$

and we obtain

$$r_\sigma(B_\alpha) = \|B_\alpha\| = \frac{M-m}{M+m} = \frac{5}{13} > \frac{1}{3}$$

which implies a slower convergence rate when using (4.48).

5. EQUATIONS OF THE FIRST KIND

In this section we address the question of iteration methods for integral equations of the first kind. Indeed, reference to Bialy's paper [7] reveals that his primary concern was with operator equations of the first kind, his purpose being to give extensions of earlier work by Wiarda [6] and Fridman [12]. Apparently Bialy was unaware of the earlier paper of Landweber [13] dealing with the case of integral equations of the first kind. More recent treatments include that of Diaz and Metcalf [14] and the paper of Ahner and Hsiao [15] in which one may find explicit calculations for the two-dimensional Dirichlet problem for the ellipse. Other references are given by Nashed [16].

Let L be a bounded linear operator on the Hilbert space H and consider the equation

$$Lw = f \quad . \tag{5.1}$$

As in Section 3, for α a real parameter we may rewrite this operator equation as

$$w - w + \alpha L^* Lw = \alpha L^* f \tag{5.2}$$

or, in operator form,

$$(I-T)w = g \tag{5.3}$$

where

$$g = \alpha L^* f \tag{5.4}$$

and

$$T = I - \alpha L^* L \quad . \tag{5.5}$$

Note that, T is not, in general, a compact operator but that it is self-adjoint.

Browder and Petryshyn [17] strengthen the results of Bialy with proofs which are independent of the spectral representation of L. It is noteworthy, moreover, that the main result of their paper is established without any assumption that the operator is compact or self-adjoint. Nor do their results concerning the convergence of the Picard iterates depend on the *a priori* assumption of unique solvability. The operator T in an equation of the form (5.3) is required, however, to be asymptotically convergent that is $T^n x$ must converge strongly as $n \to \infty$ for each x in H. If T has this property then necessarily (-1) is not an eigenvalue of T. It is remarkable and relevant to the present discussion, that this condition is also sufficient in the case that T is self-adjoint and satisfies $\|T\| \leq 1$. This means that their results guarantee the convergence of the Picard iterates in the case of the Dirichlet problem discussed immediately above.

In light of the results of [17] we may state the following theorem regarding equations of the form

$$(I-K)w = f \quad . \tag{5.6}$$

Theorem 5.1: Suppose H is any Hilbert space and K is a self-adjoint map on H with $\|K\| \leq 1$ and for which $(-1) \notin \sigma_p(K)$. Then the Picard iterates defined by

$$x_{n+1} = Kx_n + f \tag{5.7}$$

converge if and only if equation (5.6) has a solution. Moreover boundedness of the Picard iterates is sufficient to insure convergence.

The force of the last statement is that the boundedness of the sequence defined in (5.7) is sufficient to guarantee that f lies in the range of $(I-K)$. We should also remark that if we assume, as in Section 3, that the equation (5.6) is uniquely solvable, then one can assert the convergence of the Picard iterates to the solution.

In order to apply Theorem 5.1 to the equation (5.3) we must insure that -1 is not an eigenvalue of T and that $\|T\| \le 1$. Both of these conditions may be met by an appropriate choice of α.

Lemma 5.1: Let α be a real parameter satisfying

$$0 < \alpha < 2/\|L^* L\| \tag{5.8}$$

then $-1 \notin \sigma_p(T)$ and $\|T\| \le 1$.

Proof: To establish the first statement, assume the contrary. Then there exists a vector u, with $\|u\| = 1$, such that

$$(I+T)u = 0 \ .$$

This implies that

$$2u = \alpha L^* Lu$$

and taking norms on both sides and using (5.8) yields the contradiction that $2 < 2$. The second statement follows from the argument used to prove Lemma 3.2 of [18] applied to the self-adjoint operator T.

Note that since we do not assume unique solvability of the equation (5.1) we conclude only that $\|T\| \le 1$ and not the stronger inequality of [18]. In this case we can assert only that the Picard iterates converge to a least-squares solution of the problem as the following result shows (see [7], [17]):

Theorem 5.2: Let H be a Hilbert space, L a bounded linear operator on H, α a real parameter satisfying the inequality (5.8), and let $\mathbf{P}$ be the orthogonal projection of H onto the closure of the range of L. Then for any given $x_o \in H$, the sequence given by

$$x_n = Tx_{n-1} + g \tag{5.9}$$

converges to a solution of

$$Lx = \mathbf{P}g \tag{5.10}$$

if and only if (5.10) has a solution.

These results may be applied to the equation of the first kernel arising in the treatment of the exterior Dirichlet problem. The exterior Dirichlet problem may be formulated, see Table 1, as a first kind equation

$$Sw = f \quad \text{on } \partial\Omega \tag{5.11}$$

which can be rewritten as

$$(I - T)w = \alpha f \quad , \ T := I - \alpha S \ . \tag{5.12}$$

Since S is self-adjoint, T is also self-adjoint for real α. When $\partial\Omega$ is a sphere of radius a it may be shown that

$$\sigma_p(T) = \left\{1 + \frac{2\alpha a}{2n+1}\right\}_{n=0}^{\infty} \ . \tag{5.13}$$

If α is chosen to satisfy

$$0 > \alpha > -\frac{1}{\alpha} \tag{5.14}$$

then

$$r_\sigma(T) = 1 \tag{5.15}$$

and

$$-1 \notin \sigma_p(T) \ . \tag{5.16}$$

Thus (5.12) is of the form (5.3) and the Picard iterates must converge. In this case the iterates as well as the exact solution may be found explicitly. The exact solution is

$$w = -\frac{1}{2a} \sum_{|l|=0}^{\infty} (2l_1+1)Y_l \tag{5.17}$$

and the nth iterate is

$$w_n = w + \frac{1}{2a} \sum_{|l|=0}^{\infty} \frac{(2l_1+1+2a\,\alpha)^{n+1}}{(2l_1+1)^n} Y_l \tag{5.18}$$

where l_1 is the first entry in the multi-index l of the spherical harmonics. It follows that

$$\|w - w_n\|^2 = \frac{1}{4a^2} \sum_{|l|=0}^{\infty} \frac{(2l_1+1+2a\,\alpha)^{2n+2}}{(2l_1+1)^{2n}} \|Y_l\|^2 \tag{5.19}$$

and so long as (5.14) is fulfilled

$$\|w - w_n\| \to 0 \quad \text{as} \quad n \to \infty \tag{5.20}$$

however the convergence is considerably slower then that achieved using 2nd kind equations as shown in the previous section.

6. CONCLUSIONS

Bialy's method is designed to ensure that a given problem can be analysed in terms of self-adjoint operators. On the other hand the eigenvalue shifting technique can in general be used with operators having a non-real spectrum. However, both these methods require uniqueness of solvability of the original equation to guarantee convergence to the solution of the problem. The result of Browder and Petryshyn shows that uniqueness of solutions is inessential and both these authors and Bialy discuss the question of convergence to a least squares solution for equations of the first kind. For equations of the second kind, the results of Browder and Petryshyn give conditions under which the iterates converge even if the spectral radius is 1. In this case the techniques of Bialy or eigenvalue shifting may be used in an attempt to improve convergence.

On the other hand when unique solvability is not initially available it can be restored by first employing a Green's function before either of the methods mentioned. For operators which have a real spectrum but which are not necessarily self-adjoint the eigenvalue shifting technique yields particularly strong results. Indeed for problems associated with the sphere it offers the better prospects for numerical work.

REFERENCES

1. R.E. Kleinman and G.F. Roach, Boundary integral equations for the three dimensional Helmholtz equation, SIAM Rev. *16,* 214-236 (1974).

2. L.V. Kantorovich and G.P. Akhilov, Functional Analysis, Pergamon Press, Oxford, (2nd Ed.) (1982).

3. K. Yosida, Functional Analysis, Springer Berlin (1968).

4. R. Kress and G.F. Roach, On the convergence of successive approximations for an integral equation in a Green's function approach to the Dirichlet problem, Jour. Math. Anal. Applic. *55* (1), 102-111 (1976).

5. R.E. Kleinman, Iterative solutions of boundary value problems, Function Theoretic Methods for Partial Differential Equations, Springer LNM 561, 198-313 (1976).

6. G. Wiarda, Integralgleichungen unter besonderer Berücksichtigung der Anwendungen, B.G. Teubner, Leipzig und Berlin (1930).

7. H. Bialy, Iterative Behandlung linearer Funktiongleichungen, Arch. Rational Mech. Anal. *4,* 166-176 (1959).

8. R. von Mises and H. Pollaczek-Geirunger, Praktische Vefahren der Gleichungsauflosung, Z. Angew. Math. Mech. *9,* 58-77 (1929).

9. R.E. Kleinman and G.F. Roach, On modified Green's function in exterior problems for the Helmholtz equation, Proc. Roy. Soc. Lond. *A383,* 313,332 (1981).

10. R.E. Kleinman and G.F. Roach, Operators of minimal norm via modified Green's functions, Proc. Roy. Soc. Edin. *94A,* 163-178 (1983).

11. W. Magnus and F. Oberhettinger, Formulas and Theorems for the Functions of Mathematical Physics, Chelsea Publishing Co., New York (1954).

12. V. Fridman, Method of successive approximations for a Fredholm integral equation of the first kind, Upsehi Mat. Nauk, *11,* 233-34 (1956), (Russian).

13. L. Landweber, An iteration formula for Fredholm integral equations of the first kind, Amer. J. Math. *73, 615-624 (1954).*

14. J.B. Diaz and F.T. Metcalf, On iteration procedures for equations of the first kind, $Ax = y$, and Picard's criterion for the existence of a solution, Math. Comp. *24,* 923-935 (1970).

15. J. Ahner and G.C. Hsiao, Iteration solutions of boundary integral equations of the first kind, to appear.

16. M.Z. Nashed, Generalized inverses, normal solvability, and iteration for singular operator equations in Nonlinear Functional Analysis and Applications, L.B. Rall ed., Publication No. 26, Mathematics Research Center, University of Wisconsin, Academic Press, New York, London (1971).

17. F.E. Browder and W.V. Petryshyn, The solution by iteration of linear functional equations in Banach spaces, Bull. Amer. Soc. *72,* 566-570 (1966).

18. R.E. Kleinman and G.F. Roach, Iterative solutions of boundary integral equations in acoustics, to appear.

ON THE POISSON EQUATION ON THE INFINITE DIMENSIONAL TORUS

Alexander D. Bendikov and Igor V. Pavlov

Rostov Civil Engineering Institute
Sotsialisticheskaya 162, Rostov-on-Don
344022, USSR

Let $\{\mu_t\}_{t>0}$ be the Wiener semigroup on the infinite dimensional torus T^∞ (cf. Heyer,[5] Ch. 5). Let us consider the Green measure

$$G(dy) = \int_0^\infty e^{-t}\mu_t(dy)dt.$$

In this paper we get differentiability properties of the potential

$$Gf(x) = \int_{T^\infty} f(x-y)G(dy)$$

which is the weak solution of the Poisson equation $\mathcal{L}u-u = -f$.

In the sequel we shall use the following notations: R^1 is the group of real numbers; Z is the group of integers; T is the torus, i.e. the factor group $R^1/2\pi Z$; R^∞ (respectively Z^∞, T^∞) is the Cartesian product of coutably many copies of R^1 (respectively Z, T); $Z^{(\infty)}$ is the subgroup of Z^∞ consisting of all sequences in Z^∞ which are eventually zero; $dx = \times_{k=1}^{\infty} dx_k$ is the Haar measure on T^∞ where dx_k is the Haar measure on the group $T_k = T$ (dx_k coinsides with the normalized Lebesgue measure on the interval $|-\pi,\pi|$ by the natural identification $T\simeq|-\pi,\pi|$); $\hat{T}^\infty$ is the dual group of T^∞ which can be identified with $Z^{(\infty)}$ in the natural way; $\hat{\mu}(\theta)$, $\hat{f}(\theta)$, $\theta\in Z^{(\infty)}$, denote the Fourier transforms of a measure μ and a function f respectively; D is the linear space of cylindric infinitely differentiable functions on T^∞; D^* is the space of distributions on T^∞; C is the Banach space of continuous functions on

T^∞; L_p is the space of functions on T^∞ with integrable p-th power with respect to the Haar measure.

Let us consider a Gaussian measure μ on T^∞ such that $\hat{\mu}(\theta) = \exp(-\psi(\theta))$, $\theta \in Z^{(\infty)}$, where the quadratic form $\psi(\theta) = a_1\theta_1^2 + a_2\theta_2^2 + \ldots$ is constructed by the sequence $A = (a_k)_1^\infty$ of positive numbers. Hence $\mu = \times_{k=1}^{\infty} \mu_k$, where μ_k is a Gaussian measure on $T_k = T$:

$$\mu_k(dx_k) = (\pi/a_k)^{1/2} \sum_{n\in Z} e^{-(x_k-2\pi n)^2/(4a_k)} dx_k.$$

By Kakutani's theorem μ is either absolutely continuous with respect to Haar measure or singular. As shown by Berg,[2] it is absolutely continuous if and only if $\sum_{k=1}^{\infty} e^{-2a_k} < \infty$. In addition the measure μ has a continuous density $\mu(x)$ if and only if $\sum_{k=1}^{\infty} e^{-a_k} < \infty$. The function $\mu(x)$ is of the form

$$\mu(x) = \prod_{k=1}^{\infty} (\pi/a_k)^{1/2} \sum_{n\in Z} e^{-(x_k - 2\pi n)^2/(4a_k)}, \quad x = (x_k)_1^\infty.$$

For $t>0$ let us consider the Gaussian measure μ_t on T^∞ associated with the sequence $tA = (ta_k)_1^\infty$ and the linear operator $P_t\colon C \to C$, $P_t f = \mu_t * f$. The family $\{P_t\}_{t>0}$ forms a continuous semigroup of contractions on C. Let $(\mathcal{L}, \mathrm{dom}\mathcal{L})$ be the generator of $\{P_t\}_{t>0}$. Using the Fourier transform techniques we obtain that $D \subset \mathrm{dom}\,\mathcal{L}$, $\mathcal{L}\colon D \to D$ and

$$\mathcal{L}_0 := \mathcal{L}|_D = \sum_{k=1}^{\infty} a_k \partial_{kk}^2 \text{ where } \partial_k f = \frac{\partial f}{\partial x_k}.$$ The differential operator $\mathcal{L}_0$ is symmetric; we define its extention on D^* by the formula

$$\mathcal{L}_0 F(\phi) := F(\mathcal{L}_0\phi), \quad F \in D^*, \quad \phi \in D.$$

1. PROPERTIES OF THE GREEN MEASURE ON T^∞

We define now the Green measure $G(dx)$ and the potential Gf:

$$G(dx) = \int_0^\infty e^{-t}\mu_t(dx)dt, \quad Gf = G * f.$$

As shown by authors,[1] it follows that for every $f \in L_p$ $(1<p<\infty)$ and for $u = Gf$ the derivatives $\partial_i u$, $\partial_{ij}^2 u$ (in the distribution sense) belong to L_p and

$$\sum_{k=1}^{\infty} a_k \partial_k^2 u - u = -f$$

where the series converges in the norm of L_p.

From the definition of $\mathcal{L}_0$ we easy get the equality $\mathcal{L}_0 G - G = -\varepsilon_0$, i.e. the Green measure plays the role of fundamental solution. But in contrast to a finite dimensional case it may not always possess the regularity properties of the classical Green function. In general it may even be singular with respect to Haar measure. It is possible to show that the measure G(dx) is absolutely continuous with respect to Haar measure if and only if for each t>0 the measures μ_t are absolutely continuous with respect to Haar measure, i.e. if and only if for each t>0

$$\sum_{k=1}^{\infty} e^{-ta_k} < \infty.$$

But even in this case the set of terminals may be infinite.

Proposition 1 - For the absolute continuity of G(dx) with respect to Haar measure, for the continuity (in the extended sense) of its density G(x) on T^∞ and for the finiteness of G(x) on the set T^∞ {0} it is sufficient that the condition

$$N_1: \quad \lim_{n\to\infty} a_n/n = \infty$$

holds and it is necessary that $\limsup_{n \to \infty} a_n/n = \infty$.

Proof - From the condition N_1 it follows that for each t > o $\sum_{k=1}^{\infty} e^{-ta_k} < \infty$ and therefore for t > 0 the measure μ_t is absolutely continuous with respect to Haar measure and has the continuous density $\mu_t(x)$:

$$\mu_t(x) = \prod_{k=1}^{\infty} \mu_t^k(x_k), \quad x = (x_k)_1^\infty$$

$$\mu_t^k(x_k) = \left(\frac{\pi}{a_k t}\right)^{1/2} \sum_{n\in Z} \exp\left\{-\frac{(x_k - 2\pi n)^2}{4a_k t}\right\}.$$

Let us show that $\lim_{t\to 0} e^{-c/t}\mu_t(0) = 0 \quad (c > 0)$. We have (cf. Berg,[2])

$$\mu_t(0) = \prod_{k=1}^{\infty} (1 + 2\phi(a_k t)), \text{ where } \phi(t) = \sum_{n=1}^{\infty} e^{-tn^2}.$$

Hence $\mu_t(0) \leq \exp(2 \sum_{k=1}^{\infty} \phi(a_k t))$. For $\varepsilon > 0$ we find such $N \geq 1$ that for $n \geq N$ the inequality $a_n \geq n/\varepsilon$ is true. Taking into account the monotonicity of $\phi(t)$ and the inequality $\phi(t) \leq t^{-1/2}$ we obtain for some $A = A(\varepsilon) > 0$

$$\sum_{k=1}^{\infty}\phi(a_k t) = \sum_{k<N} + \sum_{k\geq N} \leq A/\sqrt{t} + \int_0^{\infty}\phi(xt/\varepsilon)dx \leq A/\sqrt{t} + 2\varepsilon/t.$$

Now let $c>0$ and $\varepsilon<c/8$. We have $e^{-c/t}\mu_t(0) \leq e^{-(c/2 - 2A\sqrt{t})/t}$ and consequently $\lim_{t\to 0} e^{-c/t}\mu_t(0) = 0$.

Let now $x \to y \neq 0$. Without loss of generality we suppose that $y_1 \neq 0$. Then for some b, $c>0$ and for each $t>0$ and x_1: $|x_1 - y_1| < |y_1|/2$ the inequality $\mu_t^1(x_1) \leq be^{-c/t}$ is true. Hence

$$\mu_t(x) = \prod_{k=1}^{\infty}\mu_t^k(x_k) \leq be^{-c/t}\prod_{k=2}^{\infty}\mu_t^k(0) \leq be^{-c/t}\mu_t(0).$$

By Lebesgue's theorem $\lim_{x\to y} G(x) = \int_0^{\infty} e^{-t}\lim_{x\to y}\mu_t(x)dt = 0$. The continuity (in the extended sense) in $x = 0$ is evident.

Let us prove the second part of the proposition. Let for some $N>0$ and for any $k = 1,2,\ldots$ the inequality $a_k < Nk$ holds. There exist constants $b,c > 0$ such that

$$\mu_t(0) \geq \prod_{k\geq 1}(1 + 2\phi(Nkt)) \geq \prod_{Nkt\leq 1}(1 + 2\phi(1)) \geq$$

$$\geq (1 + 2\phi(1))^{1/(Nt)-1} \geq b\exp(c/t).$$

We get for $x = (x_1, 0, 0,\ldots)$ and $x_1^2 < 4a_1c$:

$$G(x) \geq \int_0^1 e^{-t}\mu_t(x)dt \geq \text{const}\int_0^1 e^{-x_1^2/(4a_1t) + c/t}dt = +\infty.$$

Q.E.D.

2. BESSEL POTENTIALS ON T^{∞}

For any $\beta > 0$ we put (cf. Stein,[4] Ch.5):

$$J_\beta(dx) = \frac{1}{\Gamma(\beta/2)}\int_0^{\infty} e^{-t}t^{\beta/2 - 1}\mu_t(dx)dt.$$

This definition of Bessel measures on T^{∞} was used recently by A.V.Skorikov for estimations connected with S.L.Sobolev's lemma. We have

$$J_2(dx) = G(dx), \quad \hat{J}_\beta(\theta) = (1 + \psi(\theta))^{-\beta/2}.$$

<u>Proposition 2</u>- Suppose that the condition N_1 is satisfied. Then for every $\beta > 0$ the measure $J_\beta(dx)$ is absolutely continuous with respect to Haar measure and its density $J_\beta(x)$ is finite on the set T^{∞} {0}, $J_\beta(0) = +\infty$ and continuous on T^{∞} in the extended sense.

The proof is analogous to the proof of the proposition 1.

As in the classical case we define the Bessel potential $J_\beta f$ by the equality $J_\beta f := J_\beta * f$. We shall investigate the differentiability properties of the potentials $J_\beta f$, $f \in C$, and in particular of $J_2 f = Gf$. This potential as it was mentioned above is a solution (in the distribution sense) of the Poisson equation $\mathfrak{L} u - u = -f$. We shall prove that if some conditions on the coefficients $(a_k)_1^\infty$ and on the function f are satisfied then this solution is also the strong solution, i.e. it is twice continuous differentiable and the series $\mathfrak{L}\, Gf = \sum_{k=1}^{\infty} a_k \partial_k^2 Gf$ converges uniformly.

Let us prove two lemmas.

<u>Lemma 1</u>- If $\eta_t(x) = (\pi/t)^{1/2} \sum_{k \in Z} e^{-(x - 2\pi n)^2/(4t)}$, $x \in T$ ($\simeq |-\pi,\pi|$) and $t>0$, then the following estimate holds

$$|\partial(\ln \eta_t(x))| \leq \frac{\pi}{t}|\sin \frac{x}{2}| \, .$$

The proof is omitted.

Let $\nabla_A := (\sqrt{a_k}\,\partial_k)_1^\infty$.

<u>Lemma 2</u>- If the condition

$$N_\alpha : \lim_{k\to\infty} a_k/k^\alpha = \infty \quad (\alpha > 1)$$

holds then for each $t > 0$ and $0 < \gamma < 1 - 1/\alpha$

$$\| \, |\nabla_A \mu_t| \, \|_{L^1_\infty} \leq \text{const } t^{\gamma/2-1} .$$

<u>Proof</u>- We have $\mu_t(x) = \prod_{k=1} \eta_{a_k t}(x_k)$, $x = (x_k)_1^\infty$. Applying simple transformations and lemma 1 we obtain

$$\| \, |\nabla_A \mu_t| \, \|_{L^1} \leq \frac{\pi}{t} \int_{T^\infty} \Big(\sum_{k=1}^{\infty} \frac{1}{a_k} \sin^2 \frac{x_k}{2} \Big)^{1/2} \mu_t(dx).$$

Let $\eta_t^0(dx) = (4\pi t)^{-1/2} e^{-x^2/(4t)}$ be a measure on R^1 and $\mu_t^0(dx) := \times_{k=1}^{\infty} \eta^0_{a_k t}(dx_k)$ be a Gaussian measure on R^∞. It is easy to see that the integral on the right-hand side can be presented as the integral with respect to the measure $\mu_t^0(dx)$ (after periodic extension of integrand from $T^\infty = R^\infty/2\pi Z^\infty$ on R^∞). So we get:

$$\| \, |\nabla_A \mu_t| \, \|_{L^1} \leq \frac{\pi}{t} \int_{R^\infty} \Big(\sum^{\infty} \frac{1}{a_k} \sin^2 \frac{x_k}{2} \Big)^{1/2} \mu_t^0(dx) \leq$$

$$\leq \frac{\pi}{t}\left(\int_{R^\infty} \sum_{k=1}^{\infty} \frac{1}{a_k} \sin^2\frac{x_k}{2}\, \mu_t^0(dx)\right)^{1/2} = \frac{\pi}{t}\left(\int_{R^1} g(x)\eta_t^0(dx)\right)^{1/2}$$

where $g(x) := \sum_{k=1}^{\infty} \frac{1}{a_k} \sin^2 \frac{\sqrt{a_k}x}{2}$, $x \in R^1$. From the condition N_α it follows that $\sum_{k=1}^{\infty} 1/a_k^{1-\gamma} < \infty$. Hence

$$\frac{g(x)}{|x|^{2\gamma}} = \sum_{k=1}^{\infty} \frac{\sin^2(\sqrt{a_k}x/2)}{a_k\ |x|^{2\gamma}} =$$

$$= \sum_{k=1}^{\infty} \frac{\sin^{2\gamma}(\sqrt{a_k}x/2)}{(\sqrt{a_k}\ |x|)^{2\gamma}} \frac{\sin^{2-2\gamma}(\sqrt{a_k}x/2)}{a_k^{1-\gamma}} \leq \sum_{k=1}^{\infty} \frac{1}{2^{2\gamma}\ a_k^{1-\gamma}} < \infty.$$

From this we conclude

$$\|\ |\nabla_A\mu_t|\ \|_{L_1} \leq \text{const } t^{-1/2} \left(\int_{R^1} |x|^{2\gamma}\eta_t^0(dx)\right)^{1/2} \leq \text{const } t^{\gamma/2-1}.$$

Q.E.D.

Proposition 3 - Let the condition N_α $(\alpha > 1)$ be satisfied. Then for each $\beta > 1 + 1/\alpha$

$$\|\ |\nabla_A J_\beta|\ \|_{L_1} < \infty.$$

Proof - Choosing γ such that $\beta + \gamma > 2$ and applying the lemma 2 we get

$$\|\ |\nabla_A J_\beta|\ \|_{L_1} \leq \frac{1}{\Gamma(\beta/2)}\int_0^\infty e^{-t}t^{\beta/2-1}\|\ |\nabla_A\mu_t|\ \|_{L_1}dt \leq \text{const} \int_0^\infty e^{-t}t^{(\beta+\gamma)/2-2} < \infty.$$

Q.E.D.

Remark 1 - Suppose the condition N_1 is satisfied. Let us consider a sequence of positive numbers $B = (b_k)_1^\infty$ satisfying the condition $\sum_{k=1}^{\infty} b_k/a_k < \infty$ and let $\nabla_B := (\sqrt{b_k}\partial_k)_1^\infty$. The analysis of the proof of lemma 2 shows that the following inequality is fulfilled:

$$\|\ |\nabla_B\mu_t|\ \|_{L_1} \leq \text{const } t^{-1/2}.$$

Repeating the arguments from the proposition 3 we obtain that the inequality $\|\ |\nabla_B J_\beta|\ \|_{L_1} < \infty$ is true by this time for each $\beta>1$. This implies for each $\beta>1$ the continuous differentiability of the kernel $J_\beta(x)$ on the set T^∞ {0} and integrability of its derivatives $\partial_i J_\beta(x)$, $i=1,2,\ldots$ These facts are well compatible with the known finite dimensional results (cf. Stein,[4] Ch. 5).

Remark 2 - In some cases (if the sequence $(b_k)_1^\infty$ increases stronger than $(a_k)_1^\infty$) we have the integrability of $|\nabla_B J_\beta(x)|$. Let for example the condition N_α holds with $\alpha > 2$ and $b_k = a_k \Delta_k$, where $\Delta_k = k \ln^2 k$, $k = 2, 3, \ldots$ As in the proof of lemma 2

$$\| |\nabla_B \mu_t| \|_{L_1} \leq \text{const}\ t^{-1} \left(\int_{R^1} g(x) \eta_t^0(dx)\right)^{1/2}$$

where $g(x) = \sum_{k=1}^{\infty} \frac{\Delta_k}{a_k} \sin^2 \frac{\sqrt{a_k}\, x}{2}$. For each $0 < \gamma < 1 - 2/\alpha$

$$g(x)/|x|^{2\gamma} \leq \text{const} \sum_{k=1}^{\infty} \Delta_k / a_k^{1-\gamma}$$

and

$$\frac{\Delta_k}{a_k^{1-\gamma}} = (k^\alpha / a_k)^{1-\gamma} \frac{\ln^2 k}{k^{\alpha(1-\gamma)-1}} \leq$$

$$\leq \text{const} \frac{\ln^2 k}{k^{\alpha(1-\gamma)-1}} \leq \text{const}\ k^{-\delta}$$

where $\delta = \alpha(1 - \gamma)/2 > 1$. Therefore $\sum_{k=1}^{\infty} \Delta_k / a_k^{1-\gamma} < \infty$ and $g(x)/|x|^{2\gamma} \leq$ $\leq \text{const} < \infty$. From this as in the proposition 3 we conclude that for each $\beta > 1 + 2/\alpha$

$$\| |\nabla_B J_\beta| \|_{L_1} < \infty.$$

In particular if the condition

N_∞: $(\alpha > 1)$ $\lim_{k \to \infty} a_k / k^\alpha = \infty$

holds then $|\nabla_B J_\beta| \in L_1$ for each $\beta > 1$.

3. THE DIFFERENTIABILITY PROPERTIES OF POTENTIALS

Bessel potential J_β is a linear operator $J_\beta: C \to C$ whose norm is equal to 1. It is easy to see (for example with the help of Fourier transform) that $\text{Ker}\, J_\beta = \{0\}$ and so the inverse operator

$$\mathbf{B}^\beta := J_\beta^{-1}$$

is correctly defined. We remark simultaneously that if $\beta \leq 2$ then $\mathbf{B}^\beta = (I - \mathfrak{L})^{\beta/2}$ (see Yosida,[5]).

For each $\beta \geq 0$ we define a Banach space C_β C in the following way:

$C_0 := C$, $\|f\|_0 := \|f\|_C$, $C_\beta := \text{dom}\, \mathbf{B}^\beta$, $\|f\|_\beta := \|\mathbf{B}^\beta f\|_C$, $\beta > 0$. Since the family $\{J_\beta\}_{\beta > 0}$ forms a strongly continuous semigroup of contractions in the Banach space C then C_α C_β $(\alpha > \beta)$, $\mathbf{B}^{\alpha+\beta} f = \mathbf{B}^\alpha \mathbf{B}^\beta f$ $(f \in C_{\alpha+\beta})$ and the uniform closure of $\bigcup_{\alpha > 0} C_\alpha$ coincides with C_0. And at last we note that $D \quad C_\alpha, \mathcal{B}^\alpha(D) \quad D$ $(\alpha > 0)$.

Let e be a natural homomorphism which maps R^∞ into $T^\infty = R^\infty/2\pi Z$. For a vector $\tau = (\tau_k)_1^\infty \in R^\infty$ and for $t \in R^1$ we define $\tau_t := e(\tau \cdot t)$. It is obviously that $\{\tau_t\}_{t\in R^1}$ is one-parametric subgroup of T^∞. We shall say that the function f on T^∞ is differentiable at a point $x \in T^\infty$ in the direction of τ if the function $f_{\tau,x}(t) := f(x + \tau_t)$ is differentiable at $t = o$. The corresponding derivative we shall denote by $\partial_\tau f(x)$ and call the derivative in the direction of τ. The higher derivatives are defined analogously.

We introduce a Hilbert space $\mathcal{H}_A$:

$$\mathcal{H}_A := \{x = (x_i)_1^\infty \in R^\infty \mid \|x\|_A^2 := \sum_{k=1}^\infty x_k^2/a_k < \infty\}.$$

Proposition 4. Let the condition N_α be satisfied $(\alpha > 1)$. Then for each $u \in D$, $x,y \in \mathcal{H}_A$, $\gamma > 0$ and $\delta > \gamma + 2/\alpha$ the following inequality is fulfilled:

$$\|\partial^2_{xy} u\|_\gamma \leq \text{const } \|x\|_A \|y\|_A (\|\mathcal{L}_0 u\|_\delta + \|u\|_\delta).$$

Proof. Using the equality $u = J_\beta * \mathcal{B}^\beta u$ $(\beta > 1 + 1/\alpha,\ u \in D)$ we get

$$\mathcal{B}^\gamma \partial_x u = \partial_x J_\beta * \mathcal{B}^{\beta+\gamma} u \quad (\gamma \geq 0,\ x \in \mathcal{H}_A).$$

Then

$$\|\partial_x u\|_\gamma \leq \|\partial_x J_\beta\|_{L_1} \|u\|_{\beta+\gamma} \leq$$

$$\leq \|x\|_A \| |\nabla_A J_\beta| \|_{L_1} \|u\|_{\beta+\gamma} = \text{const } \|x\|_A \|u\|_{\beta+\gamma}$$

where by proposition 3

$$\text{const} = \| \nabla_A J_\beta | \|_{L_1} < \infty.$$

Applying the obtained inequality to the function $u := \partial_y u$ we have

$$\|\partial^2_{xy} u\|_\gamma \leq \text{const } \|x\|_A \|\partial_y u\|_{\beta+\gamma} \leq \text{const } \|x\|_A \|y\|_A \|u\|_{2\beta+\gamma}.$$

Taking into account that $\|u\|_{2\beta+\gamma} = \|\mathcal{B}^2 u\|_{2\beta-2+\gamma}$ and $\mathcal{B}^2 = I - \mathcal{L}$ we get

$$\|u\|_{2\beta+\gamma} \leq \|\mathcal{L}_0 u\|_{2\beta-2+\gamma} + \|u\|_{2\beta-2+\gamma}.$$

It remains to define $\delta := 2\beta - 2 + \gamma > \gamma + 2/\alpha$. Q.E.D.

Remark 1 (continuation). Suppose the condition N_1 is satisfied. Then (in previously introduced notations) we get

$$\|\partial^2_{xy} u\|_\gamma \leq \text{const } \|x\|_\beta \|y\|_\beta (\|\mathcal{L}_0 u\|_\delta + \|u\|_\delta),$$

where $B = (b_k)_1^\infty$, $\sum_{k=1}^\infty b_k/a_k < \infty$, $x,y \in \mathcal{H}_B$, $\delta > \gamma \geq 0$. In particular for each $i,j \geq 1$

$$\|\partial_i \partial_j u\|_\gamma \leq \text{const } (\|\mathcal{L}_0 u\|_\delta + \|u\|_\delta).$$

Remark 2 (continuation). Let the condition N_α holds ($\alpha > 2$). Then

$$\|\partial^2_{xy}u\|_\gamma \leq \text{const}(\|x\|_\beta \|y\|_\beta \|\mathcal{L}_0 u\|_\delta + \|u\|_\delta),$$

where $b_k = a_k\Delta_k$, $\Delta_k = k\ln^2 k$ ($k = 2, 3, \ldots$), $\Delta_1 = 1$ and $\gamma \geq 0$, $\delta > \gamma + 4/\alpha$. In particular if the condition N_∞ is satisfied then this inequality is true for each $\gamma \geq 0$, $\delta > \gamma$.

Theorem Suppose the condition N_∞ is satisfied. Then for every $\delta > 0$ and $f \in C_\delta$ the function $u := Gf$ is a strong solution of the equation $\mathcal{L}u - u = -f$, i.e. the function u is twice continuously differentiable (in directions of $\mathfrak{H}_A$),

$$\mathcal{L}u = \sum_{k=1}^{\infty} a_k \partial_k u$$

and the series in the right-hand side converges uniformly (and even in every C_γ for each $0 \leq \gamma < \delta$).

Proof. Let the sequence $(\phi_n)_1^\infty$ D converges weakly to ε_0 and let $f_n := f * \phi_n$, $u_n := Gf_n$. The sequences $(f_n)_1^\infty$ and $(u_n)_1^\infty$ belong to D and converge in C_δ to the functions f and u respectively. We conclude from the proposition 4 that the sequences $(\partial_x u_n)_1^\infty$ and $(\partial^2_{xy}u_n)_1^\infty$ are fundamental in C_γ. From this it follows in the usual way the continuous differentiability of u and its derivatives up to second order inclusively.

We apply now the remark 2 to the proof that the series $\sum_{k=1}^{\infty} a_k\partial_k^2 u$ converges in C_γ. We obtain

$$\left\|\sum_{k=m}^{n} a_k\partial_k^2 u\right\|_\gamma = \left\|\sum_{k=m}^{n}\frac{1}{\Delta_k} b_k\partial_k^2 u\right\|_\gamma \leq \max_{k\geq 1}\|b_k\partial_k^2 u\|_\gamma \sum_{k=m}^{n}\frac{1}{\Delta_k}.$$

Let $x = (0,\ldots,0,\sqrt{b_k}, 0,\ldots)$. Then $\|x\|_\beta = 1$ and therefore for any $k = 1, 2, \ldots$

$$\|b_k\partial_k^2 u\|_\gamma = \|\partial^2_{xx}u\|_\gamma \leq \text{const}\ \|f\|_\delta.$$

It remains to remark that the series $\sum_{k=1}^{\infty}\Delta_k^{-1}$ converges. Q.E.D.

Remark 3. The theorem remains true if the condition N_α is satisfied for $\alpha > 2$. But in addition the numbers δ and γ must be connected with the inequality $\delta > \gamma + 4/\alpha$. If the condition N_α is fulfilled for $1 \leq \alpha \leq 2$ then according to remark 1 the function $u = Gf$ is twice continuously differentiable in the direction of every finite dimensional subspace of R^∞ and the series converges in L_p for each $1 < p < \infty$ (cf. Bendikov, Pavlov,[1]).

REFERENCES

1. A.D.Bendikov, I.V.Pavlov, Diffusion processes on the group T^{∞} and elliptic equations with infinite numbers of variables, in: "Probability Theory and Mathematical Statistics," V.1: 145-169, Yu.V. Prohorov et al. ed., VNU Science Press BV, Utrecht (1987).
2. Ch.Berg, Potential theory on the infinite dimensional torus, Invent. Math. 32: 49-100 (1976).
3. H.Heyer, "Probability Measures on Locally Compact Groups," Springer-Verlag, Berlin - Heidelberg (1977).
4. E.M.Stein,"Singular Integrals and Differentiability Properties of Functions," Princeton Univ. Press, Guildford (1970).
5. K.Yosida, "Functional Analysis," Springer-Verlag, Berlin - Heidelberg (1980).

A CHARACTERIZATION OF ELEMENTARY POTENTIAL THEORY

Hedi Ben Saad

Faculté des Sciences
Dep. de Mathématiques
Campus Universitaire
Tunis-Belvédère, Tunesia

Klaus Janßen

Inst.f.Statistik u. Dokumentation
Universitätsstr. 1
D - 4000 Düsseldorf
Federal Republic of Germany

In the following (V_λ) will be a Ray resolvent on a compact metric space E with set D of non-branchpoints, $\mathcal{E}$ denotes the class of positive Borel functions on E, and $\mathcal{S}$ is the cone of functions on D, which are excessive for (V_λ). For simplification we assume $V1 := V_o 1$ to be continuous.

A resolvent $(U_\lambda)_{\lambda \geq o}$ is called subordinate to $(V_\lambda)_{\lambda \geq o}$ if $U_\lambda \leq V_\lambda$ for all λ; for short we also say that U is subordinate to V. For g in $\mathcal{E}$ the kernel M_g acts by multiplication by g on $\mathcal{E}$; obviously $(M_{g_\lambda})_{\lambda \geq o}$ is a resolvent for $g_\lambda := \frac{g}{1+\lambda g}$.

We prove that there exists a largest function h in $\mathcal{E}$ subordinate to V (i.e. M_h is subordinate to V). As a consequence, $\{h > o\}$ is the set of finely isolated points in D. We characterize elementary kernels by the representation $V = \Sigma P^n M_h$, where P is the kernel supported by D such that $V = M_h + PV$, and we show that every potential kernel associated with $\mathcal{S}$ is necessarily an elementary kernel (see theorem 8). Finally we give an example of a non-elementary potential kernel, for which the fine topology is discrete.

With every Ray resolvent there is a right process associated. Most of the results of this paper are known or follow easily, if the theory of right processes is used. We wanted to give purely analytic elementary proofs of the above results without using the machinery which is necessary for right processes.

We thank M.ITÔ for discussions on this subject.

We will start off with a simple but basic observation. In the following statement ${}^{\lambda}R^A$ will denote the kernel of sweeping on A for Borel sets A in E with respect to the resolvent $(V_{\lambda+\mu})_{\mu \geq o}$.

1. LEMMA- For every finite measurable partition $(A_1,\ldots,A_n)$ of E the kernels

$$U_\lambda := \sum_{i=1}^{n} (V_\lambda - {}^{\lambda}R^{CA_i}V_\lambda) \qquad (\lambda \geq o)$$

define a resolvent subordinate to (V_λ).

PROOF - For $1 \leq i \leq n$ the kernel $V-R^{CA_i}V$ is supported by A_i and vanishes on CA_i, hence we obtain for $1 \leq j \leq n$

$$\left[\sum_{i=1}^{n} (V-R^{CA_i}V)\right](V_\lambda - {}^{\lambda}R^{CA_j}V_\lambda) = (V-R^{CA_j}V)(V_\lambda - {}^{\lambda}R^{CA_j}V_\lambda).$$

Since the resolvent of $V-R^{CA_i}V$ ist given by $(V_\lambda - {}^{\lambda}R^{CA_i}V_\lambda)_{\lambda \geq o}$ (c.f. [1]), we obtain for $\lambda \geq o$

$$\lambda U U_\lambda = \lambda \sum_{j=1}^{n} (V-R^{CA_j}V)(V_\lambda - {}^{\lambda}R^{CA_j}V_\lambda) =$$

$$= \sum_{j=1}^{n} [(V-R^{CA_j}V) - (V_\lambda - {}^{\lambda}R^{CA_j}V_\lambda)] = U - U_\lambda .$$

Consequently, (U_λ) is a resolvent. Moreover, for bounded f in $\mathcal{E}$ and x in E we have x in some A_j, hence

$$U_\lambda f(x) = \sum_{i=1}^{n} (V_\lambda - {}^{\lambda}R^{CA_i}V) f(x) = (V_\lambda - {}^{\lambda}R^{CA_j}V_\lambda) f(x)$$

$$\leq V_\lambda f(x),$$

which proves subordination of (U_λ) to (V_λ).

2. REMARK- Defining $h_i := \infty 1_{A_i}$ for $1 \leq i \leq n$, we conclude that the resolvent of

$$\sum_{i=1}^{n} V_{h_i} = \sum_{i=1}^{n} (V-\hat{R}^{CA_i}V) = \hat{U}$$

is exactly subordinate to (V_λ) (c.f.[3] p.117).

3. THEOREM- There exists a largest function h in $\mathcal{E}$ which is subordinate to V.

PROOF - Let $(A_1^1,\dots,A_{n_1}^1)$ be a finite measurable partition of E such that for all i the diameter $\delta(A_i^1) \leqq 1$, and let $U^1 := \sum_{i=1}^{n_1} (V - R^{CA_i}V)$. Consider for $1 \leqq i \leqq n_1$ a finite measurable partition $\{A_j^2 : j \in I(i)\}$ of A_i^1 by sets of diameter smaller than $\frac{1}{2}$. Then $\sum_{j\in I(i)} (V - R^{CA_j^2} V)$ is subordinate to $V - R^{CA_i^1} V$, and consequently, $U^2 := \sum_{\text{all } j} (V - R^{CA_j^1} V)$ is subordinate to U^1. Continuing this procedure, we obtain for every $n \in \mathbb{N}$ a resolvent (U_λ^n) such that $U_\lambda^{n+1} \leqq U_\lambda^n \leqq V_\lambda$ for all $\lambda \geqq o$. Then $U_\lambda^\infty := \inf_n U_\lambda^n$ determines a resolvent with U^∞ subordinate to V. For $x \in E$ and $n \in \mathbb{N}$ the measure $U^n(x,\cdot)$ is supported by a set of diameter $\frac{1}{2^n}$ containing x, hence $U^\infty(x,\cdot) = h(x)\varepsilon_x$ for some function h on E.
This shows that $h = U^\infty 1$ is a Borel function bounded by V1.

To prove that h does not depend on a particular choice of partitions, and to prove the theorem as stated, let $g \in \mathcal{E}$ such that M_g ist subordinate to V. According to [1] we have for all i, n that

$M_g M_{1_{A_i^n}}$ is subordinate to $V - R^{CA_i^n} V$ (where $\{A_i^n : i \in I_n\}$ is the decomposition of E used above for the construction of U^n), hence M_g is subordinate to M_h, which implies $g \leqq h$.

In the following h will always be the above largest function which is subordinate to V; P will be the unique kernel on E which is supported by D and satisfies $V = M_h + PV$.

4. COROLLARY- For every x in E we have $P(x,.) = R^{C\{x\}}(x,.)$ and $h(x) = V1(x) - R^{C\{x\}}V1(x)$.

In particular, the mapping $x \to R^{C\{x\}}V1(x)$ is measurable.

PROOF - Start the above construction of h with a partition where $A_1^1 = \{x\}$. Then we have for all n in $\mathbb{N}$ and bounded f in $\mathcal{E}$

$$U^n f(x) = Vf(x) - R^{C\{x\}}Vf(x),$$

which gives the result.

5. LEMMA- $\{x\}$ is finely open if and only if $R^{C\{x\}}V1(x) < V1(x)$.

PROOF - If $\{x\}$ is finely open, then the conclusion holds. To prove the converse, consider $O := D \cap \{\hat{R}^{C\{x\}}V1 < V1\}$. Then $x \in O$ and O is finely open, hence

$$O \cap C\{x\} = \{\hat{R}^{C\{x\}}V1 < R^{C\{x\}}V1\} \cap D$$

is finely open and V-negligeable, hence empty. This proves $O=\{x\}$, hence $\{x\}$ is finely open.

6. COROLLARY- $\{h > o\}$ is the set of finely isolated points in D.

7. THEOREM- Let g be subordinate to V, and let Q be a kernel on E such that Q is supported by D and $V = M_g + QV$. The following statements are equivalent:

a) $g = h$

b) $Q1_{\{x\}}(x) = O$ whenever $h(x) > O$.

PROOF - a) $\Rightarrow$ b) Assume that $h(x) > O$ and $a := Q1_{\{x\}}(x) > O$ for some x in E. We have

$$Qf(x) \geqq f(x)Q1_{\{x\}}(x) = a\,f(x) \qquad \text{for } f \text{ in } \mathcal{E},$$

hence $Q \geqq a1_{\{x\}}$, and $Q' := Q - a1_{\{x\}}$ is a kernel on E. From $V = M_h + PV$ we conclude

$$V = M_g + Q'V + a1_{\{x\}}V = M_g + Q'V + a1_{\{x\}}(M_h + PV)$$

$$= M_{(g+a1_{\{x\}}h)} + (Q' + a1_{\{x\}}P)\,V.$$

$Q' + a1_{\{x\}}P$ is a kernel; since every function in $\mathcal{E}$ is $M_{g+a1_{\{x\}}h}$ - supermedian, we conclude from [1] that the function $g+a1_{\{x\}}h$ is subordinate to V. Consequently $g+a1_{\{x\}}h \leqq h$, which implies $g(x) < h(x)$.

b)$\Rightarrow$ a) We always have $g \leqq h$. To prove equality, let x in E such that $h(x) > o$. Since g is subordinate to V, we conclude for $\varphi_n := n1_{C\{x\}}$ from [1]

$$V_{\varphi_n} = (M_g)_{\varphi_n} + [I-(M_g)_{\varphi_n} M_{\varphi_n}]\,QV_{\varphi_n},$$

and $V_{\varphi_n}1 \to V1-\hat{R}^{C\{x\}}V1 = h(x)1_{\{x\}}$ for $n \to \infty$;

in particular, we have $QV_{\varphi_n}1(x) \to h(x)Q1_{\{x\}}(x) = O$.

Now $(M_g)_{\varphi_n} = M_{\frac{g}{1+g\varphi_n}} \to M_{g1_{\{x\}}}$, and $I-(M_g)_{\varphi_n} M_{\varphi_n} = M_{\frac{1}{1+g\varphi_n}} \to M_{1_{\{g=o\}\cup\{x\}}}$ implies $g(x) = h(x)$.

DEFINITION- V is called an *elementary kernel* if $V = \sum_{n\geq o} Q^n M_g$ for a kernel Q on E and a function g in $\mathcal{E}$.

For $g = 1$, this notion coincides with the classical notion of an elementary kernel.

Remember that $\mathcal{S}$ denotes the cone of functions on D which are excessive with respect to (V_λ).

8. THEOREM- Let h be the largest function subordinate to V and let P be the unique kernel supported by D such that $V = M_h + PV$.

The following properties are equivalent:

a) V is an elementary kernel.

b) $V = \sum_{n\geq o} P^n M_h$.

c) $\lim_n P^n V1 = 0$.

d) $\lim_n P^n s = 0$ for some strictly positive s in $\mathcal{S}$.

e) Let $(W_\lambda)_{\lambda\geq o}$ be any resolvent on D such that the initial kernel W is proper. If $\mathcal{S}$ is the cone of excessive functions with respect to (W_λ), then W is an elementary kernel. Moreover, $W = \sum_{n\geq o} P^n M_g$ for g defined by

$g(x) := W1_{\{x\}}(x) - PW1_{\{x\}}(x)$ for x in D.

PROOF- a) ⇒ b) Assume that

$V = \sum_{n\geq o} Q^n M_g$ for some g in $\mathcal{E}$ and a kernel Q on E.

Then we have for every n in $\mathbb{N}$

$$V = \sum_{k<n} Q^k M_g + Q^n V = \sum_{k<n} P^k M_h + P^n V .$$

Since $Q^n V1 \to 0$ by assumption, it is enough to prove $P^n V1 \leq Q^n V1$ for all n. This will be accomplished in the

following way: We show by induction, that for every V- excessive function u the function $P^n u$ is V-supermedian and below $Q^n u$. To prove this for $n=1$ we observe that g is subordinate to V, hence $g \leq h$ by theorem 3, and consequently $M_g + QV = M_h + PV$ implies $PV \leq QV$. This gives $Pu \leq Qu$ for every excessive function u. Moreover, since h is subordinate to V, the function PVf is supermedian for every f in E (c.f. proposition III.4.7 in [3]), hence Pu is supermedian for every excessive function u.
If the above statement is true for some n in $\mathbb{N}$, then we obtain for every excessive function u

$$P^{n+1}u = P(P^n u) = P(\widehat{P^n u}),$$

since P is supported by D, and since the fine topology is discrete. Consequently $P^{n+1}u$ is supermedian, and the above statement follows for $n+1$ from

$$P^{n+1}u = P(\widehat{P^n u}) \leq Q(\widehat{P^n u}) \leq Q(Q^n u) = Q^{n+1}u.$$

b) ⇒ c) ⇒ d) are obvious.

d) ⇒ e) obviously we have $W = M_g + PW$ for g defined by $g(x) = W1_{\{x\}}(x) - R^{C\{x\}}\, W1_{\{x\}}(x)$ $(x \in D)$, hence $W = \sum_{k<n} P^k M_g + P^n W$ for all n. In particular, we have $\sum_{k \geq 0} P^k M_g \leq W$, and we have equality of these two kernels if for some strictly positive f on D the equality $\lim P^n Wf = 0$ holds. Using the domination principle for W, we construct for s of the assumption d) a strictly positive f such that $Wf \leq s$ and thus obtain e):

For $A_n := \{n\, s \geq 1\}$ we have $W1_{A_n} \leq n\, s$ on A_n, hence on D by the domination principle.
Obviously, $f := \sum_{n \geq 1} \frac{1}{n2^n} 1_{A_n}$ is strictly positive and $Wf \leq s$.
e) ⇒ a) is obvious.

9. EXAMPLE - There exists a Ray resolvent (V_λ) such that the fine topology is discrete and V is not an elementary kernel.

Therefor let $E := \mathbb{N} \,\bar{\cup}\, \{\infty\} \cup \{a\}$, where a is an isolated extra point. Choose h in $\mathcal{E}$ such that $h(\infty) = o$, h is strictly positive on $E \setminus \{\infty\}$ and $\sum_{x \in E} h(x)$ is finite.

Let P be the kernel on E defined by $Pf(n) = f(n+1)$ for n in $\mathbb{N}$, $Pf(\infty) = f(a)$, and $Pf(a) = 0$ for f in $\mathcal{E}$.

Define the kernel Q on E by $Qf(n) = f(\infty)$ for n in $\mathbb{N}$ and $Qf(\infty) = o = Qf(a)$ for f in $\mathcal{E}$. Then it is easily verified that $V := (I+Q) \sum_{n \geq o} P^n M_h$ is the initial kernel of a Ray resolvent on E.

Moreover, $PQ=Q$ and $QM_h = 0$ imply $V = M_h + PV$, hence h is the largest function which is subordinate to V according to theorem 7. According to theorem 8 V is not an elementary kernel. Since the excessive functions are exactly those functions, which "decrease on E", the fine topology is discrete.

10. REMARK - a) It can be shown, that in the above example there exists no elementary kernel which has the same excessive functions as V.

b) If the fine topology on D is discrete, then

$$V = \sum P^\alpha M_h$$

where the sum is taken over all countable ordinal numbers, h denotes the greatest function subordinate to V, and the kernels P^α are defined in the following way:

P^1 is the kernel supported by D satisfying

$$V = M_h + P^1 V;$$

if $\alpha = \beta + 1$ for some countable ordinal number, then $P^\alpha := P^\beta P$; if $\alpha = \sup \{\beta : \beta < \alpha\}$, then there exists a unique kernel P^α supported by D such that $P^\alpha V f = \lim_{\beta \uparrow \alpha} P^\beta V f$ for f in $\mathcal{E}$.

It follows for each x in E

$$V1(x) = \sum_{\alpha<\alpha(x)} P^{\alpha}h(x)$$

for a countable ordinal number $\alpha(x)$, but it is not clear, if $\sup\{\alpha(x) : x \in E\}$ is a countable ordinal number. In the above example we are exactly in this situation.

REFERENCES

1. H. Ben Saad: Généralisation des noyaux V_h et applications, Lecture Notes Math. 1061, Springer, Berlin (1984), p. 14-39.

2. H. Ben Saad: Exemples de noyaux admettant des résolvantes Math.Ann. 265 (1983), p. 149-154.

3. R.M. Blumenthal, R.K. Getoor: Markov processes and potential theory, Academic Press, New-York and London (1968).

4. P.A. Meyer: Probabilités et potentiel, Hermann, Paris (1966)

5. G. Mokobodzki: Ensembles compacts de fonctions fortement surmédianes, Lecture Notes Math. 713, Springer, Berlin (1979), p. 178-193.

WIENER ESTIMATES FOR SOLUTIONS OF ELLIPTIC EQUATIONS IN NONDIVERGENCE FORM

Marco Biroli

Department of Mathematics
Politecnico di Milano
Milano, Italy

1. INTRODUCTION AND RESULTS

In this paper we deal with a Wiener estimate of the modulus of continuity at the boundary for the solutions of elliptic problems in nondivergence form. It is well known that the Wiener criterion for the continuity of the Laplace equation at a point of the boundary was given at first by Wiener (1923) and was extended to general linear elliptic (second order) equations in divergence form by Littman, Stampacchia, Weinberger (1963) (for the intermediate stages see the references of this last paper).

The proof of Wiener criterion given in those papers is a qualitative one and gives no estimates on the modulus of continuity (in particular on the dependence of the modulus of continuity on the rate of divergence of the Wiener integral).

A sharp estimate on the modulus of continuity at the boundary by the Wiener integral was firstly given by Maz'ja (1967) and subsequently extended to some nonlinear second order elliptic equations (Maz'ja, 1976) and to the case of bilaplacian (Maz'ja, 1977). Such an estimate has been proved for general second order elliptic equations (in divergence form) by Gariepy, Ziemer (1977).

It is well known that the "regular" points of the boundary for a second order elliptic equation in nondivergence form with Hölder continuous coefficients are the same as for the Laplace equation (see for ex. Ancona, 1978). In the case of continuous coefficients such a result is false (Miller, 1970) and sufficient conditions for the regularity of a point at the boundary has been given by Landis (1968).

Recently Bauman (1985) has proved a Wiener criterion (necessary and sufficient condition) for the regularity of a point at the boundary. Bauman's proof is a qualitative one, then no estmates on the modulus of continuity in terms of the rate of divergence of the Wiener integral are obtained.

The goal of this paper is to prove such an estimate.

Now we state precisely the framework of the paper.

Let $\Omega \subset R^N$ be a bounded open set with boundary Γ and let a_{ij} be continuous functions such that:

(1.1.) $$\sum_{i,j=1}^{N} a_{ij}(x)\,\xi_i\xi_j \geq \alpha\,|\xi|^2 \ (\alpha>0), \quad a_{ij}=a_{ji}$$

We consider the problem

(1.2.) $$\begin{aligned} &Lu = 0 \ \text{in } \Omega \qquad \Big(Lv = \sum_{i,j=1}^{N} a_{ij} D^2_{ij} v \Big) \\ &u = g \ \text{on } \Gamma, \quad g \in H^{2,p}(\Omega) \qquad p \in (N,+\infty) \end{aligned}$$

We recall at first some properties of the Green function of the problem (1.2.) in a ball proved by Bauman (1985). Let B be a ball in R^N and denote by $G(x,y)=G^x(y)$ the Green function of the problem (1.2.) in B with singularity in x (i.e. if Lu=f in B , $f \in L^p(B)$ p>N and u=0 on B , then $u(x) = \int_B G(x,y)f(y)\,dy$) ; G(x,y) is a nonegative function and has a norm in the space $L^{N/N-1}(B)$ with respect to the variable y , which is uniformly bounded with respect to x in B . We define the normalized Green function relative to an open set ω, $\bar\omega \subset B$ (G_ω Green function of ω) by

$$\tilde G_\omega(x,y) = \tilde G^x_\omega(y) = G_\omega(x,y)\,(\,G(\bar x,y)\,)^{-1} \qquad (\,\tilde G = \tilde G_B\,)$$

where $\bar x$ is a fixed point in B . Let now B' be a ball with closure $\bar B'$ contained in B and suppose $\bar x \in B-\bar B'$; then $\tilde G(x,y)$ is locally Hölder continuous in $B - \{x,\bar x\}$ and if $x,y \in B'$ the following estimate holds

$$\tilde G(x,y) \sim \int_{|x-y|}^{1} s\,(\,w(B(y,s))\,)^{-1}\,ds$$

where $w(.) = w(\bar x,.) = G(\bar x,y)$ and $w(E) = \int_E w\,dx$ (B(y,s) denotes the ball of center y and radius s) . Moreover w is an A_∞ weight (for the definition of Muckenhoupt's class A_p $p \in (1,+\infty]$ see for ex. Muckenhoupt (1974)). Finally the capacity relative to the operator L and B can be defined by the relation

(1.3.) $$\operatorname{cap}(K) = \inf \{\, \nu(K)\ ;\ \nu \text{ positive measure with } \int_B G(x,y)\,d\nu_y \geq 1 \,\}$$

where K is a compact set in B.
We come back now to the problem (1.2.) ; it is well known that , if Γ is smooth enough there is a unique solution to (1.2.) in $C(\bar\Omega) \cap H^{2,p}(\Omega)$, $p \in (N,+\infty)$.

We define now the solution of the problem (1.2.) in the case of a general boundary Γ. Let Ω_n be an increasing sequence of open sets with smooth boundary Γ_n , such that $\Omega_n \uparrow \Omega$. We denote by u_n the solution of the problem

(1.2$_n$.) $$Lu_n = 0 \ \text{in } \Omega_n, \quad u = g \ \text{on } \Gamma_n$$

<u>Definition 1</u> – <u>The function</u> u <u>is a solution of</u> (1.2.) <u>if there exists a a sequence</u> $\Omega_n \uparrow \Omega$ <u>such that</u>

$$\lim_{n\to\infty} u_n = u$$

<u>in</u> $H^{2,p}_{loc}(\Omega)$ <u>and in the weak * topology of</u> $L^\infty(\Omega)$ (we denote here by u_n also the extension of u_n to Ω by g).

Remark 1- We are able to prove that u_n converges to u in $L^q(\Omega)$, $q \in (1,+\infty)$; moreover, taking into account the Caccioppoli's inequality of Section 3. we can suppose that Du_n converges to Du in $L^2(\Omega;w)$ (w is as defined before with $\bar{x} \in B - \bar{B}'$, $\bar{\Omega}$ B') in the weak topology and using the boundary condition in (1.2.)(1.2_n) we are able to replace the weak by the strong convergence.

Definition 2. - A point x in Γ is a regular point iff every solution of (1.2.) is continuous at x.

P. Bauman (1985) has proved the following Wiener criterion :

Theorem 1. - Let $\bar{\Omega} \cap B'$; $x \in \Gamma$ is a regular point iff

(a) cap $\{x\} > 0$

(b) $\int_0^1 r\,(w(B(x,r)))^{-1}\,\mathrm{cap}(\Omega^c \cap B(x,r))\,dr = +\infty$

We observe that in the case (a) the condition (b) holds again, then we can restrict ourself to condition (b) ; moreover if $N \geq 3$ there are no points of positive capacity.. in the following we consider only the case $N \geq 3$, but the results can be easily extended to the case N=2.

The result we will prove is the following :

Theorem 2. - Let u be a solution of the problem (1.2.) (according to definition 1.) and $x \in \Gamma$; then the following estimate holds

$$\mathrm{osc}_{B(x,r)} u \leq C_1 \exp(-C_2 \int_r^R s(w(B(x,s)))^{-1} \mathrm{cap}(\Omega^c \cap B(x,s))\,ds)\, \mathrm{osc}_{B(x,R)} u + \mathrm{osc}_{B(x,R)}\, g .$$

We observe that our definition of solution of (1.2.) is different from Bauman's definition. Our definition follows the framework of Wiener (1923), Bauman's definition comes from potential theory and Perron's method.

In the second section of the paper we prove a suitable Poincaré's inequality and in the third section we sketch the proof of Theorem 2.

2. THE POINCARE'S INEQUALITY

We will give now the Poincaré's inequality, which we use in section 3. for the proof of Theorem 1. We observe at first that wLv can be defined in distribution sense for functions $v \in L^1_{loc}(\omega)$ with $wv^2 \in L^1(\omega)$ and $w|Dv|^2$ in $L^1(\omega)$ (where is a bounded open set with smooth boundary , $\bar{\omega} \subset B'$ and Dv denotes the distributional gradient of v) by means of the following relation

$$(2.1.) \quad \langle wLv, \phi\rangle = -\langle wv, L\phi\rangle + \int_\omega wDvD\phi\,dx \quad , \quad \phi \in C_0^\infty(\omega) .$$

In the following of this section we use the notations $B(0,r) = B(r)$, $S(r_1,r_2) = B(r_1)-B(r_2)$; given a function v in $\bar{v}$ is a constant such that

$$\mathrm{meas}\{x \in \omega ;\ (v-\bar{v})^+ = 0\} ,\quad \mathrm{meas}\{x \in \omega ;\ (v-\bar{v})^- = 0\} \geq 4^{-1}\mathrm{meas}(\omega)$$

Proposition 1 - Let ω be the ball B(2) (the anulus S(4,1)) and v be a function in $L^1_{loc}(\omega)$ with wv^2, $w|Dv|^2 \in L^1(\omega)$,$w\{L[(v-\bar{v})^\pm]^2\}^- \in L^1(\omega)$; then

$$(2.2) \qquad \int_{\tilde{\omega}} w\,(v-\bar{v})^{\pm\,2}\,dx \leq C[\int_\omega w\{L[(v-\bar{v})^\pm]^2\}^-\,dx + \int_\omega w|Dv|^2 dx\,]$$

where $\tilde{\omega} = B(1)$ ($\tilde{\omega} = S(3,2)$).
We give the proof in the case of the ball and for $(v-\bar{v})^+$; the other cases ($(v-\bar{v})^-$ or the case of the anulus are substantially analogous.

If the result of (2.2.) does not hold there exists a sequence v_n such that

(2.3.) $$\int_{B(1)} w[(v_n-\bar{v}_n)^+]^2\,dx = 1$$

(2.4.) $$\int_{B(2)} w\,\{L\,[(v_n-\bar{v}_n)^+]^2\}^-\,dx \to 0$$

(2.5.) $$\int_{B(2)} w\,|Dv_n|^2\,dx \to 0\ .$$

Let ψ be a function in $H^{2,p}(B(2))\cap H_0^1(B(2))$, $p>N$, we have

(2.6.) $$w\,L\{[(v_n-\bar{v}_n)^+]^2\,\psi\}\geq w[(v_n-\bar{v}_n)^+]^2\,L\psi - w\psi\{L[(v_n-\bar{v}_n)^+]^2\}^- -$$
$$-\sum_{i,j=1}^N a_{ij}\,D_i[(v_n-\bar{v}_n)^+]^2\,D_j\psi\ . \qquad (\psi\geq 0)$$

Choosing $\psi=\sigma$, where $L\sigma=\chi(B(2))$ ($\chi(B(2))$ = carachteristic function of $B(2)$), we obtain

(2.7.) $$w\,L\,\{[(v_n-\bar{v}_n)^+]^2\sigma\} + w\sigma\{\,L\,[(v_n-\bar{v}_n)^+]^2\}^- + C_1\,w|Dv|^2 \geq$$
$$\geq w\,[(v_n-\bar{v}_n)^+]^2\ .$$

We multiply (2.7.) by $\tilde{G}_{B(2)}$ and we rewrite (2.7.) as

(2.8.) $$G_{B(2)}(x,.)\,L\{[(v_n-\bar{v}_n)^+]^2\sigma\}+C_2\,G(x,.)\{\{L\,[(v_n-\bar{v}_n)^+]^2\}^- + G(x,.)|Dv|^2\}$$
$$\geq G_{B(2)}(x,.)[(v_n-\bar{v}_n)^+]^2\ .$$

From the lemma 2.6 in Bauman (1985) we have

(2.9.) $$\int_{B(2)} G(x,.)\,\{L\,[(v_n-\bar{v}_n)^+]^2\}^-dy \to 0$$ a.e. , then a. uniformly

(2.10.) $$\int_{B(2)} G(x,.)|Dv|^2\,dy \to 0$$ a.e. , then a. uniformly.

From the definition of $\bar{v}$ there exists a sequence $x_n \in B(1)$ such that the following relations hold

(2.11.) the integral in (2.9.)(2.10) converges uniformly for $x=x_n$

(2.12.) $$(v_n-\bar{v}_n)^+(x_n) = 0\ .$$

Integrating (2.8.) on $B(2)$ we obtain

(2.13.) $$\int_{B(2)} G_{B(2)}(x_n,.)\,[(v_n-\bar{v}_n)^+]^2\,dy \leq$$
$$\leq \int_{B(2)} G(x_n,.)\{\,\{L[(v_n-\bar{v}_n)^+]^2\}^- + |Dv|^2\,\}\,dy\ .$$

Then $(v_n-\bar{v}_n)^+$ converges to $\chi=0$ in $L^2(B(1))$ strongly.

From (2.3.) we obtain

$$\int_{B(1)} w\, \chi\, dx = 1 \quad ;$$

then a contradiction arises and the result is proved.

Corollary 1. - Let $v \in L^\infty(\omega)$ be as in Proposition 1. and suppose $w\{L[(v-\bar{v})^{\pm}]^2\}^- \geq -cw\,|Dv|^2$; then

$$\int_{\tilde{\omega}} w\, (v-\bar{v})^{\pm\, 2}\, dx \leq C \int_\omega w|Dv|^2\, dx \quad .$$

To prove the result is enough to estimate $w\, \{L[(v-\bar{v})^+]^2\}^-$.

A usual trick gives now the following refinement of the Poincaré's inequality :

Proposition 2. - Let v as in the Corollary 1. ; then

$$\int_{\tilde{\omega}} wv^2\, dx \leq C\, (\, cap(N)\,)^{-1} \int_\omega w|Dv|^2 dx \ ,$$

where N is the set $\{\, x \in \omega |\ \ v(x)=0\ \}$.

3. PROOF OF THEOREM 1

By a double approximation (with regular coefficients in Ω_n and as in the Definition 1. by smooth sets) we obtain the following Caccioppoli's inequality

$$(3.1.) \qquad \int_{B(\bar{x}\ ,3r)} |Dv|^2\ G^{\bar{x}}\ dx + v^2(\bar{x}) \leq C\ r^{-2} \int_{B(\bar{x},4r)-B(\bar{x},3r)} v^2 G^{\bar{x}}\ dx$$

where $v = (u-k)^{\pm}$, $\bar{x} \in B(x_o,r)$ $x_o \in \Gamma$, $\delta > 0$; here and in the following we suppose δ choosen such that v=0 on the intersection of $B(x_o,5r)$ with the boundary Γ.

Taking the supremum for $\bar{x} \in B(x_o,r)$, we obtain

$$(3.2\ ) \qquad \int_{B(x_o,r)} |Dv|^2 G^{x_o} dx + \sup_{B(x_o,r)} v^2 \leq$$

$$\leq C\ r^{-2} (\ \inf_{B(x_o,5r)-B(x_o,r)} \tilde{G}^{x_o}\ ) \ .$$

$$\cdot \int_{B(x_o,5r)-B(x_o,r)} w\ v^2\ dx$$

(To prove (3.2.) Corollary 1.6. and Theorem 2.3. of Bauman (1985) are also used).

Now we apply to (3.2.) the Poincaré's inequality of Proposition 2. and we obtain

$$(3.3.) \qquad \int_{B(x_o,r)} |Dv|^2\ G^{x_o}\ dx + \sup_{B(x_o,r)} v^2 \leq$$

$$\leq C\ r^{-2}\ w(B(x_o,r)\ (\ \inf_{B(x_o,5r)-B(x_o,r)} \tilde{G}^{x_o}\ ) .$$

$$\cdot \int_{B(x_o,\ 10r)\ -\ B(x_o,r/2)} |Dv|^2 w\ dx$$

(To obtain (3.3.) the estimate $|wL[(v-k)^{\pm}]^2 \geq -c\, w|Dv|^2$ is also used).

We can write (3.3) as

(3.4.) $$\int_{B(x_o,r)} |Dv|^2 G^{x_o} dx + \sup_{B(x_o,r)} v^2 \leq$$

$$\leq C\, \delta(r)^{-1} \int_{B(x_o,\, 20r)-B(x_o,r)} |Dv|^2 G^{x_o} dx$$

where

$$\delta(r) = r^2 (w(B(x_o,r))^{-1} \mathrm{cap}(\Omega^c \cap (B(x_o,5r)-B(x_o,r)))$$

(To obtain (3.4.) the estimates on the normalized Green function of Bauman (1985) are also used).

From (3.4.) the result follows as in Biroli, Marchi (1986).

Acknowledgment - The most part of the results in this paper has been obtained jointly with S. Marchi.

REFERENCES

Ancona, A., Principe de Harnack à la fontière et théorème de Fatou pour un opérateur elliptique dans un domaine lipscitzien, Ann. Inst. Fourier, 28, 4, 169-213.

Bauman, P., 1985, A Wiener test for nondivergence structure, second order elliptic equations, Indiana Un. Math. J., 34, 4, 825-844.

Biroli, M., and Marchi, S., 1986, Wiener estimates at boundary points for degenerate elliptic equations, Boll. U.M.I., 6, 5(B), 689-706.

Gariepy, R., and Ziemer, E., 1977 A regularity condition at the boundary for solutions of quasilinear elliptic equations, Arch. Rat. Mech. An., 67, 25-39.

Landis, E.M., 1968, s-capacity and the behavior of a solution of a second order elliptic equation with discontinuous coefficients in the neighborhood of a boundary point, Sov. Math. Dokl., 9, 582-586.

Littman, W., Stampacchia, G., and Weinberger, H., 1963, Regular points for elliptic equations with discontinuous coefficients. Ann. Sc. Norm. Sup. Pisa. 17, 45-79.

Maz'ja, V.G., 1967, Behavior near the boundary of solutions for a second order elliptic equation in divergence form, Math. Notes, 2, 610-617.

Maz'ja, V.G., 1976, On the continuity at a boundary point of the solution of quasilinear elliptic equations, Vestnik Leningrad Un. Math., 3, 225-242.

Maz'ja, V.G., 1977, On the behavior near the boundary of solutions of the Dirichlet problem for the biharmonic operator, Dokl. Akad. Nauk SSR, 235, 6, 1263-1266.

Wiener, N., 1924, The Dirichlet problem, J. Math. and Phys., 3, 127-146.

Wiener, N., 1924, Certain notions in potential theory, J. Math. and Phys., 3, 24-51.

*During this conference Prof. Ibragimov has informed me that he has obtained by different methods similar results in the case of Dini-continuous coeff.

APPROXIMATION BY CONTINUOUS POTENTIALS

Jürgen Bliedtner

Fachbereich Mathematik
Universität Frankfurt
D-6000 Frankfurt, FRG

Wolfhard Hansen

Fakultät für Mathematik
Universität Bielefeld
D-4800 Bielefeld, FRG

INTRODUCTION

In this note we improve theorems in [1] and [2] dealing with approximation of (super)harmonic functions by continuous potentials. That is, we intend to show that for every finely open set G of a balayage space $(X,\mathcal{W})$ there exists a continuous potential $q \in \mathcal{P}$ such that

$$S(G) = \overline{\mathcal{P} + \mathbb{R}q} \quad , \quad H(G) = \overline{H(q)} \; .$$

Here,

$S(G) = \{s \in C_{\mathcal{P}}(X) : s \text{ finely superharmonic on } G\}$,
$H(G) = S(G) \cap (- S(G))$,
$H(q) =$ linear space generated by q_K , $K \subset \beta(\complement G)$ compact.

To obtain the above identities we prove the simpliciality of the function cone $\mathcal{P} + \mathbb{R}q$, determine its representing measures, and apply methods developed in [2] .

1. PRELIMINARIES

In the sequel let $(X,\mathcal{W})$ be a balayage space. We shall use the notations of [2]; in particular, $\mathcal{P}$ denotes the convex cone of all continuous potentials, $C_{\mathcal{P}}(X)$ the space of all continuous real functions on X which are $\mathcal{P}$-bounded.

Let $\mathcal{T}$ be a family of $\mathcal{W}$-dilations and define

$$S(\mathcal{T}) = \{s \in C_{\mathcal{P}}(X) : Ts \leq s \quad \text{for every} \quad T \in \mathcal{T}\} ,$$
$$H(\mathcal{T}) = \{h \in C_{\mathcal{P}}(X) : Th = h \quad \text{for every} \quad T \in \mathcal{T}\} .$$

By [2] (section VII.5) $S(\mathcal{T})$ is a min-stable simplicial function cone such that $Ch_{S(\mathcal{T})}X = \beta(b(\mathcal{T}))$, where $b(\mathcal{T}) = \bigcap_{T \in \mathcal{T}} b(T)$, in particular $Ch_{S(\mathcal{T})}X$ is a basic set. The following case is of special interest: $G \subset X$ finely open, $\mathcal{T} = \{(\varepsilon_x^{\complement V})_{x \in X} : V \text{ finely open}, \ \overline{V} \subset G\}$. Then

$$S(\mathcal{T}) = S(G) \ , \ H(\mathcal{T}) = H(G) \quad \text{and} \quad Ch_{S(G)}X = \beta(\complement G) .$$

For a precise statement we shall recall the following topological definition: For every subset $\mathcal{F}$ of $C_{\mathcal{P}}(X)$ let $\overline{\mathcal{F}}$ denote the set of all $g \in C_{\mathcal{P}}(X)$ such that there exists $p_0 \in \mathcal{P}$ such that for every $\varepsilon > 0$ there exists a function $f \in \mathcal{F}$ satisfying $|f - g| \leq \varepsilon p_0$. Note that

$$\overline{S(\mathcal{T})} = S(\mathcal{T}) \quad \text{and} \quad \overline{H(\mathcal{T})} = H(\mathcal{T}) \ ;$$

furthermore by the general approximation theorem $\overline{\mathcal{P} - \mathcal{P}} = C_{\mathcal{P}}(X)$.

2. THE SPACE H(q)

For every continuous potential $q \in \mathcal{P}$ let $H(q)$ the linear space generated by the set $\{q_K : K \subset \delta(q) \text{ compact}\}$, where $\delta(q) = Ch_{\mathcal{P}+\mathbb{R}q}X$ is the fine support of q .

Let $\mathcal{T}$ be a family of $\mathcal{P}$-dilations. Since $Ch_{S(\mathcal{T})}X = \beta(b(\mathcal{T}))$ is basic, there exists by [2] (VI.8.6) a $q \in \mathcal{P}$ such that $\delta(q) = Ch_{S(\mathcal{T})}X$.

THEOREM– Let $q \in \mathcal{P}$ such that $\delta(q) = Ch_{S(\mathcal{T})}X$. Then

$$\overline{H(q)} = H(\mathcal{T}) .$$

Proof. Let $K \subset \delta(q)$ be compact. Since $R_{q_K}^{\delta(q)} = q_K$ we have $q_K \in H(\mathcal{T})$ by [2] (VII.5.2), hence $H(q) \subset H(\mathcal{T})$.

Suppose now conversely that $h \in H(\mathcal{T})$. Take $p_0 \in \mathcal{P}$ such that $p_0 > 0$ and $|h| \leq p_0$, choose $q_0 \in \mathcal{P}$ such that $q_0 > 0$ and $p_0 \in o(q_0)$, and fix $\varepsilon > 0$. Then by [2] (I.1.3) there exist $p', q' \in \mathcal{P}$ such that

$$(*) \qquad |h - (p' - q')| \leq \varepsilon q_o .$$

As in the proof of [2] (VII.9.1) we may assume that

$$p' + q' \leq a\, p_o$$

for some $a \geq 1$.

We fix a compact subset K of X such that

$$a\, p_o \leq \varepsilon q_o \quad \text{on} \quad \complement K .$$

Since by [2] (VII.5.7) $\varepsilon_x^{\delta(q)}(h) = h(x)$ for every $x \in X$, the inequality (*) implies that

$$|h - (R_{p'}^{\delta(q)} - R_{q'}^{\delta(q)})| \leq \varepsilon \cdot R_{q_o}^{\delta(q)} \leq \varepsilon q_o .$$

Let V denote the potential kernel associated with q . By [2] (II.7.12) there exist sequences $(f_n),(g_n)$ in $\mathcal{B}_b^+(X)$ such that

$$V f_n \uparrow R_{p'}^{\delta(q)} , \quad V g_n \uparrow R_{q'}^{\delta(q)}$$

and such that $S(f_n) \cup S(g_n)$ is a compact subset of $\delta(q)$ for all $n \in \mathbb{N}$.

By [2] (I.1.11) there exist $\alpha_1,\dots,\alpha_n,\beta_1,\dots,\beta_m \in \mathbb{R}_+$ such that $\sum_{i=1}^{n} \alpha_i = 1$, $\sum_{j=1}^{m} \beta_j = 1$ and $f = \sum_{i=1}^{n} \alpha_i f_i$, $g = \sum_{j=1}^{m} \beta_j g_j$ satisfy

$$|h - (V f - V g)| < 2\, \varepsilon q_o \quad \text{on} \quad K .$$

Clearly, $f,g \in \mathcal{B}_b^+(X)$ such that $V f \leq p'$, $V g \leq q'$.

Define

$$\mathcal{F} = \{ \sum_{i=1}^{m} a_i 1_{K_i} : m \in \mathbb{N} , \alpha_i \in \mathbb{R}_+ , K_i \subset \delta(q) \text{ compact}, i=1,\dots,m\} .$$

Then there exist increasing sequences $(\varphi_n),(\psi_n)$ in $\mathcal{F}$ such that $\varphi_n \leq f$, $\psi_n \leq g$ for all $n \in \mathbb{N}$ and

$$V \varphi_n \uparrow V f , \quad V \psi_n \uparrow V g .$$

Indeed, if (x_k) is dense in X then for every $k \in \mathbb{N}$ there exists $(\varphi_{kn})_n \subset F$ such that $(\varphi_{kn})_n$ is increasing, $\varphi_{kn} \leq f$ and $V\varphi_{kn}(x_k) \uparrow_n Vf(x_k)$. Define $\varphi_n = \sup\{\varphi_{kn} : k \leq n\}$. Then (φ_n) is increasing to $\varphi \leq f$ and $V\varphi(x_k) = Vf(x_k)$ for every $k \in \mathbb{N}$, hence $V\varphi = Vf$.

Applying [2] (I.1.11) again we obtain functions $\varphi, \psi \in F$ such that $\varphi \leq f$, $\psi \leq g$ and

$$|h - (V\varphi - V\psi)| < 2\varepsilon q_0 \quad \text{on} \quad K .$$

Since $|h| \leq p_0 \leq \varepsilon q_0$ on $\complement K$ and $V\varphi + V\psi \leq Vf + Vg \leq p' + q' \leq a p_0 \leq$ $\leq \varepsilon q_0$ on $\complement K$ we conclude that

$$|h - (V\varphi - V\psi)| \leq 2\varepsilon q_0 \quad \text{on} \quad X .$$

The proof is finished since obviously $V\varphi - V\psi \in H(q)$.

COROLLARY- Let $G \subset X$ be finely open. Then there exists a continuous potential $q \in P$ such that

$$\overline{H(q)} = H(G) .$$

REMARK- Let $q \in P$ be strict. Then the above theorem implies

$$\overline{H(q)} = C_P(X) .$$

3. THE CONVEX CONE $P + \mathbb{R}q$

Let $q \in P$. Then $P + \mathbb{R}q$ is a function cone. Furthermore, $P + \mathbb{R}q$ is min-stable. Indeed, let $p_1, p_2 \in P$ and $\alpha, \beta \in \mathbb{R}$. We may assume $\alpha \geq \beta$. Then

$$\inf(p_q + \alpha q \,,\, p_2 + \beta q) = \inf(p_1 + (\alpha - \beta)q_1, p_2) + \beta q \in P + \mathbb{R}q .$$

The following lemma generalizes the characteristic property of a strict potential to arbitrary continuous potentials.

LEMMA – Let $q \in P$ and $\mu, \nu \in M_+(X)$ such that $\mu(p) \leq \nu(p)$ for every $p \in P$, $\mu(q) = \nu(q) < +\infty$ and $\nu(\complement\, \delta(q)) = 0$. Then $\mu = \nu$.

Proof. Let $p \in P$ and define

$$F = \{p' \in P : p' \leq p \text{ and } p' \prec \lambda q \text{ for some } \lambda \geq 0\} .$$

By [2] (II.7.10) F is increasingly filtered such that $\sup F = R_p^{\delta(q)}$. For every $p' \in F$ we have $\mu(p') = \nu(p') < +\infty$ by assumption, hence

$$\nu(p) = \int R_p^{\delta(q)}\, d\nu = \sup \{\nu(p') : p' \in F\}$$

$$= \sup \{\mu(p') : p' \in F\} = \int R_p^{\delta(q)}\, d\mu \leq \mu(p) .$$

Thus $\nu(p) = \mu(p)$ and therefore $\mu = \nu$.

PROPOSITION– Let $q \in P$. Then $P + \mathbb{R}q$ is a simplicial cone such that for every $x \in X$, $\varepsilon_x^{\delta(q)}$ is the unique measure in $M_x(P + \mathbb{R}q)$ such that $\varepsilon_x^{\delta(q)}(\complement\, \delta(q)) = 0$. Furthermore,

$$M_x(P + \mathbb{R}q) = \{\mu \in M_x(P) : \mu^{\delta(q)} = \varepsilon_x^{\delta(q)}\} .$$

Proof. Let $x \in X$. Since $\varepsilon_x^{\delta(q)}(p) \leq p(x)$ for every $p \in P$ and $\varepsilon_x^{\delta(q)}(q) = R_q^{\delta(q)}(x) = q(x)$ we have $\varepsilon_x^{\delta(q)} \in M_x(P + \mathbb{R}q)$ and clearly $\varepsilon_x^{\delta(q)}(\complement\, \delta(q)) = 0$.

Now let $\mu \in M_x(P + \mathbb{R}q)$. Then $\mu \in M_x(P)$ and therefore $\mu^{\delta(q)}(p) \leq \varepsilon_x^{\delta(q)}(p)$ for every $p \in P$. Furthermore,

$$\mu^{\delta(q)}(q) = \int R_q^{\delta(q)}\, d\mu = \mu(q) = q(x) = \varepsilon_x^{\delta(q)}(q) < +\infty ,$$

hence $\mu^{\delta(q)} = \varepsilon_x^{\delta(q)}$ by the above lemma. If moreover $\mu(\complement\, \delta(q)) = 0$ then $\mu = \mu^{\delta(q)}$, thus $\mu = \varepsilon_x^{\delta(q)}$.

Let finally $\mu \in M_x(P)$ such that $\mu^{\delta(q)} = \varepsilon_x^{\delta(q)}$. Then

$$\mu(q) = \int R_q^{\delta(q)}\, d\mu = \mu^{\delta(q)}(q) = \varepsilon_x^{\delta(q)}(q) = q(x) ,$$

thus $\mu \in M_x(P + \mathbb{R}q)$.

The following lemma can be proved in the same way as (4.5) of [1] .

LEMMA. Let S_1, S_2 be min-stable function cones such that $P \subset S_i \subset C_P(X)$, $i = 1,2$. Then the following statements are equivalent:

(1) $\overline{S_1} = \overline{S_2}$.

(2) $M_x(S_1) = M_x(S_2)$ for every $x \in X$.

THEOREM Let $G \subset X$ be finely open and $q \in P$ such that $\delta(q) = \beta(\complement G)$. Then

$$\overline{P + \mathbb{R}q} = S(G) .$$

Proof. Let $x \in X$. By the above theorem and [2] (VII.9.5) we have $M_x(P + \mathbb{R}q) = M_x(S(G))$. Hence the assertion follows from the lemma.

REMARK If T is an arbitrary family of W-dilations then in general there does not exist a $q \in P$ such that $S(T) = \overline{P + \mathbb{R}q}$. Indeed, consider the harmonic space $(\mathbb{R}, W)$ of translation on $\mathbb{R}$ and let $T = \{T\}$, where $T(x,\cdot) = \varepsilon_x$ if $x \leq 0$ and $T(x,\cdot) = \varepsilon_0$ if $x \geq 0$. Then for every $q \in P$ such that $\delta(q) = \mathrm{Ch}_{S(T)}\mathbb{R} =]-\infty,0]$ we have $S(T) \underset{\neq}{\supset} S(]0,+\infty[) = \overline{P + \mathbb{R}q}$.

References

J. BLIEDTNER and W. HANSEN:

[1] Simplicial Cones in Potential Theory II (Approximation Theorems). Inventiones Math., 46 (1978), 255 - 275.

[2] Potential Theory - An Analytic and Probabilistic Approach to Balayage. Universitext. Springer-Verlag (1986).

THE SCHRÖDINGER EQUATION $\Delta u = \mu u$ IN A NEIGHBOURHOOD OF AN ISOLATED SINGULARITY OF μ

Abderrahman Boukricha

Département de Mathématiques
Faculté des Sciences de Tunis
1060 Tunis, Tunisia

In this paper we present new results about the Picard principle and the behaviour of the continuous solutions of the Schrödinger equation $\Delta u = \mu u$ at an isolated singularity of the measure μ. Using the axiomatic potential theory, a criterion for the integrability of μ is given.

Some of the results presented here were obtained in [3].

Let $U = \{x \in \mathbf{R}^n;\ 0 < |x| < 1\}$, $n \geq 2$, and $M(U)$ be the set of positive Radon measures μ on $\overline{U} \setminus \{0\}$ with the following property: For every open set $V \subset \overline{V} \subset \overline{U} \setminus \{0\}$ the function

$$P_V = \int G_t^V \mu(dt)$$

is continuous and real on V.

This property was introduced by B. Walsh for the definition of perturbation of harmonic spaces; see [7].

For every $\mu \in M(U)$ and $V \subset U$ we consider the following sets:

$${}^{\mu}H(V) = \{u \in C(V);\ \Delta u = \mu u \text{ in the distributional sense}\},$$

$${}^{\mu}H_0 = \{h \in {}^{\mu}H^{+}(U);\ \lim_{x \to z} h(x) = 0 \quad \forall z \in \partial\overline{U}\}.$$

1. Definition: Let $\mu \in M(U)$. We say that the Picard principle (in notation P.P.) is valid for μ, if the cone ${}^{\mu}H_0$ is generated by exactly one function.

The notion of P.P. was introduced by G. Bouligand and adopted by M. Nakai in the case where μ has a Hölder continuous density relatively to the Lebesgue measure λ in $\overline{U} \setminus \{0\}$.

We will denote by $M_p(U)$ the set of measures in $M(U)$ for which the P.P. is valid.

2. Definition: Let $\mu \in M_p(U)$. A subset $E \subset U$ is said to be μ-thin at zero iff E is minimally thin in the point $x = 0$ as an isolated point of the Martin boundary for the harmonic space $(U, {}^{\mu}H)$ (i.e. ${}^{\mu}\hat{R}_h^E \neq h$ for every $h \in {}^{\mu}H_0 \setminus \{0\}$; ${}^{\mu}\hat{R}_h^E$ is the reduced of h on the set E).

3. Proposition: Let $\mu \in M_p(U)$ be rotation free (i.e. $\mu(f \circ \sigma) = \mu(f)$ $\forall \sigma \in SO(n)$) and $E \subset U$. If there exists an open cone C with vertex at 0, a hyperplane H containing 0, and $r > 0$ such that:

$$\emptyset \neq C \cap B \cap B_r(0) \subset E,$$

then E is not μ-thin at zero; ($B_r(0) = \{x \in R^n; |x| < r\}$).

4. Theorem: Let $\lambda \in M(U)$ be rotation free. The following properties are equivalent:

1) $\lambda(U) < +\infty$

2) $\exists \alpha > 0$ and $E \subset U$ $\alpha/|t|^2$ -thin at zero such that $\lambda^*(U \setminus E) < +\infty$.

5. Lemma: Let $\mu \in M(U)$. Then

a) There exists exactly one bounded μ-harmonic function e_μ such that $\lim_{|x| \to 1} e_\mu(x) = 1$.

If μ is rotation free, then

b) There exists exactly one rotation free function $h_\mu \in {}^\mu H_0$ with $h_\mu(x) = 1$ for $|x| = 1/2$; furthermore $\lim_{x \to 0} h_\mu(x) = \infty$.

c) Let μ' be another rotation free measure in $M(U)$ such that $\mu \leq \mu'$. Then

$$\lim_{x \to 0} \frac{e_{\mu'}}{e_\mu}(x) \text{ and } \lim_{x \to 0} \frac{h_\mu}{h_{\mu'}}(x)$$

exist and both are in [0,1].

Proof: For a) see [2], b) and c) see [4].

6. Theorem of M. Brelot (cf. [5]): Let c be a positive continuous function on $\overline{U} \setminus \{0\}$ such that

$$\lim_{x \to 0} |x|^2 c(x) = +\infty$$

then $\lim |x|^{-\alpha} u(x) = 0$ for every $\alpha > 0$ and every bounded continuous solution of the equation $\Delta u = cu$.

7. Theorem (cf. [3]): Let $\mu \in M(U)$, $n > 2$. Then $\lim_{x \to 0} u(x) = 0$ for every bounded continuous function u such that $\Delta u = \mu u$ in the distributional sense iff

$$\int_V |t|^{2-n} \mu(dt) = +\infty$$

for every nonvoid fine neighbourhood V of the origin. If $n = 2$, we replace $|t|^{2-n}$ by $\log(1/|t|)$.

The following theorem gives a converse of the theorem of M. Brelot (Th. 6).

8. Theorem: Let $\mu \in M(U)$. If

$$\lim_{x \to 0} |x|^{-\alpha} e_\mu(x) = 0 \text{ for every } \alpha \geq 0;$$

then for every $\alpha \geq 0$ and $E \subset U$ $\alpha/|t|^2$-thin at the origin we have:

$$\int^*_{U \setminus E} \frac{1}{|t|^{n-2}} \left(\mu - \frac{\alpha}{|t|^2} \lambda\right)^+ (dt) = +\infty .$$

9. Theorem: Let $\mu, \mu' \in M(U)$ be rotation free such that $\mu \leq \mu'$. The following properties are equivalent:

1) $\lim_{x \to 0} \frac{e_{\mu'}}{e_\mu}(x) = 0;$

2) $\lim_{x \to 0} \frac{h_\mu}{h_{\mu'}}(x) = 0$;

3) $\int_U e_\mu(t) h_\mu(t)\,(\mu' - \mu)(dt) = +\infty$.

10. Theorem: Let $\mu \in M(U)$ be rotation free. Then the following properties are equivalent:

1) The Picard principle is valid for μ;

2) $\lim_{x \to 0} (h_\mu / h_{(\mu + (\alpha/|t|^2)\lambda)})(x) = 0$ for a suitable $\alpha \geq 0$;

3) $\lim_{x \to 0} (e_{(\mu + (\alpha/|t|^2)\lambda)} / e_\mu)(x) = 0$ for a suitable $\alpha \geq 0$;

4) $\int e_\mu(t)\, h_\mu(t)\, (1/|t|^2)\, dt = +\infty$.

Remark: 1) $\Longleftrightarrow$ 2). The same proof as in the case where μ is absolutely continuous relatively to the Lebesgue measure; see [4]. 1) $\Longleftrightarrow$ 3) is done by [6] in the case $n = 2$ and μ has a locally Hölder continuous density relatively to the Lebesgue measure on the set $\overline{U} \setminus \{0\}$.

11. Theorem: Let $\mu \in M(U)$ be rotation free. If the Picard principle is not valid for μ, then:

$\lim_{x \to 0} |x|^{-\alpha} e_\mu(x) = 0$ and $\lim_{x \to 0} |x|^{\alpha} h_\mu(x) = \infty$

for every $\alpha \geq 0$. Furthermore, $\liminf_{t \to 0} e_\mu(t)\, h_\mu(t) = 0$.

References

1. Bliedtner J., Hansen W.: Potential theory. An analytic and probabilistic approach to balayage. Universitext, Springer, Berlin-Heidelberg-New York-Tokyo, 1986.
2. Boukricha A.: Das Picard-Prinzip und verwandte Fragen bei Störung von harmonischen Räumen. Math. Ann. 239, 247-270(1979).
3. Boukricha A.: Principe de Picard et comportement des solutions continues de l´équation de Schrödinger, au voisinage d´une singularité isolée. Research Center Bielefeld-Bochum Stochastics. University of Bielefeld,FRG, Nr. 247(1987).
4. Boukricha A., Hueber H.: The Poisson space ${}^cP_X$ for $\Delta u = cu$ with rotation free c. Académie Royale de Belgique. Bulletin de la Classe des Sciences, 5-e série, Tome LXIV, 1978 - 10.
5. Brelot M.: Etude de l´équation de la chaleur $\Delta u = c(M)\, u(M), c(M) \geq 0$, au voisinage d´un point singulier du coefficient. Ann. Sci. Ecole Norm. Sup. 48, 153-246 (1931).
6. Nakai M.: A test for Picard principle for rotation free densities, J. Math. Soc. Japan 27, 412-431(1975).
7. Walsh B.: Perturbation of harmonic structures and an index-zero theorem. Ann. Inst. Fourier 20, 317-359(1970).

A CHARACTERIZATION OF THE FINE SHEAF PROPERTY

Aurel Cornea

Katholische Universität Eichstätt

INTRODUCTION

Throughout this article S will be a standard H-cone of functions on a set X. For definitions and notations see Springer Lecture Notes in Mathematics 853.

We call a specific band in S any subset $\mathcal{B}$ which is 1. a subcone, 2. specifically solid, and 3. specifically complete, i.e. for any family $\mathcal{F}$ in $\mathcal{B}$ the specific l.u.b. $\curlyvee\mathcal{F}$ belongs to $\mathcal{B}$ if it exists. If $u \in S$ put $\mathcal{B}u$ the specific band generated by u.

An element $u \in S$ is called a strongly charged unit if

$$u := \sum_{m,n} \alpha_{mn}(u_m \wedge u_n)$$

where α_{mn} is real strictly positive and $(u_n)_{n\in\mathbb{N}}$ is dense in order from below in S. Denote $\mathcal{C}(S) = \mathcal{C} := \bigcap_u \mathcal{B}u$ where u runs through the set of all strongly charged units of S. In this article we study this band $\mathcal{C}$ and show that the fine sheaf property is characterized by the fact that $\mathcal{C}$ reduces to the null element. If S is the cone of positive superharmonic functions on a harmonic space X, $A\subset X$ is a basic subset, and $T = \{s\in S : s = B_s^A\}$, then $\mathcal{C}(T)$ is a band generated of all elements $h\in T$ such that $h = B_s^A$ with $s \curlywedge B_s^A = 0$. Roughly speaking, s is "harmonic" on A. In the general case a set of generators for $\mathcal{C}$ is given by means of balayages.

1. THE OPERATOR O_B

Throughout this paragraph B will be a fixed balayage on S. We remember that b(B) denotes the base of B and d(B) denotes the domain of B.

For any $s\in S$ we denote $O_B s = s - s \curlywedge Bs$. It is easy to see that $O_B s$ is the projection of s onto the specific band $\{t\in S : t \curlywedge Bv = 0$ for any $v \in S\}$.

If s is nearly continuous then $O_B s = s_{d(B)}$ using the notation of 3.4. of SLN 853. The following proposition follows immediately from the additivity of O_B.

Proposition 1.1 - Let $s,t,u,v \in S$ and put $p = s + u \wedge v$, $q = u + s \wedge t$, then:

$$O_B(s \wedge t) + O_B u + O_B p = O_B(u \wedge v) + O_B s + O_B q.$$

Proposition 1.2 - Let $s,t \in S$ be such that $\{ x \in X : t(x) < s(x) \} \subset b(B)$, then

(a) $O_B(s \wedge t) = O_B s + O_B(t \wedge Bs)$

(b) if moreover $t \leq s$ then $O_B t = O_B s + O_B(t \wedge Bs)$

Proof - Take in proposition 1.1. $u = Bs$ and $v = t$. Then we have $p = q$ and $O_B u = 0$.

Proposition 1.3 - Let $s,t,u,v \in S$ be such that $\{ x \in X : t(x) < s(x) \} \subset b(B)$ and $(u-v)_+ \leq (s-t)_+$, then

(a) $O_B(s \wedge t) \succ O_B s + O_B(v \wedge Bu)$

(b) $O_B(t \wedge Bs) \succ O_B(v \wedge Bu)$.

Proof - Take p,q as in proposition 1.1. Then we have $p-q = (s-t)_+ - (u-v)_+$ hence $q \leq p$, $\{ x \in X : q(x) < p(x) \} \subset b(B)$, and $\{ x \in X : v(x) < u(x) \} \subset b(B)$. The assertion follows now applying the above proposition 1.2. (b).

For a sequence $(s_n)_{n \in \mathbb{N}}$ denote

$$\operatorname*{liminfspec}_{n \to \infty} s_n = \bigvee_n \left(\bigwedge_{j \geq n} s_j \right)$$

and

$$\operatorname*{limsupspec}_{n \to \infty} s_n = \bigwedge_n \left(\bigvee_{j \geq n} s_j \right)$$

whenever they make sense. If they are equal write $\operatorname*{limspec}_{n \to \infty} s_n$ for their common value.

Proposition 1.4 - Let $s_n \uparrow s$, $t_n \uparrow t$ be such that $\{x \in X : t_1(x) < s(x)\} \subset b(B)$, then we have $O_B(t \wedge Bs) = \operatorname*{limspec}_{n \to \infty} O_B(t_n \wedge Bs_n)$.

Proof - Denote $u_n = t_n \wedge Bs$, $v_n = t \wedge Bs_n$, $u = t \wedge Bs$. Obviously $u_n \uparrow u$, $v_n \uparrow u$, $\{x \in X : u_n(x) < u(x)\} \subset b(B)$. From proposition 1.2. (b) $O_B u_n \succ O_B u$. Replacing in this reasoning s with s_n we get $O_B(t_n \wedge Bs_n) \succ O_B v_n$. Further from $\{x \in X : t(x) < s(x)\} \subset b(B)$ and $s_n - t \leq s - t$ we get using proposition 1.3. (b) $O_B v_n \prec O_B u$ and analogously $O_B(t_n \wedge Bs_n) \prec O_B u_n$. From the following two lemmas we shall have $O_B u = \bigvee O_B v_n = \bigwedge O_B u_n$ which completes the proof.

Lemma 1 - $w_n \uparrow w$, $O_B w_n \prec O_B w \Rightarrow O_B w = \bigvee O_B w_n$.

Proof - Putting $v = \bigvee O_B w_n$ we have $v \prec O_B w$. From the definition of O_B,

$O_B q - B(O_B q) = q - Bq$ for any $q \in S$. Further $w - Bw = \lim_{n\to\infty}(w_n - Bw_n) = \lim_{n\to\infty}(O_B w_n - B(O_B w_n)) \leq v - Bv$. Hence $O_B w - B(O_B w) \leq v - Bv$ and from SLN 853, proposition 5.1.2., $O_B w \leq v$.

Lemma 2 — $w_n \uparrow w$, $O_B w_n \succ O_B w \Rightarrow O_B w = \curlywedge O_B w_n$.

Proof — Putting $v = \curlywedge O_B w_n$ we have $v \succ O_B w$ and $O_B v = v$. From $w_n \succ O_B w_n \succ v$ we get $w \succ v$, $O_B w \succ v$.

Proposition 1.5 — Let $(B_n)_{n \in \mathbb{N}}$ be a decreasing sequence of balayages on S such that $b(B) = \bigcap b(B_n)$. Then for any nearly continuous element $s \in S$ we have $O_B s = \curlyvee O_{B_n} s$.

Proof — Put $v_n = O_B s - O_{B_n} s$ and $v = \curlywedge v_n$. Since v is nearly continuous we have carr $v \subset b(B_n)$, carr $v \subset b(B)$. Thus $v = Bv$. Since $v \curlywedge Bv = 0$ we have $v = 0$.

2. THE OPERATOR Θ_B

Fix B a balayage on S. We remember that B' denotes the balayage on the fine open set $d(B)$.

For any $s \in S$ denote $\Theta_B s = O_B(Bs \wedge B's)$. From the relation $s + Bs \wedge B's = Bs + B's$ we see that $s \mapsto \Theta_B s$ is additive.

Proposition 2.1

(a) Let $s,t \in S$ be such that $s - B's \leq t - B't$, then $\Theta_B s \prec \Theta_B t$.

(b) If $s_n - B's_n \uparrow s - B's$ then $\Theta_B s = \curlyvee \Theta_B s_n$.

Proof — (a) From $s + B't \leq t + B's$ using SLN 853, theorem 5.1.6. (d), there exists $f \in \hat{S}_B$ such that on $d(B)$ we have $Bs + B(B't) + f = Bt + B(B's)$, hence $Bs - B(B's) + f = Bt - B(B't)$.

It is easy to see that for any $v \in S$ we have $Bv - B(B'v) = Bv \wedge B'v - B(Bv \wedge B'v)$ on $d(B)$. Hence $Bs \wedge B's - B(Bs \wedge B's) + f = Bt \wedge B't - B(Bt \wedge B't)$. Then we get that $f = u - Bu$ with $u \in S$ and $\Theta_B s + u = \Theta_B t$.

(b) Construct similarly u_n such that $\Theta_B s_n + u_n = \Theta_B s$. From the condition stated in (b) we must have $u_n - Bu_n \downarrow 0$ and therefore $u_n \downarrow 0$.

Proposition 2.2 — If $s_n \uparrow s$ then $\Theta_B s \prec \operatorname{liminfspec}_{n\to\infty} \Theta_B s_n$.

Proof — Denote $f_n = \bigwedge_{k\geq n}(s_k - B's_k) \in \hat{S}_{B'}$. Then there exists $t_n \in S$ such that $f_n = t_n - B't_n \uparrow s - B's$. The assertion follows from the preceding proposition.

Proposition 2.3 – Let p be generator in S. Then for any $s \in S$, $\theta_B s \in \mathcal{B}(\theta_B p)$.

Proof – Take $t_n \in S$ such that $t_n \curlywedge B't_n = 0$ and $t_n - B't_n = (s-B's) \wedge n(p-B'p)$. Since p is a generator we have $t_n - B't_n \uparrow s - B's$. Hence from proposition 2.1. (b): $\theta_B s = \curlyvee \theta_B t_n$. From $t_n - B't_n \leq n(p-B'p)$ and proposition 2.1. (a) $\theta_B t_n \prec n\theta_B p \in \mathcal{B}(\theta_B p)$ and therefore $\theta_B s = \curlyvee \theta_B t_n \in \mathcal{B}(\theta_B p)$.

Corollary 1 – If p and q are generators in S we have

$$\mathcal{B}(\theta_B p) = \mathcal{B}(\theta_B q).$$

We call quasi bounded any element s of S such that $s = \sum s_n$ where any $s_n \in S$ is dominated by a universally continuous element of S. This property is equivalent with the following one: for any unit $u \in S$ we have $\bigwedge R(s-nu) = 0$.

Corollary 2 – For any $s \in S$, $\theta_B s$ is quasi bounded.

Proof – Let $p = \Sigma s_n$ where $(s_n)_{n \in \mathbb{N}}$ is dense in order from below and any s_n is universally continuous. Then $\theta_B p = \Sigma \theta_B s_n$ is quasi bounded. It is easy to see that if u is quasi bounded then $\mathcal{B}u$ consists only of quasi bounded elements.

Remark – There are examples where $\theta_B s$ is not nearly continuous.

3. THE CHARGED ELEMENTS

Denote by $\Gamma(S) = \Gamma$ the specific band generated by the elements of the form $\theta_B s$ where B is a balayage on S and $s \in S$. The elements of Γ will be called charged elements. From proposition 2.3. we may take as generators of Γ all elements of the form $\theta_B p$ with B balayage and p a fixed generator of S. In this paragraph we investigate the relation between $\Gamma(S)$ and $\mathcal{C}(S)$.

Proposition 3.1 – Let $s \in S$ be nearly continuous and such that for any $t \in \Gamma$ we have $s \curlywedge t = 0$. Then if X has no isolated (for the natural topology) points one may construct a strongly charged unit u such that $s \curlywedge u = 0$.

We do the proof in three steps.

(i) Take $v, w \in S$ and denote

$$B_1 = B^{\beta(\{x \in X \,:\, v(x) \leq w(x),\ w(x) \neq 0\})}$$

$$B_2 = B^{\beta(\{x \in X \,:\, v(x) \geq w(x),\ v(x) \neq 0\})}$$

where $\beta(A)$ is the essential base of A, see SLN 853, § 5.3. We shall show that if $s \quad (v + w + B_1 s \wedge B_2 s) = 0$ then $s \quad (v \wedge w) = 0$.

From $\{x \in X : v(x) < w(x)\} \subset b(B_1)$ and proposition 1.2. (a) we have

$$O_B(v \wedge w) = O_{B_1} w + O_{B_1}(v \wedge B_1 w) .$$

Further again from proposition 1.2. (b) applied to the pair $v \wedge B_1 w$, $B_1 w \wedge B_1^{\dagger} w$ we get $O_B(v \wedge B_1 w) \prec O_{B_1}(B_1 w \wedge B_1^{\dagger} w) = \Theta_{B_1} w$. Now we conclude $s \curlywedge O_{B_1}(v \wedge w) = 0$ and analogously $s \curlywedge O_{B_2}(v \wedge w) = 0$.

It is easy to see that $v \wedge w = O_{B_1}(v \wedge w) + O_{B_2}(v \wedge w) + h$ where $h \in S$ is such that $h = B_1 h = B_2 h$. From this property of h and from the condition stated above we have $s \curlywedge h = 0$ and therefore $s \curlywedge (v \wedge w) = 0$.

(ii) Let v, w be such that $s \curlywedge (v+w) = 0$. Then except for countably many positive real numbers α we have $s \curlywedge (v \wedge \alpha w) = 0$.

In order to get this denote for any positive real α

$$B_{1,\alpha} = B^{\beta(\{x \in X : v(x) \leq \alpha w(x),\ w(x) \neq 0\})}$$

$$B_{2,\alpha} = B^{\beta(\{x \in X : v(x) \geq \alpha w(x),\ v(x) \neq 0\})}$$

$$s_\alpha = s \curlywedge B_{1,\alpha} s \curlywedge B_{2,\alpha} s .$$

We show first that if $\alpha \neq \gamma$ then $s_\alpha \curlywedge s_\gamma = 0$. Indeed take $h = s_\alpha \curlywedge s_\gamma$ then from $h \prec s$ we have that h is nearly continuous. From $h = B_{j,\alpha} h$ for $j = 1,2$ we get $\operatorname{carr} h \subset b(B_{1,\alpha} \cap B_{2,\alpha})$. Since this last set is emty we have $\operatorname{carr} h = \emptyset$, and therefore $h = 0$. We can now conclude that $s_\alpha + s_\gamma = s_\alpha \curlyvee s_\gamma \prec s$. Therefore there exist only countably many α with $s_\alpha \neq 0$. The assertion for step (ii) now follows from step (i).

(iii) We construct now the required unit. Take $(s_n)_{n \in \mathbb{N}}$ universally continuous and dense in order from below. Since X has no isolated points one may chose them so that $s \curlywedge s_n = 0$ for any $n \in \mathbb{N}$. Construct now inductively a sequence $(\beta_n)_{n \in \mathbb{N}}$ of real numbers with $\frac{1}{n} < \beta_n < 1$ and $s \curlywedge (\beta_j s_j \wedge \beta_n s_n) = 0$ for any $j \leq n$. If we put $u_n = \beta_n s_n$ we see that $(u_n)_{n \in \mathbb{N}}$ is dense in order from below. By a convenient choice of strictly positive real numbers $(\alpha_{nm})_{n,m \in \mathbb{N}}$ the series $\Sigma \alpha_{nm} u_n \wedge u_m$ is convergent in S and is the required strongly charged unit.

Remark — The step (ii) in the above proof may be done with other techniques for s quasi bounded provided that all nonzero elements of S are units in S. One may show then that under the above restriction $C \subset \Gamma$. For the proof of this inclusion see the proof of the following theorem 3.3.

Proposition 3.. — If the axiom of nearly continuity holds then $\Gamma \subset C$.

Proof — Let p be a quasi bounded generator in S, let $u = \Sigma \alpha_{mn} u_m \wedge u_n$ be a strongly charged unit, and let B be a balayage. We shall show that $\Theta_B p \in \mathcal{B}u$ which will complete the proof according to proposition 2.3.

Chose two subsequences $(s_n)_{n \in \mathbb{N}}$, $(t_n)_{n \in \mathbb{N}}$ of the sequence

$(u_n)_{n \in \mathbb{N}}$ such that $s_n \uparrow Bp$, $t_n \uparrow B'p$, fix $j \in \mathbb{N}$, and denote

$$B_j = B^{\{x \in X \,:\, t_j(x) < Bp(x)\}}$$

From proposition 1.2. (a) $O_{B_j}(t_n \wedge B_j s_n) \prec O_{B_j}(s_n \wedge t_n) \prec s_n \wedge t_n \in \mathcal{B}u$ for $n \geq j$. Using now proposition 1.4. and the fact that $B_j B = B$ we get $O_{B_j}(Bp \curlywedge B'p) \in \mathcal{B}u$. From proposition 1.5. we have $O_B(Bp \wedge B'p) = \curlyvee_j O_{B_j}(Bp \wedge B'p) \in \mathcal{B}u$. Hence $\theta_B p \in \mathcal{B}u$.

Theorem 3.3 – If X has no isolated points and the axiom of nearly continuity holds then $\Gamma(S) = \mathcal{C}(S)$.

Proof – According to the preceding proposition it is enough to show $\mathcal{C} \subset \Gamma$. Take $s \in \mathcal{C}$ and $t = \curlyvee\{v \in S : v \prec s,\ v \curlywedge w = 0$ for any $w \in \Gamma\}$. Since $t \in \mathcal{C}$ it is quasi bounded and since the axiom of nearly continuity holds it is nearly continuous. From proposition 3.1. there exists a strongly charged unit u with $t \curlywedge u = 0$. Thus $t = 0$ since $t \in \mathcal{B}u$. Hence $s \in \Gamma$.

4. THE FINE SHEAF PROPERTY

Theorem 4.1 – The following assertions are equivalent:

(a) The fine sheaf property holds in S.

(b) (i) For any universally continuous element $s \in S$ there exists a strongly charged unit u with $s \curlywedge u = 0$.

(ii) The axiom of nearly continuity holds in S.

(c) $\Gamma(S) = \{0\}$.

Proof – We have by SLN 853, theorem 5.6.10.:

(a) $\Leftrightarrow$ for any balayage B and any $x \in d(B)$ the measure $B^*(\varepsilon_x)$ charges only the fine boundary of $b(B)$

$\Leftrightarrow$ for any balayage B and any $s \in S$ we have $Bs - B(B's) = 0$ on $d(B)$

$\Leftrightarrow$ for any balayage B and any $s \in S$ we have $Bs \wedge B's - B(Bs \wedge B's) = 0$

$\Leftrightarrow$ for any balayage B we have $\theta_B = 0$

$\Leftrightarrow$ (c) .

The assertion (b) (i) follows from (c) using proposition 3.1. and (b) (ii) follows from (a) using SLN 853, proposition 5.6.8. Conversely from (b) (ii) and the fact that the elements of Γ are quasi bounded we see that they are nearly continuous. If $\Gamma \neq \{0\}$ then there exists a universally continuous $s \in \Gamma$, $s \neq 0$. Using now the unit u from (b) (i) and proposition 3.2. we get the contradictory relations $s \in \mathcal{C} \subset \mathcal{B}u$, $s \curlywedge u = 0$.

Let B be a balayage on S. We denote by $B(S)$ the H-cone of all elements $t \in S$ for which $t = Bt$. Remark that the g.l.b. operation in $B(S)$ denoted by $\wedge_B$ is given by $s \wedge_B t := B(s \wedge t)$. Using this computation of g.l.b., one can see that for any strongly charged unit u of S,

Bu is a strongly charged unit of $B(S)$. We denote by $\mathcal{L}(B(S))$ the specific band in $B(S)$ generated by all elements $h \in B(S)$ such that $h = Bt$ with $t \in S$ and with $t \curlywedge Bt = 0$. If we are in the classical potential theory, $t \curlywedge Bt = 0$ means roughly speaking that t is "harmonic on $b(B)$". Thus $\mathcal{L}(B(S))$ is the specific band generated by balayages of potentials which are "harmonic on $b(B)$". We shall show that $\mathcal{L}(B(S))$ and $\mathcal{C}(B(S))$ are equal.

Proposition 4.2 — Let B be a balayage such that for any $x \in d(B)$, $B^*(\varepsilon_x) \neq \alpha\varepsilon_y$. Then $\mathcal{L}(B(S)) \subset \mathcal{C}(B(S))$.

Proof — Let $u = \Sigma\, \alpha_{ij}\, u_i \wedge_B u_j$ be a strongly charged unit in $B(S)$ and denote $v = \Sigma\, \alpha_{ij}\, u_i \wedge u_j$. Further take $x \in d(B)$ and denote $\mu = B^*(\varepsilon_x)$. We show first that $v(x) > u(x)$. Indeed, if $v(x) = u(x)$ then we have for any i and j, $(u_i \wedge u_j)(x) = (u_i \wedge_B u_j)(x)$, then we get

$$\inf(\mu(u_i),\ \mu(u_j)) = \mu(u_i \wedge u_j).$$

Since $(u_i)_{i \in \mathbb{N}}$ is dense in order from below in $B(S)$, we get for $s,t \in B(S)$

$$\inf(\mu(s),\ \mu(t)) = \mu(s \wedge t)$$

which is characteristic for μ being an extremal H-integral on $B(S)$ and thus of the form $\alpha \cdot \varepsilon_y$.

Take now $h = Bt$ with $t \curlywedge Bt = 0$. For any $n \in \mathbb{N}$ there exists t_n such that

$$(t-Bt) \wedge n(v-Bv) = t_n - Bt_n.$$

We have $Bt_n \prec nBv = nu$ and $Bt = \curlyvee Bt_n$. It follows $h \in \mathcal{B}u$.

Theorem 4.3 — Assume that the fine sheaf property holds in S and let B be a balayage on S. Then we have $\Gamma(B(S)) \subset \mathcal{L}(B(S))$. If moreover the base of B has no isolated points (in the natural topology) then $\mathcal{C}(B(S)) = \mathcal{L}(B(S))$.

Proof — We remark first that for $s,t \in B(S)$, $s \curlywedge t \in B(S)$. Hence this element is also the specific g.l.b. in $B(S)$. Fix now B_1 a balayage on $B(S)$, that means a balayage on S with $b(B_1) \subset b(B)$ and denote $B_2 = B^{[b(B) \setminus b(B_1)]}$. Take $p \in B(S)$ a generator and denote $q = B_1 p \wedge B_2 p$. Denote $w = Bq - Bq \curlywedge B_1 q$. One can see that $w = \theta_{B_1} p$ where the operator θ_{B_1} is considered in $B(S)$. Further denote

$$B_3 = B^{\beta(\{x \in X\ :\ B_1 p(x) \geq B_2 p(x)\})}.$$

It is easy to see that $q = B_1 q$ on $d(B_3)$ and $b(B_1) \subset b(B_3)$, hence $B_3(B_1 p) = B_1 p$. From these and from the fine sheaf property we deduce that $q = B_3 q$. Putting now $t = q \curlywedge Bq$ and $s = q - t$ we get $t = B_3 t = Bt$. Since p is a generator in $B(S)$ we have $b(B_3) \cap b(B) = b(B_1)$ and using again the fine sheaf property we get $t = B_1 t$. Finally $w = Bq - Bq \curlywedge B_1 q = Bs - Bs \curlywedge B_1 s \prec Bs$. Obviously $Bs \in \mathcal{L}(B(S))$.

The last assertion of the theorem is a direct consequence of theorem 3.3.

HARMONIC MORPHISMS AND RAY PROCESSES

Laszlo Csink

Faculty of Arts
Eötvös University
P.O.Box 107
H-1364 Budapest
Hungary

Bernt Øksendal

Department of Mathematics
University of Oslo
P.O.Box 1053 Blindern
N-0316 Oslo 3
Norway

INTRODUCTION

In the following we will let $(X,\mathcal{U})$ and $(Y,\mathcal{V})$ denote $\mathcal{P}$-harmonic spaces in the sense of Constantinescu and Cornea [3]. (X,Y are locally compact Hausdorff spaces with countable bases and $\mathcal{U},\mathcal{V}$ are hyperharmonic sheaves on X,Y, respectively.) We will assume that the constant function 1 is hyperharmonic. It is now well known that there exist Hunt processes X_t, Y_t on X,Y with continuous paths (i.e. diffusions) such that the family of X_t-excessive, resp. Y_t-excessive, functions coincide with the family of non-negative $\mathcal{U}$-hyperharmonic, resp. $\mathcal{V}$-hyperharmonic, functions. The construction of the process X_t goes as follows:

Let $p>0$ be a bounded, hyperstrict continuous potential on X. If $g \geqq 0$ is bounded and continuous we define the specific multiplication $g*p$ by (see p. 189 in [3]):

$$g*p = \bigvee_{\delta\in\Delta_p} \delta_*(g) = \bigwedge_{\delta\in\Delta_p} \delta^*(g) \tag{0.1}$$

where

$$\delta^*(g) = \sum_{i\in I} (\sup_{S(p_i)} g)\cdot p_i \;, \quad \delta_*(g) = \sum_{i\in I} (\inf_{S(p_i)} g)\cdot p_i \;.$$

Here Δ_p is the set of all partitions $\delta=(p_i)_{i\in I}$ of p, $S(p_i)=X\setminus\{x;\ p_i$ is harmonic in a neighbourhood of $x\}$ is the set of non-harmonicity of p_i and $\bigwedge,\bigvee$ denotes the infimum and supremum relative to the partial order $\precsim$ on the family of nonnegative superharmonic functions $\mathcal{S}$ given by

$$u \precsim v \Leftrightarrow \exists w\in\mathcal{S} \quad \text{s.t.} \quad v=u+w$$

By Theorem 10.2.1 p. 252 in [3] there exists a unique sub-Markov semigroup $P_t;\ t\geq 0$ such that

$$(f*p)(x) = \int_0^\infty (P_t f)(x)dt \tag{0.2}$$

for all $x \in X$ and all non-negative continuous functions f with compact support. Then by results of Meyer, Boboc-Constantinescu-Cornea and Hansen (see p. 271 in [3]) there exists a diffusion X_t having the properties stated above and such that P_t coincides with the transition semigroup of X_t, i.e.

$$(P_t f)(x) = E^x[f(X_t); t<\zeta] , \tag{0.3}$$

where E^x denotes expectation wrt. the probability law P^x of X_t^* (i.e. of X_t with $X_0=x$) and $\zeta \leqq \infty$ denotes the life time of X_t.

1. RAY PROCESSES ASSOCIATED TO SUPERHARMONIC FUNCTIONS

If the hyperstrict potential p above is replaced by a continuous non-negative superharmonic function u, can we still associate a stochastic process to u in some sense? The answer is yes. We outline the construction in the following (details can be found in [4]):

a) First consider a compact $K \subset X$. Define the kernel $V=V_{u,K}(\cdot,\cdot): K \times B_b(K) \to \mathbb{R}$ by

$$(V_f)(x) = V_{u,K}(x,f) = ((f\cdot\chi_K)*u)(x) \;;\quad x\in K,\ f\in B_b(K) \tag{1.1}$$

where $B_b(K)$ denotes the bounded Borel functions on K. In [3], Theorem 8.2.1 it is shown how to extend the definition of $g*p$ to the case when $g \in B_b(K)$. A similar type of specific multiplication and a similar extension can be carried out to define $g*u$ for all $g \in B_b(K)$. This makes the term $(f\cdot\chi_K)*u$ in (1.1) meaningful.

b) There exists a unique resolvent $\{U_\alpha\}_{\alpha\geqq 0} = \{U_\alpha^{(u,K)}\}_{\alpha\geqq 0}$ such that

$$U_0 = V \tag{1.2}$$

To obtain this it suffices to prove that V satisfies the complete maximum principle and apply Theorem 7.7 in [2].

c) $\{U_\alpha\}_{\alpha\geqq 0}$ is a <u>Ray resolvent</u> and therefore there exists a <u>Ray process</u> $Z_t = Z_t^{(K)}$ on K whose resolvent coincide with U_α, i.e.

$$(U_\alpha f)(x) = E^x[\int_0^\zeta e^{-\alpha t} f(Z_t)dt] \;;\quad \alpha \geqq 0 \tag{1.3}$$

where $\zeta = \zeta^{(K)}$ denotes the life time of Z_t. (See [5] for information about Ray processes).

d) Finally we let $\{G_n\}$ be an increasing sequence of relatively compact open subsets of X such that

$$X = \bigcup_{n=1}^{\infty} G_n$$

Then in a canonical way we obtain a process Z_t on X (with life time $\zeta = \lim \zeta^{(\bar{G}_n)}$) as a limit of the Ray processes $Z_t^{(n)} = Z_t^{(\bar{G}_n)}$, $n=1,2,\ldots$ We call Z_t <u>the extended Ray process associated with</u> u.

In general the Z_t-excessive functions will always be $\mathcal{U}$-hyperharmonic, but there may be $\mathcal{U}$-hyperharmonic functions which are not Z_t-excessive.

2. HARMONIC MORPHISMS AND RAY PROCESSES

We now apply this general construction to a given harmonic morphism $\varphi:\mathcal{X}\to\mathcal{Y}$: Let $q>0$ be the bounded, hyperstrict continuous potential on $\mathcal{Y}$ from which the diffusion Y_t is constructed. Then $u=q\circ\varphi$ is superharmonic on $\mathcal{X}$ (but not necessarily a potential). Let Z_t be the extended Ray process associated to u as desribed in §1, with probability law $\tilde{P}^x$. Then one can prove that

$$(f*q)\circ\varphi = (f\circ\varphi)*u \tag{2.1}$$

for all bounded continuous functions f on $\mathcal{Y}$. The stochastic interpretation of this is the following: Define

$$W_t = \begin{cases} \varphi(Z_t) & ; \quad t<\zeta \\ Y^y_{t-\zeta} & ; \quad t\geq\zeta \end{cases},$$

with the natural probability law, where $y=\lim_{t\to\zeta}\varphi(Z_t)$ exists a.s. on $\{\zeta<\infty\}$. Then W_t coincide in law with Y_t. In other words, (2.1) implies that <u>φ is Z_t-Y_t pathpreserving without time change</u>. The converse is also true, in the following sense:

<u>Theorem 1</u> ([4], Theorem 3.1) – Let $\varphi:\mathcal{X}\to\mathcal{Y}$ be a continuous function. Then the following are equivalent:

(1) φ is a harmonic morphism

(2) There exists an extended Ray process Z_t on $\mathcal{X}$ such that

(i) $\zeta=0$ a.s. $\tilde{P}^x \Rightarrow \varphi$ is finely locally constant at x

(ii) g Z_t-excessive $\Rightarrow$ g $\mathcal{U}$-hyperharmonic

and

(iii) φ is Z_t-Y_t pathpreserving (without time change).

The <u>branch set</u> B of Z_t can be described by

$$B = \mathcal{X}\setminus\{x; Z_0=x \text{ a.s. } \tilde{P}^x\} \tag{2.2}$$

It turns out that B can be given the following characterization in terms of φ:

<u>Theorem 2</u> ([4], Theorem 4.4) – The following are equivalent:

(i) $x\in B$

(ii) φ is finely locally constant at x

(iii) There exists a fine neighbourhood U of x such that $\varphi(x)$ does not belong to the fine interior of $\varphi(U)$ (in $\mathcal{Y}$).

In particular, this gives that for any $x \in X$ either φ is finely locally constant at x or φ maps every fine neighbourhood of x onto a fine neighbourhood of $\varphi(x)$ ([4], Corollary 4.5).

Moreover, if we apply the main result of Blumenthal, Getoor and McKean in [1] we get the following consequence of Theorems 1 & 2:

Theorem 3 ([4], Theorem 4.6) – Let $\psi: X \to Y$ be a continuous, finely locally non-constant function. Then the following are equivalent:

a) ψ is a harmonic morphism

b) ψ is X_t-Y_t pathpreserving, with a continuous time change.

The easy part of the proof is b) ⇒ a). To obtain the converse we use that the branch set B of the Ray process Z_t associated to ψ must be empty by Theorem 2. It follows that X_t and Z_t have the same excessive functions, hence the same hitting distributions and therefore Z_t is a continuous time change of X_t by [1].

For detailed proofs and other applications (about value distribution and boundary behaviour of harmonic morphisms) we refer the reader to [4].

ACKNOWLEDGEMENTS

We wish to thank A. Cornea and G. Michaletzky for their comments. And we are very grateful to B. Fuglede for helpful communications throughout this project.

REFERENCES

[1] Blumenthal, R.M., Getoor, R.K., and McKean, H.P., Markov processes with identical hitting distributions, Illinois J. Math., 6 (1962), 402-420 and 7 (1963), 540-542.
[2] Bliedtner, J., and Hansen, W., 1986, "Potential Theory", Springer-Verlag.
[3] Constantinescu, C., and Cornea, A., 1972, "Potential Theory on Harmonic Spaces", Springer-Verlag.
[4] Csink, L., and Øksendal, B., A stochastic characterization of harmonic morphisms. (To appear).
[5] Getoor, R.K., 1975, "Markov Processes: Ray Processes and Right Processes", Springer LNM 440.

ON DIRICHLET'S BOUNDARY VALUE PROBLEM FOR CERTAIN ANISOTROPIC DIFFERENTIAL AND PSEUDO-DIFFERENTIAL OPERATORS

Karl Doppel — FU Berlin, Arnimallee 3, D - 1000 Berlin 33

Niels Jacob — FAU Erlangen, Bismarckstr.1 1/2, D - 8520 Erlangen

Introduction

In our previous paper [1] a calculus for pseudo-differential operators was used to find periodic solutions for a large class of not necessarily elliptic pseudo-differential equations with constant coefficients. In this paper we consider a generalized homogeneous Dirichlet-problem for pseudo - differential operators with constant coefficients and prove Fredholm's alternative theorem. We give two examples how to apply this result. The first one is a differential operator arising in stochastics. In the second example we deal with an anisotropic pseudo-differential operator and using the theory of Dirichlet forms (see [4]) it follows that this operator generates a symmetric Hunt process. Some of our results had been already used in [6].

1. Sobolev-space generating functions and related operators

Troughout this paper $\Omega \subset \mathbb{R}^n$ is a bounded open set with smooth boundary $\partial\Omega$ and $c > 0$ is a generic constant which may change from line to line. We say that $a : \mathbb{R}^n \to \mathbb{R}$ is a Sobolev-space generating function if it fulfills the two following conditions :

A 1 : Let $a : \mathbb{R}^n \to \mathbb{R}^+ \cup \{0\}$ be a continuous function and if $\lambda^{(n)}$ denotes the Lebesgue measure in $\mathbb{R}^n$ assume that

$$\lambda^{(n)}(\{ \xi \in \mathbb{R}^n, a(\xi) = 0\}) = 0.$$

A 2 : There exists a number $m \geq 0$ such that

$$a(\xi) \leq c (1 + |\xi|)^m \tag{1.1}$$

holds for all $\xi \in \mathbb{R}^n$.

The set of all Sobolev-space generating functions is denoted by $\mathcal{A}$. Let $a \in \mathcal{A}$ and $s \in \mathbb{R}$, $s \geq 0$. Then the space $H_a^s(\mathbb{R}^n)$ is defined as follows

$$H_a^s(\mathbb{R}^n) := \{ u \in L^2(\mathbb{R}^n) , \int_{\mathbb{R}^n} (1 + a(\xi)^{2s}) \mid Fu(\xi) \mid^2 d\xi < \infty \} ,$$

where $L^2(\mathbb{R}^n)$ is the usual Hilbert-space consisting of all Lebesgue-measurable functions $u : \mathbb{R}^n \to \mathbb{C}$ such that $|u|^2$ is integrable. Furthermore Fu denotes the Fourier-transform of u.The norm in $H_a^s(\mathbb{R}^n)$ is given by

$$|| u ||_{s,a}^2 = \int_{\mathbb{R}^n} (1 + a(\xi)^{2s}) \mid Fu(\xi) \mid^2 d\xi . \tag{1.2}$$

Obviously, for $s = 0$ we have $H_a^0(\mathbb{R}^n) = L^2(\mathbb{R}^n)$ .The space $\tilde{C}_c^\infty(\Omega)$ consists of all arbitrarily often differentiable functions $\varphi : \mathbb{R}^n \to \mathbb{C}$ with compact support $\operatorname{supp} \varphi \subset \Omega$; and $\tilde{L}^2(\Omega)$ is the space of all functions $u \in L^2(\mathbb{R}^n)$ with $\operatorname{supp} u \subset \Omega$ .Trivially we have a continuous imbedding of $\tilde{L}^2(\Omega)$ into $L^2(\mathbb{R}^n)$. By definition $H^s_{a,0}(\Omega)$ is the closure of $\tilde{C}_c^\infty(\Omega)$ in $H_a^s(\mathbb{R}^n)$.

Take $a \in \mathcal{A}$ and let $p : \mathbb{R}^n \to \mathbb{C}$ be a continuous function such that

$$\mid p(\xi) \mid \leq c (1 + a(\xi)^s) \tag{1.3}$$

is valid for $\xi \in \mathbb{R}^n$ and $s \in \mathbb{R}$, $s \geq 0$. The set of all continuous functions satifying (1.3) is denoted by $\Sigma(a,s)$. Take $p \in \Sigma(a,s)$ and define for $\varphi \in \tilde{C}_c^\infty(\Omega)$ the operator $P(D)$ by

$$P(D)\varphi(x) = \int_{\mathbb{R}^n} e^{ix\xi} p(\xi) F\varphi(\xi) d\xi , x \in \mathbb{R}^n. \tag{1.4}$$

The function p is called the symbol of the operator $P(D)$.

Proposition 1.1. - The operator $P(D)$ has a continuous extension $\hat{P}(D) : H^s_{a,0}(\Omega) \to L^2(\mathbb{R}^n)$.

Proof - We have to show that the estimate

$$(1.5) \qquad \| P(D)\varphi \|^2_{L^2(\mathbb{R}^n)} \leq c \| \varphi \|^2_{s,a}$$

holds for all $\varphi \in \tilde{C}^\infty_c(\Omega)$. Now, for $\varphi \in \tilde{C}^\infty_c(\Omega)$ we have

$$\begin{aligned} \| P(D)\varphi \|^2_{L^2(\mathbb{R}^n)} &= c \int_{\mathbb{R}^n} | p(\xi) |^2 \, | F\varphi(\xi) |^2 \, d\xi \\ &\leq c \int_{\mathbb{R}^n} (1 + a(\xi)^s)^2 \, | F\varphi(\xi) |^2 \, d\xi \\ &\leq c \int_{\mathbb{R}^n} (1 + a(\xi)^{2s}) \, | F\varphi(\xi) |^2 \, d\xi \\ &= c \| \varphi \|^2_{s,a} . \end{aligned}$$

As an immediate consequence we have

Corollary 1.2 - Let $a \in \mathcal{A}$. Then the operator $\hat{A}(D) : H^s_{a,0}(\Omega) \to L^2(\mathbb{R}^n)$, $s \geq 0$, defined by

$$\hat{A}^s(D)u(x) = \int_{\mathbb{R}^n} e^{ix\xi} \, a(\xi)^s \, Fu(\xi) \, d\xi$$

is continuous.

Furthermore we have

Proposition 1.3 - Let $a \in \mathcal{A}$ and $\hat{A}^s(D)$ as in Corollary 1.2. Then $\hat{A}^s(D)u = 0$ implies $u = 0$ $\lambda^{(n)}$-almost everywhere.

Proof - Since $\hat{A}^s(D)u = 0$, it follows that

$$\| \hat{A}^s(D)u \|^2_0 = c \int_{\mathbb{R}^n} a(\xi)^{2s} \, | Fu(\xi) |^2 \, d\xi = 0 ,$$

hence

$$a(\xi)^{2s} \, | Fu(\xi) |^2 = 0 \quad \lambda^{(n)}\text{-almost everywhere.}$$

But $Fu \in L^2(\mathbb{R}^n)$ and by A 1 we have

$$\lambda^{(n)}(\{\ \xi \in \mathbb{R}^n,\ a(\xi) = 0\ \}) = 0 ,$$

therefore it follows that $Fu(\xi) = 0$ $\lambda^{(n)}$-almost everywhere, which implies $u = 0$ $\lambda^{(n)}$-almost everywhere.

The last result leads us to

<u>Proposition 1.4</u> – (Poincaré's inequality) Take $a \in \mathcal{A}$ and let $\hat{A}^s(D)$ be given as in Corollary 1.2. If in addition the imbedding of $H^s_{a,0}(\Omega)$ into $\tilde{L}^2(\Omega)$ is compact, then the estimate

$$(1.6) \qquad \| u \|_0^2 \leq c \| \hat{A}^s(D)u \|_0^2$$

holds for all $u \in H^s_{a,0}(\Omega)$.

<u>Proof</u> – Suppose that (1.6) is false. Then there exists a sequence $(u_\nu)_{\nu\in\mathbb{N}}$, $u_\nu \in H^s_{a,0}(\Omega)$, such that $\| u_\nu \|_{s,a} = 1$ and

$$(1.7) \qquad \| u_\nu \|_{s,a}^2 \geq \nu \| \hat{A}^s(D)u_\nu \|_0^2$$

holds. From (1.7) we get that $\| \hat{A}^s(D)u \|_0 \to 0$ for $\nu \to \infty$, and hence

$$(1.8) \qquad \hat{A}^s(D)u_\nu \to .0 \text{ in } L^2(\mathbb{R}^n) .$$

Since $H^s_{a,0}(\Omega)$ is reflexive, there exists a subsequence of $(u_\nu)_{\nu\in\mathbb{N}}$ denoted by $(u_\nu)_{\nu\in\mathbb{N}}$, such that u_ν converges weakly in $H^s_{a,0}(\Omega)$ to an element $u \in H^s_{a,0}(\Omega)$. By the compactness of the imbedding of $H^s_{a,0}(\Omega)$ into $\tilde{L}^2(\Omega)$ it follows that u_ν converges strongly to $u \in \tilde{L}^2(\Omega)$, hence $\|u_\nu - u\|_0 \to 0$. Now let $\varphi \in C_c^\infty(\mathbb{R}^n)$, then we have

$$\int_{\mathbb{R}^n} \overline{\hat{A}^s(D)(u_\nu - u)}\ \varphi\ dx = c \int_{\mathbb{R}^n} a(\xi)^s\ \overline{F(u_\nu - u)(\xi)}\ F\varphi(\xi)\ d\xi$$

$$= c \int_{\mathbb{R}^n} \overline{(u_\nu - u)}\ \hat{A}^s(D)\varphi\ dx,$$

This gives $(\hat{A}^s(D)u_\nu,\varphi)_0 \to (\hat{A}^s(D)u,\varphi)_0$ for all $\varphi \in C_c^\infty(\mathbb{R}^n)$ and by (1.8) it follows that $\hat{A}^s(D)u_\nu \to \hat{A}^s(D)u = 0$. By Proposition 1.3 we find $u = 0$ $\lambda^{(n)}$-almost everywhere and therefore

$$\| u_\nu \|_{s,a}^2 \leq c \left[\| u_\nu \|_0^2 + \| \hat{A}^s(D)u_\nu \|_0^2 \right] \to 0 ,$$

which is a contradiction to our assumption $\| u_\nu \|_{s,a} = 1$.

Remark 1.5 – In the proof of Proposition 1.4 we mainly followed the ideas of [9], p.83.

2. On sesquilinearforms associated with symbols

Let $p \in \Sigma(a,2s)$ and consider the sesquilinear form B_p defined by

$$B_p(\varphi,\psi) = \int_{\mathbb{R}^n} \overline{p(\xi)}\ \overline{F\varphi(\xi)}\ F\psi(\xi)\ d\xi \tag{2.1}$$

for $\varphi,\psi \in \tilde{C}_c^{\infty}(\Omega)$. We get

Proposition 2.1 – The sesquilinear form B_p has a continuous extension $\hat{B}_p$ on $H^s_{a,0}(\Omega)$, i.e. there exists a constant $c > 0$ such that

$$|\ B_p(u,v)| \leq c\ ||\ u\ ||_{s,a}\ ||\ v\ ||_{s,a} \tag{2.2}$$

holds for all $u,v \in H^s_{a,0}(\Omega)$.

Proof – Since $\tilde{C}_c^{\infty}(\Omega)$ is dense in $H^s_{a,0}(\Omega)$ it is sufficient to prove (2.2) for all $\varphi,\psi \in \tilde{C}_c^{\infty}(\Omega)$. Now we have

$$|\ B_p(\varphi,\psi)\ | = |\int_{\mathbb{R}^n} \overline{p(\xi)}\ \overline{F\varphi(\xi)}\ F\psi(\xi)\ d\xi\ |$$

$$\leq c \int_{\mathbb{R}^n} (\ 1 + a(\xi)^{2s}\ )\ |\ F\varphi(\xi)\ |\ |\ F\psi(\xi)\ |\ d\xi$$

$$\leq c \left[\int_{\mathbb{R}^n} (1+a(\xi)^{2s})\,|F\varphi(\xi)|^2 d\xi \right]^{1/2} \left[\int_{\mathbb{R}^n} (1+a(\xi)^{2s}\,|F\psi(\xi)|^2 d\xi \right]^{1/2}$$

$$= c\ ||\ \varphi\ ||_{s,a}\ ||\ \psi\ ||_{s,a}\ .$$

Furthermore we have

Proposition 2.2 – (Gårding's inequality) Let $p \in \Sigma(a,2s)$ and in addition suppose that there is a constant $c > 0$ and a number $\rho \geq 0$ such that

$$\operatorname{Re} p(\xi) \geq c\ a(\xi)^{2s} \tag{2.3}$$

holds for all $\xi \in \mathbb{R}^n$, $|\ \xi\ | \geq \rho$. Then there exists two constants $c_1 > 0$

and $c_2 \geq 0$ such that

(2.4) $$\operatorname{Re} B_p(\varphi,\varphi) \geq c_1 \| \varphi \|^2_{s,a} - c_2 \| \varphi \|^2_0$$

holds for all $\varphi \in \tilde{C}^\infty_c(\Omega)$.

Proof – Let $\varphi \in \tilde{C}^\infty_c(\Omega)$, then we find

$$\operatorname{Re} B_p(\varphi,\varphi) = \int_{\mathbb{R}^n} \operatorname{Re} p(\xi) \, | F\varphi(\xi) |^2 \, d\xi$$

$$= \int_{|\xi| \leq \rho} \operatorname{Re} p(\xi) \, | F\varphi(\xi) |^2 \, d\xi + \int_{|\xi| \geq \rho} \operatorname{Re} p(\xi) \, | F\varphi(\xi) |^2 \, d\xi$$

$$\geq c \int_{|\xi| \geq \rho} a(\xi)^{2s} \, | F\varphi(\xi) |^2 \, d\xi - \int_{|\xi| \leq \rho} \operatorname{Re} p(\xi) \, | F\varphi(\xi) |^2 \, d\xi$$

$$= c \int_{\mathbb{R}^n} a(\xi)^{2s} |F\varphi(\xi)|^2 \, d\xi - \int_{|\xi| \leq \rho} \left[|p(\xi)| + a(\xi)^{2s} \right] |F\varphi(\xi)|^2 \, d\xi$$

$$\geq c_1 \| \varphi \|^2_{s,a} - c_2 \| \varphi \|^2_0 .$$

Using Proposition 2.1 and Proposition 2.2 we get the following theorem, the proof of which is standard.

Theorem 2.3 – Suppose that the assumptions of Proposition 2.1 and 2.2 are fulfilled. In addition suppose that the imbedding of $H^s_{a,0}(\Omega)$ into $\tilde{L}^2(\Omega)$ is compact. Let $f \in \tilde{L}^2(\Omega)$ be a given function. Then for the re - presentation problem, to find all $u \in H^s_{a,0}(\Omega)$ such that

(2.5) $$\hat{B}_p(u,\varphi) = (f,\varphi)_0$$

holds for all $\varphi \in \tilde{C}^\infty_c(\Omega)$, Fredholm's alternative theorem is valid.

3. Applications to generalized Dirichlet-problems

In this section we want to discuss two examples of operators the results of the previous section can be applied to. The first example is a differential operator which symbol is a Sobolev-space generating function itself.

Example 3.1 – Consider the differential operator $(-1)^k \Delta_1 \cdots \Delta_k$, $k \in \mathbb{N}$, introduced by E.B.Dynkin [3], which has the symbol

$$a(\xi) = \prod_{j=1}^{k} |\xi^j|^2 , \tag{3.1}$$

where $\xi = (\xi^1,\dots,\xi^k) \in \mathbb{R}^n$ and $\xi^j \in \mathbb{R}^{n_j}$, $\sum_{j=1}^{k} n_j = n$. It was shown in our paper [2] how to solve a generalized Dirichlet-problem for this non-hypoelliptic operator. Moreover, there we discussed why the representation problem of Theorem 2.3 generalizes the Dirichlet-problem posed for the operator $(-1)^k \Delta_1 \dots \Delta_k$.

The next example is a pseudo-differential operator which generates an anisotropic regular Dirichlet-form (for the definition see [4], p.5).

Example 3.2 – Let $\mathbb{R}^n = \mathbb{R}^{n_1} \times \mathbb{R}^{n_2}$, $\xi = (\eta,\varsigma) \in \mathbb{R}^{n_1} \times \mathbb{R}^{n_2}$, and r_1, r_2 be two numbers such that $0 < r_1 \le r_2 \le 1$. Now consider the function $\Lambda^{r_1,r_2} : \mathbb{R}^{n_1} \times \mathbb{R}^{n_2} \to \mathbb{R}$ defined by $\Lambda^{r_1,r_2}(\eta,\varsigma) = |\eta|^{2r_1} + |\varsigma|^{2r_2}$ and set $\lambda^{r_1,r_2}(\eta,\varsigma) = [\Lambda^{r_1,r_2}(\eta,\varsigma)]^{1/2}$. Clearly λ^{r_1,r_2} is a Sobolev-space generating function and $\Lambda^{r_1,r_2} \in \Sigma(\lambda^{r_1,r_2},2)$. Moreover, the imbedding $\iota : H^1_{\lambda^{r_1,r_2},0}(\Omega_1 \times \Omega_2) \to \tilde{L}^2(\Omega_1 \times \Omega_2)$ is for all bounded open sets $\Omega_1 \subset \mathbb{R}^{n_1}$ and $\Omega_2 \subset \mathbb{R}^{n_2}$ compact, since $H^1_{\lambda^{r_1,r_2},0}(\Omega_1 \times \Omega_2)$ is continuously imbedded into the space $H_0^{r_1}(\Omega_1 \times \Omega_2)$, the classical Sobolev-space (for the definition and for properties of the space $H_0^{r_1}(\Omega_1 \times \Omega_2)$ see [8], chap. 1). The sesquilinear form associated with Λ^{r_1,r_2} is given by

$$B_{r_1,r_2}(u,v) = \int_{\mathbb{R}^{n_1}} \int_{\mathbb{R}^{n_2}} \left[|\eta|^{2r_1} + |\varsigma|^{2r_2} \right] \overline{F_{x\to\eta} F_{y\to\varsigma} u(\eta,\varsigma)}\, F_{x\to\eta} F_{y\to\varsigma} v(\eta,\varsigma)\, d\varsigma\, d\eta$$

and by Proposition 1.4 we find that B_{r_1,r_2} gives an equivalent norm to $\|\cdot\|_{1,\lambda^{r_1,r_2}}$. Since B_{r_1,r_2} is obviously continuous in $H^1_{\lambda^{r_1,r_2}}(\Omega_1 \times \Omega_2)$, we can apply Theorem 2.3. Moreover, as in [10], p.97, one can prove for real valued functions

$$B_{r_1,r_2}(u,v) = c_1 \int_{\mathbb{R}^{n_2}} \left[\int_{\mathbb{R}^{n_1}} \int_{\mathbb{R}^{n_1}} \frac{(u(x+\hat{x},y)-u(x,y))(v(x+\hat{x},y)-v(x,y)}{|\hat{x}|^{n_1+2r_1}} \, dx \, d\hat{x} \right] dy$$

$$+ c_2 \int_{\mathbb{R}^{n_1}} \left[\int_{\mathbb{R}^{n_2}} \int_{\mathbb{R}^{n_2}} \frac{(u(x,y+\hat{y})-u(x,y))(v(x,y+\hat{y})-v(x,y)}{|\hat{y}|^{n_2+2r_2}} \, dy \, d\hat{y} \right] dx.$$

Hence, if we restrict our considerations to real valued functions, B_{r_1,r_2} is a regular Dirichlet-form. This example shows that the pseudo-differential operator can be treated almost in the same way as the operator $\Lambda^{2r}(D)$ with symbol $|\xi|^{2r}$ was treated in [6]. In particular potential theoretical techniques and techniques used in probability theory do apply to this operator. Conversly, by Theorem 6.2.1 in [4] there corresponds a symmetric Hunt-process M_{r_1,r_2} to the Dirichlet-form B_{r_1,r_2}.

The problem to classify the set of Sobolev-space generating functions for which the imbedding of $H^s_{a,0}(\Omega)$ into $\tilde{L}^2(\Omega)$ is compact for each bounded open set $\Omega \subset \mathbb{R}^n$ is still open. However, there exist some sufficient conditions; see [5], Theorem 10.1.10 or [7], Theorem 4.

References

[1] Doppel, K., and Jacob, N. : Zur Konstruktion periodischer Lösungen von Pseudodifferentialgleichungen mit Hilfe von Operatorenalgebren. Ann.Acad.Sci.Fenn.Ser.A.I.Math.8(1983)193 - 217.

[2] Doppel, K., and Jacob, N. : A non-hypoelliptic Dirichlet-problem from stochastics. Ann.Acad.Sci.Fenn.Ser.A.I.Math.8(1983)375 - 389.

[3] Dynkin, E.B. : Harmonic functions associated with several Markov processes. Adv.Appl.Math.2(1981)260 -283.

[4] Fukushima, M. : Dirichlet forms and Markov processes. North-Holland Publishing Company, Amsterdam Oxford New York, (1980).

[5] Hörmander, L. : The analysis of lineare partial differential op - erators II. Springer Verlag, Berlin Heidelberg New York Tokyo, (1983).

[6] Jacob, N. : Sur les fonctions 2r-harmoniques de N.S.Landkof. C.R. Acad. Sc. Paris 304(1987)169 - 171.

[7] Jacob, N., and Schomburg, B. : On Gårding's inequality. Aequationes Math. 31(1986)7 - 17.

[8] Lions, J.L., and Magenes, E. : Non-homogeneous boundary value problems and applications I. Springer Verlag, Berlin Heidelberg New York, (1972).

[9] Morrey, C.B. jr. : Multiple integrals in the calculus of variations. Springer Verlag, Berlin Heidelberg New York, (1966).

[10] Wloka, J. : Partielle Differentialgleichungen. Teubner Verlag, Stuttgart, (1982).

HYPERHARMONIC CONES

Sirkka-Liisa Eriksson-Bique

Department of Mathematics
University of Joensuu
SF-80101 Joensuu, Finland

INTRODUCTION

The theory of H-cones ([3]) covers the superharmonic case in potential theory. This work continues the algebraic axiomatization of the hyperharmonic case. Our basic structure is a hyperharmonic structure defined by M. Arsove and H. Leutwiler in [1]. The cancellation law does not hold in hyperharmonic structures. In order to characterize cancellable elements we assume that a hyperharmonic structure is a convex cone and the greatest lower bound of $(u/n)_{n\in\mathbb{N}}$ exists for all u.

Hyperharmonic and σ-hyperharmonic cones are ordered convex cones possessing order properties similar to those of positive hyperharmonic functions on harmonic spaces. Hyperharmonic cones are studied in [7]. The theory of σ-hyperharmonic cones is very similar to the theory of hyperharmonic cones. Some proofs are more difficult since the greatest lower or least upper bounds do not necessarily exist. We show that σ-hyperharmonic (resp. hyperharmonic) cones are also σ-hyperharmonic (resp. hyperharmonic) cones with respect to specific order. Both σ-hyperharmonic and hyperharmonic cones can be represented as a subcone of positive continuous extended real valued functions on a locally compact space.

Additive order continuous functions between σ-hyperharmonic cones are called σ-hyperharmonic morphisms. The set of extended real valued σ-hyperharmonic morphisms called the dual is not necessarily a σ-hyperharmonic cone. Under certain conditions the dual is a σ-hyperharmonic cone and the second dual is the natural imbedding cone. The corresponding result also holds for hyperharmonic cones ([7]).

1. PREHYPERHARMONIC CONES

Our basic structure is a partially ordered abelian semigroup $(W,+,\leqq)$ with a neutral element 0 and satisfying the following properties:

(1.1) $\quad u \geqq 0$ for all $u \in W$,

(1.2) $\quad x \geqq y \Rightarrow u + x \geqq u + y$ for all $u, x, y \in W$.

Specific order in W is defined by

$$u \preccurlyeq v \quad \text{if} \quad v = u + u' \quad \text{for some} \quad u' \in W.$$

An element $u \in W$ is called *cancellable* in W if

(1.3) $\quad u + x \leqq u + y \Rightarrow x \leqq y$

for all $x, y \in W$.

DEFINITION 1.1 – A partially ordered abelian semigroup $(W,+,\leqq)$ satisfying (1.1) and (1.2) is called a *hyperharmonic structure* if for $x, u \in W$ the set $\{w \in W : u \leqq w + x\}$ has the least element denoted by $S_x u$ with $S_x u \preccurlyeq u$. A hyperharmonic structure W is called a *superharmonic structure* if all elements are cancellable in W.

An important property of hyperharmonic structures is

(1.4) $\quad x \preccurlyeq u \Rightarrow x + S_x u = u.$

Superharmonic structures have the following equivalent characterization proved by M. Arsove and H. Leutwiler ([1], Theorem 2.5).

THEOREM 1.2 – *A partially ordered abelian semigroup $(W,+,\leqq)$ satisfying (1.1) and (1.3) is a superharmonic structure if and only if for all u and v in W there exist both the mixed lower envelope*

$$u \curlywedge v = \max \{ w \in W : w \preccurlyeq u, w \leqq v\}$$

and the mixed upper envelope

$$u \curlyvee v = \min \{ w \in W : w \succcurlyeq u, w \geqq v\}$$

satisfying

$$u \curlywedge v + v \curlyvee u = u + v.$$

Mixed upper envelopes always exist in hyperharmonic structures ([1], Theorem 17.2). However mixed lower envelopes do not necessarily exist in hyperharmonic structures ([7], p.29). We shall prove that a prehyperharmonic cone defined next has mixed lower envelopes.

DEFINITION 1.3 – A hyperharmonic structure $(W,+,\leqq)$ is called a *prehyperharmonic cone* if it satisfies the following axioms:

AXIOM 1 – Multiplication by strictly positive real numbers is defined in W such that

$$r(x + y) = rx + ry, \quad (r + q)x = rx + qx,$$

$$(rq)x = r(qx),$$

$$x \leqq y \Rightarrow rx \leqq ry$$

for all $r, q \in \mathbb{R}_+ \setminus \{0\}$ and $x, y \in W$.

AXIOM 2 – *For any element* $u \in W$, *the set* $\{u/n : n \in \mathbb{N}\}$ *has the greatest lower bound denoted by* $\underline{u}$ *with* $\underline{u} \preccurlyeq u$.

AXIOM 3 – *For all* u *and* v *in* W *there exists the greatest lower bound of* u *and* v *denoted by* $u \wedge v$.

An element $\underline{u}$ plays an important role in prehyperharmonic cones. Briefly, it behaves like a function which is zero when u is finite and infinite when u is infinite. An essential property is

(1.5) $\quad \underline{u} \leqq w \Longleftrightarrow \underline{u} + w = w$

([7], Lemma 3.8; [6], p.5).

THEOREM 1.4 – *Let* $(W,+,\leqq)$ *be a prehyperharmonic cone. Then the following assertions are mutually equivalent for* $u \in W$:

(i) $\quad u$ *is cancellable*,

(ii) $\quad S_{x+u}(y + u) = S_x y$ *for all* $x, y \in W$,

(iii) *if* $w \succcurlyeq u$ *for* $w \in W$ *then there exists a unique* $u' \in W$ *satisfying* $w = u + u'$,

(iv) $\quad x + u \preccurlyeq y + u \Rightarrow x \preccurlyeq y$ *for all* $x, y \in W$,

(v) $\quad \underline{u} = 0$.

The proof is the same as in [7], Theorem 3.9.

COROLLARY 1.5 – *Let* $(W,+,\leqq)$ *be a prehyperharmonic cone. If an element* $u \in W$ *is cancellable, then* $v \in W$ *with* $v \leqq u$ *is also cancellable*.

An important tool for handling uncancellable elements is

THEOREM 1.6.– *If* $(W,+.\leqq)$ *is a prehyperharmonic cone and* u *is an element of* W *then* $(\underline{u}+W,+,\leqq)$ *is also a prehyperharmonic cone and* u *is cancellable in* $\underline{u} + W$.

PROOF – Let x and y be arbitrary elements of W. Using the same ideas as in the proof of [7], Lemma 4.11, we see that

(1.6) $\quad \min\{w \in \underline{u}+W : y + \underline{u} \leqq w + x + \underline{u}\} = \underline{u} + S_x y.$

Axiom 2 obviously holds in $\underline{u} + W$ and Axiom 3 follows from the equality

(1.7) $\quad a + x \wedge y = (a + x) \wedge (a + y)$

for all $a \in W$ ([1], (17.38)). This completes the proof.

COROLLARY 1.7 – *Let* W *be a prehyperharmonic cone and* u, x, y *be elements of* W. *Then*

(a) $\quad x + u \leqq y + u \Longleftrightarrow x + \underline{u} \leqq y + \underline{u}$,

(b) $x + u \preccurlyeq y + u \Longleftrightarrow x + \underline{u} \preccurlyeq y + \underline{u}$,

(c) $S_{x+u}(y+u) = S_{x+\underline{u}}(y + \underline{u})$.

We shall prove that for all u and v in a prehyperharnonic cone the least upper bound $u \vee v$ exists. In addition it has a similar form as in H-cones ([3], Corollary 2.1.3).

PROPOSITION 1.8 – Let W be a prehyperharmonic cone. Then for all elements u and v in W the least upper bound $u \vee v$ exists and

$$u \vee v = \underline{u} + \underline{v} + S_{u\wedge v}(u + v).$$

PROOF – Let u and v be elements of W. Applying Corollary 1.7 (a) to the inequality

$$u + S_{u\wedge v}(u + v) \geqq u \wedge v + S_{u\wedge v}(u + v) \geqq u + v$$

we obtain

$$\underline{u} + S_{u\wedge v}(u + v) \geqq \underline{u} + v \geqq v.$$

Hence we have

$$\underline{u} + \underline{v} + S_{u\wedge v}(u + v) \geqq v + \underline{v} = v.$$

Similarly we see that

$$\underline{u} + \underline{v} + S_{u\wedge v}(u + v) \geqq u.$$

Now let $w \in W$ be an arbitrary upper bound of u and v. Then (1.7) yields

$$w + u \wedge v = (w + u) \wedge (w + v) \geqq u + v.$$

This inequality leads to $w \geqq S_{u\wedge v}(u + v)$; whence

$$w + \underline{u} + \underline{v} \geqq \underline{u} + \underline{v} + S_{u\wedge v}(u + v).$$

From $w \geqq u$ and $w \geqq v$ it follows that $\underline{w} \geqq \underline{u}$ and $\underline{w} \geqq \underline{v}$. Therefore

$$w = w + \underline{w} = w + \underline{w} + \underline{w} \geqq w + \underline{u} + \underline{v} \geqq w.$$

Consequently, $w \geqq \underline{u} + \underline{v} + S_{u\wedge v}(u + v)$ completing the proof.

Next we give a weak version of Boboc-Cornea theorem [4], Théorème 1, for prehyperharmonic cones. A stronger version of Boboc-Cornea theorem is proved later.

PROPOSITION 1.9 – Let W be a prehyperharmonic cone. If u is a specific upper bound of x and y in W then $u \succcurlyeq x \vee y$.

PROOF – Let x and y be elements of W and $u \in W$ be a specific upper bound of x and y. Assume first that u is cancellable. According to Corollary 1.5 x and y are cancellable. Then the preceding proposition shows that $x \vee y = S_{x\wedge y}(x + y)$. Since $u \succcurlyeq x$ and $u \succcurlyeq y$, there exist elements x' and y' in W such that $u = x + x' = y + y'$. By Corollary 1.5 x' and y' are cancellable. Using Theorem 1.4 we find that

$$S_{x\wedge y}(x + y) = S_{(x+x'+y')\wedge(y+y'+x')}2u = S_{u+y'\wedge x'}2u = S_{y'\wedge x'}u.$$

Hence $x \vee y = S_{y'\wedge x'}u \preccurlyeq u$.

Suppose now that u is an arbitrary element of W. Then u is cancellable in $\underline{u} + W$ by Theorem 1.6. According to the first part of this proof we have $(x + \underline{u}) \vee (y + \underline{u}) \preccurlyeq u + \underline{u} = u$. The inequality $\underline{u} + x \vee y \geqq (\underline{u} + x) \vee (\underline{u} + y)$ is trivial. Applying (1.5) to $(\underline{u} + x) \vee (\underline{u} + y) \geqq \underline{u}$ we obtain

$$\underline{u} + (\underline{u} + x) \vee (\underline{u} + y) = (\underline{u} + x) \vee (\underline{u} + y).$$

This result implies that

$$(\underline{u} + x) \vee (\underline{u} + y) \geqq \underline{u} + x \vee y.$$

Thus we conclude

$$u \succcurlyeq (\underline{u} + x) \vee (\underline{u} + y) = \underline{u} + x \vee y \succcurlyeq x \vee y,$$

which is the desired result.

The preceding result can be used to show that mixed lower envelopes exist in prehyperharmonic cones.

THEOREM 1.10 – Let $(W,+,\leqq)$ be a prehyperharmonic cone. For any elements u and x in W, the mixed lower envelope $u \curlywedge x$ exists.

PROOF – Let u be an arbitrary element of W. Assume first that $x \in W$ is cancellable in W. We show that the mixed lower envelope $(w + \underline{u})$ x exists for all cancellable elements w in W. Suppose w is cancellable in W. Since cancellable elements of W form a superharmonic structure by Theorem 1.2 and [1], Theorem 19.1, the mixed lower envelope $(w + \underline{u}\wedge x)\curlywedge x$ exists. According to (1.5) $\underline{u} \wedge x \preccurlyeq u$ and so

$$(w + \underline{u} \wedge x) \curlywedge x \preccurlyeq w + \underline{u} \wedge x \preccurlyeq w + \underline{u}.$$

Certainly, the inequality $(w + \underline{u} \wedge x)\curlywedge x \leqq x$ also holds. Hence the element $(w + \underline{u} \wedge x) \curlywedge x$ is a candidate for $(w + \underline{u})\curlywedge x$. Now let y be an arbitrary element of W satisfying $y \preccurlyeq \underline{u} + w$ and $y \leqq x$. Using (1.7) we obtain

$$y \preccurlyeq y + w \wedge S_y(w + \underline{u}) = (y + w) \wedge (y + S_y(w + \underline{u}) = w + \underline{u} \wedge y.$$

This result implies that $y \preccurlyeq w + \underline{u} \wedge y + \underline{u} \wedge x$. Since $y \leqq x$, Corollary 1.5 states that y is cancellable. Thus by Theorem 1.2 the mixed lower envelope $(w + \underline{u} \wedge x + \underline{u} \wedge y) \curlywedge x$ exists and

$$y \leqq (w + \underline{u} \wedge x + \underline{u} \wedge y) \curlywedge x.$$

Noticing the equality $\underline{u} \wedge x + \underline{u} \wedge y = \underline{u} \wedge (x + y)$ and referring to [2], (4.24), we see that

$$y \leqq ((w \curlywedge y + \underline{u} \wedge (x + y)) \curlywedge y) \curlywedge y \leqq (w + \underline{u} \wedge x) \curlywedge x.$$

Consequently, the result $(w + \underline{u} \wedge x) \curlywedge x = (w + \underline{u}) \curlywedge x$ holds.

Next note that the mixed lower envelope $u \curlywedge (\underline{u} + x)$ exists. Indeed, the elements u and $\underline{u} + x$ are cancellable in $\underline{u} + W$. Therefore the mixed lower envelope of u and $\underline{u} + x$ exists in $\underline{u} + W$ by Theorem 1.2. Clearly the mixed lower envelopes of u and $\underline{u} + x$ in W and in $\underline{u} + W$ are equal.

Choose now $w = S_{\underline{u}}(u \curlywedge (x + \underline{u}))$. From $\underline{u} \leqq u \curlywedge (x + \underline{u})$ it follows by (1.5) $\underline{u} \preccurlyeq u \curlywedge (x + \underline{u})$ and further by (1.4) $w + \underline{u} = u \curlywedge (x + \underline{u})$. On account of the first part of the proof the mixed lower envelope $(w + \underline{u}) \curlywedge x$ exists and $(w + \underline{u}) \curlywedge x = (w + \underline{u} \wedge x) \curlywedge x$. Hence

$$(u \curlywedge (x + \underline{u})) \curlywedge x = (w + \underline{u}) \curlywedge x = (w + \underline{u} \wedge x) \curlywedge x.$$

Certainly the element $(u \curlywedge (x + \underline{u})) \curlywedge x$ is a candidate for the mixed lower envelope $u \curlywedge x$. Assume that t is an arbitrary element of W satisfying $t \preccurlyeq u$ and $t \leqq x$. Since $w \leqq x$ and

$$w \preccurlyeq \underline{u} + w = u \curlywedge (x + \underline{u}) \preccurlyeq u,$$

Proposition 1.9 gives $w \vee t \leqq x$ and $w \vee t \preccurlyeq u$. Thus $w \vee t + \underline{u} \preccurlyeq u + \underline{u} = u$ and $w \vee t + \underline{u} \leqq x + \underline{u}$. These inequalities imply that

$$u \curlywedge (\underline{u} + x) \geqq w \vee t + \underline{u} \geqq w + \underline{u} = u \curlywedge (x + \underline{u}).$$

Hence $w \vee t \preccurlyeq u \curlywedge (x + \underline{u})$. Since also $w \vee t \leqq x$, we obtain

$$t \leqq w \vee t \leqq (u \curlywedge (x + \underline{u})) \curlywedge x.$$

Thus $(u \curlywedge (x + \underline{u})) \curlywedge x$ is the mixed lower envelope $u \curlywedge x$ provided that x is cancellable.

Finally, assume that x is arbitrary. It is enough to prove that $u \curlywedge (u \wedge x))$ exists, since $u \curlywedge x = u \curlywedge (u \wedge x)$. We may assume that $x \leqq u$. The element x is cancellable in $\underline{x} + W$ by Theorem 1.6 and therefore the mixed lower envelope $(u + \underline{x}) \curlywedge (x + \underline{x})$ exists in $\underline{x} + W$. It follows that $(u + \underline{x}) \curlywedge (x + \underline{x})$ also exists in W. But then $u \geqq x \geqq \underline{x}$ leads by (1.5) to $u + \underline{x} = u$ and $x + \underline{x} = x$. Consequently $u \curlywedge x$ exists in W, which completes the proof.

Mixed lower and upper envelopes can be used to prove the generalization of Proposition 1.9. The proof of M. Arsove and H. Leutwiler [1], Theorem 17.9, applies also in prehyperharmonic cones.

THEOREM 1.11 – Let E be a subset of a prehyperharmonic cone W.

If E *admits the least upper bound* $\bigvee E$ *and* $w \in W$ *is a specific upper bound of* E, *then* $\bigvee E \preccurlyeq w$. *Similarly, if* E *admits the greatest lower bound* $\bigwedge E$ *and* w *is a specific lower bound of* E, *then* $\bigwedge E \succcurlyeq w$.

In order to show that mixed lower envelopes are translation invariant we need the following result.

LEMMA 1.12 – *Let* W *be a prehyperharmonic cone and* u *be an element of* W. *If* $w \in W$ *satisfies* $w \geqq \underline{u}$, *then* $S_{\underline{u}} w$ *is the specifically smallest element of the set* $V = \{ x \in W : w \preccurlyeq x + \underline{u} \}$.

PROOF – Note first that $S_{\underline{u}} w$ belongs to V. Indeed, (1.4) and (1.5) yields $w = \underline{u} + S_{\underline{u}} w$. Assume next that $x \in V$ is cancellable. Then the property $x \geqq S_{\underline{u}} w$ implies that by Corollary 1.5 that $S_{\underline{u}} w$ is cancellable. From $x + \underline{u} \succcurlyeq w \succcurlyeq S_{\underline{u}} w$ it follows that

$$x + \underline{u} \leqq \underline{u} + S_{\underline{u}} w \curlyvee x \leqq \underline{u} + x.$$

Hence $S_{\underline{u}} w + x + \underline{u} = x \curlywedge S_{\underline{u}} w + x + \underline{u}$ by Theorem 1.2. This leads to

$$S_{\underline{u}} w + \underline{u} = x \curlywedge S_{\underline{u}} w + \underline{u}.$$

Thus $x \curlywedge S_{\underline{u}} w$ belongs to V and $S_{\underline{u}} w \leqq x \curlywedge S_{\underline{u}} w \leqq S_{\underline{u}} w$. Therefore we have that $x \succcurlyeq x \curlywedge S_{\underline{u}} w = S_{\underline{u}} w$ provided $x \in V$ is cancellable.

Lastly, suppose $x \in V$ is arbitrary. Then x is cancellable in $\underline{x} + W$ and $S_{\underline{u}} w + \underline{x} + \underline{u} \preccurlyeq x + \underline{x} + \underline{u}$. Our previous work shows that

$$\min \{ w \in \underline{x} + W : S_{\underline{u}} w + \underline{x} + \underline{u} \preccurlyeq w + \underline{x} + \underline{u} \} \preccurlyeq x.$$

Using (1.6) and the equality $w = \underline{u} + S_{\underline{u}} w$ we notice that

$$x \succcurlyeq \underline{x} + S_{\underline{u}}(\underline{u} + S_{\underline{u}} w) = \underline{x} + S_{\underline{u}} w \succcurlyeq S_{\underline{u}} w.$$

Consequently, $S_{\underline{u}} w$ is the specifically smallest element of V.

THEOREM 1.13 – *The following properties hold for all* u, v, x *and* y *in a prehyperharmonic cone* W:

(a) $(u + x) \curlyvee (u + y) = u + x \curlyvee y$,

(b) if $x \preccurlyeq y$ then $v \curlyvee x \preccurlyeq v \curlyvee y$,

(c) if $v \preccurlyeq y$ and $x \preccurlyeq y$, then $v \curlyvee x \preccurlyeq y$,

(d) $(u + x) \curlywedge (u + y) = u + x \curlywedge y$,

(e) if $x \preccurlyeq y$ and $x \preccurlyeq y$, then $x \preccurlyeq v \curlywedge y$,

(f) if $x \preccurlyeq y$ then $v \curlywedge x \preccurlyeq v \curlywedge y$,

(g) $x \curlywedge y + y \curlyvee x = x + y$.

The proof of these assertions is the same as in hyperharmonic cones (see [7], Theorem 5.3).

Calculation rules of mixed lower and upper envelopes in prehyperharmonic cones are exactly like those in hyperharmonic cones (see [7], p.34-37).

EXAMPLE – None of Axioms 1, 2 or 3 is sufficient alone to imply the existence of mixed lower envelopes. We first consider an example similar to [7], Remark 2.6 (b). Let us denote

$$S = \{(x,0,0) : x \in \mathbb{R}_+\} \cup \{(\infty,\infty,0), (\infty,0,\infty), (\infty,\infty,\infty)\}.$$

With the usual partial ordering, addition and multiplication by $\mathbb{R}_+ \setminus \{0\}$ the set S is a hyperharmonic structure satisfying Axioms 1 and 2. However, the mixed lower envelope $(\infty,\infty,0) \curlywedge (\infty,0,\infty)$ does not exist.

Secondly, we set

$$W = \mathbb{R}_+^2 \cup \{ (\infty,x) : x > 0\} \cup \{(\infty,x,\infty) : x \geqq 0\}.$$

Addition is defined in W as usual for elements having the same number of coordinates. Otherwise, it is defined as follows

$$(\infty,x,\infty) + (y,z) = (\infty,x + y,\infty).$$

Multiplication by $\mathbb{R}_+ \setminus \{0\}$ is defined in W as usual and partial ordering is alphabetical ordering. The cone $(W,+,\leqq)$ is a hyperharmonic structure satisfying Axioms 1 and 3. However, for $u = (\infty,1)$ the mixed lower envelope $u \curlywedge \underline{u}$ does not exist. The preceding example also shows that mixed lower envelopes do not necessarily exist without the assumption $\underline{u} \preccurlyeq u$.

2. HYPERHARMONIC CONES AND HYPERHRAMONIC MORPHISMS

In this section prehyperhamonic cones satisfying convergence axioms are studied.

DEFINITION 2.1 – A prehyperharmonic cone $(W,+,\leqq)$ is called a <u>σ-hyperharmonic</u> (resp. <u>hyperharmonic</u>) <u>cone,</u> if the following axioms hold:

AXIOM 4 – Any increasing sequence (resp. upward directed family) F has the least upper bound $\bigvee F$ such that $\bigvee(a + F) = a + \bigvee F$ for all $a \in W$;

AXIOM 5 – Any decreasing sequence (resp. downward directed family) F has the greatest lower bound $\bigwedge F$.

Note that the greatest lower bound is always translation invariant in σ-hyperharmonic cones and hyperharmonic cones ([1], (17.38)).

Hyperharmonic cones have already been studied in [7]. Their theory with natural changes carries also over to σ-hyperharmonic cones.

EXAMPLES – (1) Positive hyperharmonic functions on a harmonic space form a hyperharmonic cone.

(2) Positive measurable functions form a σ-hyperharmonic cone.

(3) If V is a kernel on $(E,\mathcal{B})$, then V-supermedian functions form a σ-hyperharmonic cone.

(4) Every H-cone can be extended to a hyperharmonic cone.

PROPOSITION 2.2 – Let W be a σ-hyperharmonic cone. For any u and v in W there exist the greatest specific lower bound $u \curlywedge v$ and the least specific upper bound $u \curlyvee v$ having the properties

$$u \curlywedge v + a = (u + a) \curlywedge (v + a)$$

and

$$u \curlyvee v + a = (u + a) \curlyvee (v + a).$$

PROOF – Theorem 1.10 enables us to apply the same methods as in [1], Theorem 4.8. Let u and v be arbitrary elements of a σ-hyperharmonic cone W. Define inductively

$$u_1 = u,\ v_1 = v,\ u_{n+1} = u_n \curlyvee v_n,\ v_{n+1} = v_n \curlyvee u_n.$$

Then the sequences (u_n) and (v_n) are specifically increasing. Moreover, we have $u_{n+1} \geqq v_n$ and $v_{n+1} \geqq u_n$ for all $n \in \mathbb{N}$. Hence the sequences (u_n) and (v_n) admit the common least upper bound denoted by w. By (1.4) we have $u_n = u_1 + S_{u_1} u_n$ for all $n \in \mathbb{N}$. This implies

$$w = \bigvee u_n = u_1 + \bigvee S_{u_1} u_n,$$

since $(S_{u_1} u_n)$ is increasing. Therefore $w \succcurlyeq u_1 = u$. Similarly, we obtain $w \succcurlyeq v$. Assuming $x \succcurlyeq u$ and $x \succcurlyeq v$ for $x \in W$ we find that $x \succcurlyeq u_n$ by Theorem 1.13 (c) and so $x \succcurlyeq \bigvee u_n = v$. Consequently, $w = u \curlyvee v$.

Analogously, we can prove that the greatest specific lower bound $u \curlywedge v$ exists. The last assertions follow from Theorem 1.13 (a), (b) and the methods used above.

The preceding propositon is applied to prove the following theorem.

THEOREM 2.3 – Let $(W,+,\leqq)$ be a σ-hyperharmonic (resp. hyperharmonic) cone. Then $(W,+,\preccurlyeq)$ is also a σ-hyperharmonic (resp. hyperharmonic) cone.

The proof is the same as in [7], Theorem 4.16 (b).

COROLLARY 2.4 – If $(W,+,\leqq)$ is a σ-hyperharmonic cone and $u \preccurlyeq v_1 + v_2$ for $u, v_1, v_2 \in W$, then there exist elements u_1 and u_2 such that $u = u_1 + u_2$ and $u_1 \preccurlyeq v_1$, $u_2 \preccurlyeq v_2$.

Both σ-hyperharmonic and hyperharmonic cones have a representation theorem which can be proved similarly to [7], Theorem 7.4.

THEOREM 2.5 — *Let* $(W,+,\leqq)$ *be a* σ*-hyperharmonic (resp. hyperharmonic) cone. Then there exists a* σ*-hyperharmonic (resp. hyperharmonic) cone* V *and a bijection* $\varphi: W \to V$ *satisfying the following conditions:*

(a) V *is a convex subcone of the set* $\overline{C}^{+}(X)$ *formed by continuous, positive extended real valued functions on a locally comapct Hausdorff space* X;

(b) φ *is additive and positively homogeneous*;

(c) $s \leqq t \Longleftrightarrow \varphi(s) \leqq \varphi(t)$ *for all* $s, t \in W$;

(d) s *is cancellable if and only if* $\varphi(s)$ *is finite on a dense set*.

An additive map φ from a σ-hyperharmonic (resp. hyperharmonic) cone W into a σ-hyperharmonic (resp. hyperharmonic) cone V is called a *σ-hyperharmonic* (resp. *hyperharmonic*) *morphism* if

(2.1) $\varphi(\bigvee F) = \bigvee_{f \in F} \varphi(f)$ for any increasing sequence (resp. upward directed family) F.

THEOREM 2.6 — *The set* W^* *of* σ*-hyperharmonic (resp. hyperharmonic) morphisms satisfying* $\varphi(0) = 0$ *from a* σ*-hyperharmonic (resp. hyperharmonic) cone* W *into* $\overline{\mathbb{R}}_{+}$ *is a* σ*-hyperharmonic (resp. hyperharmonic) cone, if* W *possesses the property*

(H4) $s \wedge (\bigvee F) = \bigvee_{f \in F}(s \wedge t)$ *for any increasing sequences (resp. upward directed families)* F.

The proof is similar to [7], Corollary 6.16.

Note that every H-cone can be extended to a hyperharmonic cone satisfying (H4) ([8], Proposition 2.2).

Define $\tilde{s}: W^* \to \overline{\mathbb{R}}_{+}$ by $\tilde{s}(\mu) = \mu(s)$. Then $\tilde{s}$ is a hyperharmonic morphism.

THEOREM 2.7 — *Let* W *be a* σ*-hyperharmonic (resp. hyperharmonic) cone satisfying (H4) and the separation property*

(S) *for any* $u, v \in W$ *such that* $u \neq \underline{u}$ *and* $u < v$, *there exists a* σ*-hyperharmonic (resp. hyperharmonic) morphism* $\mu: W \to \overline{\mathbb{R}}_{+}$ *satisfying* $0 < \mu(u) < \mu(v)$.

Then the evaluation mapping $s \to \tilde{s}$ *has the following properties:*

(a) $S_{\tilde{t}}\tilde{s} = \widetilde{S_t s}$,

(b) $\tilde{s} \wedge \tilde{t} = \widetilde{s \wedge t}$,

(c) $s \leqq t \Longleftrightarrow \tilde{s} \leqq \tilde{t}$,

(d) $(\underline{\tilde{s}}) = \underline{(\tilde{s})}$,

(e) $$\widetilde{\bigvee F} = \bigvee_{f \in F} \tilde{f}$$

for any $s, t \in W$ and $F \subset W$ provided that $\bigvee F$ exists.

The proof is similar to the proof of [7], Theorem 3.7, with obvious changes. Note that (a) does not necessarily hold without (S). For counter-example see [9], p.14.

ACKNOWLEDGMENT

The author is grateful to the Academy of Finland for their support of the present research.

REFERENCES

1. Arsove M.G. and H. Leutwiler: Algebraic potential theory. Mem. Amer. Math. Soc. 226 (1980).
2. Arsove M.G. and H. Leutwiler: Quasi-units in mixed lattices structures. Lecture Notes in Math. 789, Springer-Verlag, Berlin-Heidelberg-New York (1979), 35-54.
3. Boboc N., Gh. Bucur and A. Cornea: Order and convexity in potential theory: H-cones. Lecture Notes in Math. 853, Springer-Verlag, Berlin-Heidelberg-New York (1981).
4. Boboc, N. and A. Cornea: H-cônes et biadjoints des H-cônes. C. R. Acad. Sci. Paris 270 (1970) 1679-1682.
5. Cornea, A. and S.-L. Eriksson: Order continuity of the greatest lower bound of two functionals. Analysis 7 (1987) 173-184.
6. Cornea, A. and G. Licea: Order and potential resolvent families of kernels. Lecture Notes in Math. 494, Springer-Verlag, Berlin-Heidelberg-New York (1975).
7. Eriksson, S.-L.: Hyperharmonic cones and hyperharmonic morphims. Ann. Acad. Sci. Fenn. Ser. AI Math. Dissertationes 49 (1984).
8. Eriksson. S.-L.: Hyperharmonic cones and cones of hyperharmonics. An. Stiint. Univ. "Al.I.Cuza" Iasi Sect Ia Mat. (1985), 109-116.
9. Eriksson, S.-L.: Representations of hyperharmonic cones. To appear in Trans. Amer. Math. Soc.

JUMP RELATIONS FOR SURFACE POTENTIALS IN CASE OF CONTINUOUS DENSITIES

Alexander Hornberg

Mathematisches Institut II
Universität Karlsruhe
Karlsruhe, FRG

ABSTRACT

It is well-known that the single layer surface potential with uniform HÖLDER continuous density is continuously differentiable with respect to each side of the surface and fulfills jump relations for the first order derivatives. In this paper it is shown that the jump relations still hold in the CAUCHY sense if the density is supposed to be only continuous.

Let $S \in \mathbb{R}^3$ be a C^2-regular closed surface orientated by its unit normal vector $n(x)$, $x \in S$. Furthermore with respect to the orientation of S, let S divide the $\mathbb{R}^3$ into the open connected sets $D_{\pm} \subset \mathbb{R}^3$ and let $\cdot|_{\pm}$ denote the values of a function on the positive or negative side of S, respectively. Then as it is well-known, the gradient of the single layer surface potential

$$\Phi(x) = \iint_S \frac{\rho(y)}{|x-y|} \, ds(y) \quad , \quad x \in \mathbb{R}^3 \quad , \tag{1}$$

with uniform HÖLDER continuous densitiy $\rho(x)$, $x \in S$, can be uniformly extended in a HÖLDER continuous way from D_+ into $\overline{D_+}$ as well as from D_- into $\overline{D_-}$ satisfying the vector jump relation [1]

$$\operatorname{grad} \Phi|_+(x) - \operatorname{grad} \Phi|_-(x) = -4\pi n(x)\rho(x) \quad , \quad x \in S \quad . \tag{2}$$

If the density $\rho(x)$, $x \in S$, is supposed to be only continuous on S, then the normal components of (2) remain still valid in the sense of

$$\lim_{h \to +0} \frac{\partial \Phi}{\partial n}(x + hn(x)) - \lim_{h \to +0} \frac{\partial \Phi}{\partial n}(x - hn(x)) = -4\pi\rho(x) \quad , \quad x \in S \quad , \tag{3}$$

where the limits exist in the sense of uniform convergence on S [1].

Furthermore, in case of a merely C^1-regular surface S with continuous density $\rho(x)$, $x \in S$, the following CAUCHY version of (3) holds pointwise on S [4]:

$$\lim_{h\to+0} \left\{\frac{\partial\Phi}{\partial n}(x + hn(x)) - \frac{\partial\Phi}{\partial n}(x - hn(x))\right\} = -4\pi\rho(x) \quad , \quad x\in S \quad . \tag{4}$$

In the following we refer to a C^2-regular closed surface $S \subset \mathbb{R}^3$. For this it is well-known that there exists a constant $L > 0$ depending on S only and satisfying the inequality [1, p.35]

$$|(n(x),x-y)| \leqq L|x-y|^2 \quad , \quad x,y\in S \quad . \tag{5}$$

If we define now

$$x_{\pm} := x \pm hn(x) \quad , \quad x\in S \quad , \tag{6}$$

then for all $0 < h < \frac{1}{2L}$, the parallel surfaces $\{x_{\pm} \mid x\in S\}$ are C^1-regular and disjoint [3, p.70].

Lemma 1 – For all $0 < h \leqq \frac{1}{4L}$ it holds the inequality

$$|x_{\pm}-y|^2 \geqq \frac{1}{2}|x-y|^2 + h^2 \quad , \quad x,y\in S \quad . \tag{7}$$

Proof – For $0 < h \leqq \frac{1}{4L}$, $x,y\in S$, together with (5) and (6), we obtain

$$2|x_{\pm}-y|^2 - |x-y|^2 - 2h^2 = 2|x-y \pm hn(x)|^2 - |x-y|^2 - 2h^2$$
$$= |x-y|^2 \pm 4h(n(x),x-y) \geqq |x-y|^2 - 4hL|x-y|^2 \geqq 0 \quad .$$

Theorem 1 - Let S be a C^2-regular closed surface and $\rho(x)$, $x\in S$, a continuous density. Then for the corresponding single layer surface potential $\Phi(x)$, $x\in\mathbb{R}^3$, the vector jump relation in the CAUCHY sense

$$\lim_{h\to+0} \{\operatorname{grad}\Phi(x + hn(x)) - \operatorname{grad}\Phi(x - hn(x))\} = -4\pi n(x)\rho(x) \quad , \quad x\in S \quad , \tag{8}$$

holds uniformly on S.

Proof – Let $0 < h \leqq \frac{1}{4L}$. Using the GRASSMANN identity we get [1)]

$$\operatorname{grad}\Phi(x_+) - \operatorname{grad}\Phi(x_-) + 4\pi n(x)\rho(x)$$
$$= \{(n(x),\operatorname{grad}\Phi(x_+) - \operatorname{grad}\Phi(x_-)) + 4\pi\rho(x)\}n(x) \tag{9}$$
$$+ [[n(x),\operatorname{grad}\Phi(x_+) - \operatorname{grad}\Phi(x_-)],n(x)] \quad , \quad x\in S \quad .$$

Take now any real number $\varepsilon > 0$. Then because of (3), there exists a real number $0 < \delta_1 \leqq \frac{1}{4L}$ such that

$$|(n(x),\operatorname{grad}\Phi(x_+) - \operatorname{grad}\Phi(x_-)) + 4\pi\rho(x)| < \frac{\varepsilon}{2} \tag{10}$$

for all $0 < h \leqq \delta_1$ and all $x\in S$. From (6) we get the elementary vector identity

1) The scalar product and the vector product are denoted by $(\cdot,\cdot)$ and $[\cdot,\cdot]$, respectively.

$$[n(x),\ \frac{x_+-y}{|x_+-y|^3} - \frac{x_--y}{|x_--y|^3}] = [n(x),\ \frac{x-y}{|x_+-y|^3} - \frac{x-y}{|x_--y|^3}]$$

$$= - \frac{4h(n(x),x-y)[n(x),x-y]}{|x_+-y||x_--y|(|x_+-y|+|x_--y|)} \left\{\frac{1}{|x_+-y|^2} + \frac{1}{|x_+-y||x_--y|} + \frac{1}{|x_--y|^2}\right\} , \tag{11}$$

valid for all $0 < h \leqq \frac{1}{4L}$ and $x,y \in S$. From (1) and (11) it follows for all $0 < h \leqq \frac{1}{4L}$ and all $x \in S$ that

$$[n(x),\ \mathrm{grad}\,\Phi(x_+) - \mathrm{grad}\,\Phi(x_-)]$$

$$= \iint_S \rho(y) \frac{4h(n(x),x-y)[n(x),x-y)]}{|x_+-y||x_--y|(|x_+-y|+|x_--y|)} \left\{\frac{1}{|x_+-y|^2} + \frac{1}{|x_+-y||x_--y|} + \frac{1}{|x_--y|^2}\right\} ds(y) . \tag{12}$$

By repeated application of Lemma 1, together with (5) and a bound $M > \max_{x\in S} |\rho(x)|$, we obtain for all $0 < h \leqq \frac{1}{4L}$ the estimation

$$|[n(x),\mathrm{grad}\,\Phi(x_+) - \mathrm{grad}\,\Phi(x_-))]| \leqq \frac{24}{\sqrt{2}} ML \iint_S \frac{h\,ds(y)}{|x_+-y||x_--y|}$$

$$\leqq \frac{24}{\sqrt{2}} ML \iint_S \frac{\sqrt{h}\,ds(y)}{\sqrt[4]{|x_+-y|}^3\ \sqrt[4]{|x_--y|}^3} \leqq 24\ \sqrt[4]{2}\ ML\ \sqrt{h} \iint_S \frac{ds(y)}{\sqrt{|x-y|}^3} , \quad x \in S . \tag{13}$$

Since there exists a constant $C > 0$ depending only on S such that [2, p.149]

$$\iint_S \frac{1}{\sqrt{|x-y|}^3}\, ds(y) \leqq C \quad , \quad x \in S \quad , \tag{14}$$

we get with $\delta_2 = \min\left\{\frac{1}{4L}, \frac{1}{\sqrt{2}} (\frac{\varepsilon}{48MLC})^2\right\} > 0$ from (13) and (14)

$$|[n(x),\mathrm{grad}\,\Phi(x_+) - \mathrm{grad}\,\Phi(x_-)]| \leqq \frac{\varepsilon}{2} \tag{15}$$

for all $0 < h \leqq \delta_2$ and all $x \in S$. Taking finally $\delta = \min\{\delta_1,\delta_2\} > 0$ and referring to (9),(10), and (15) we conclude

$$|\mathrm{grad}\,\Phi(x_+) - \mathrm{grad}\,\Phi(x_-) + 4\pi n(x)\rho(x)| < \varepsilon$$

for all $0 < h \leqq \delta$ and all $x \in S$.

Writing (8) in components we are led to the following

Corollary 1 – For the vector potential

$$A(x) = \iint_S \frac{j(y)}{|x-y|}\, ds(y) \quad , \quad x \in \mathbb{R}^3 \quad , \tag{16}$$

with continuous current density $j(x)$, $x \in S$, the jump relations

$$\lim_{h\to+0} \{\mathrm{div}\,A(x+hn(x)) - \mathrm{div}\,A(x-hn(x))\} = -4\pi(n(x),j(x)) \quad , \quad x \in S \ , \tag{17}$$

$$\lim_{h\to+0} \{\mathrm{rot}\,A(x+hn(x)) - \mathrm{rot}\,A(x-hn(x))\} = -4\pi[n(x),j(x)] \quad , \quad x \in S \ , \tag{18}$$

hold uniformly on S.

For an application of the foregoing results let $D \subset \mathbb{R}^3$ be a connected open set for which the boundary $S \subset \mathbb{R}^3$ consists of a finite number of C^2-regular closed surfaces. An exterior boundary surface and interior boundary surfaces are denoted by S^o and $S^1,\ldots,S^p$, respectively. In every case the unit normal vector n on S is orientated to the outside of D. Obviously the foregoing results are also valid for this case.

Next we consider the physical situation that the complement $\mathbb{R}^3 \setminus D$ is filled out by a conductor or supra conductor and that a charge or current is brought into $\mathbb{R}^3 \setminus D$, respectively. Then in a well-known way charge or current densities being in equilibrium are established on the boundary S, respectively. Here we shall show by Theorem 1 and Corollary 1 that this equilibrium densities can be introduced and discussed without using HÖLDER continuity.

<u>Definition 1</u> – A continuous charge density ρ on S is said to be in equilibrium on S if the vector field $v = -\text{grad}\,\Phi$, Φ being the scalar potential (1), vanishes in $\mathbb{R}^3\setminus\overline{D}$.

<u>Theorem 2</u> – If ρ is a continuous charge density being in equilibrium on S, then $v = -\text{grad}\,\Phi$ forms a DIRICHLET vector field in $\overline{D}$ 2).

<u>Proof</u> – Immediately, v is a harmonic vector field in D and if there is no exterior boundary S^o, then in the infinity v converges uniformly to zero. Together with $v = 0$ in $\mathbb{R}^3 \setminus \overline{D}$, we conclude from Theorem 1 that v can be extended continuously from D to $\overline{D}$ with normal boundary values

$$v|_- = -\text{grad}\,\Phi|_- = -4\pi n\rho \quad \text{on} \quad S \quad .$$

<u>Definition 2</u> – A continuous tangential current density j on S is said to be in equilibrium on S if the vector field $v = \text{rot}\,A$, A being the vector potential (16), vanishes in $\mathbb{R}^3\setminus\overline{D}$.

<u>Theorem 3</u> – If j is a continuous tangential current density being in equilibrium on S, then $v = \text{rot}\,A$ forms a NEUMANN vector field in $\overline{D}$ 3).

<u>Proof</u> – First we get $\text{div}\,v = 0$ in D and if there is no exterior boundary S^o, then in the infinity v converges uniformly to zero. Next together with $v = 0$ in $\mathbb{R}^3 \setminus \overline{D}$, Corollary 1 yields the continuous extension of v from D to $\overline{D}$ with tangential boundary values

---

2) A vector field v in $\overline{D}$ is called a DIRICHLET vector field if it is continuous in $\overline{D}$, harmonic in D, and normal on S; if D is unbounded, v tends uniformly to zero in the infinity.

3) A vector field v in $\overline{D}$ is called a NEUMANN vector field if it is continuous in $\overline{D}$, harmonic in D, and tangential on S; if D is unbounded, v tends uniformly to zero in the infinity.

$$v|_- = \text{rot}\,A|_- = 4\pi\,[n,j] \quad \text{on} \quad S \quad .$$

From

$$\text{rot}\,v = \text{rot}\,\text{rot}\,A = \text{grad}\,\text{div}\,A = 0 \quad \text{in} \quad \mathbb{R}^3\setminus\overline{D} \quad , \quad (n,j) = 0 \quad \text{on} \quad S \quad ,$$

and Corollary 1 it follows that div A can be continuously extended from D to $\overline{D}$ with boundary values

$$\text{div}\,A|_- = \text{const} \quad \text{on} \quad S \quad .$$

Furthermore we get from (16)

$$\Delta\,\text{div}\,A = \text{div}\,\Delta A = 0 \quad \text{in} \quad D \quad .$$

If D is unbounded, then in the infinity div A tends uniformly to zero; in case of bounded D the constant values of div A outside of S^o must be 0 by the same argument like before so that

$$\text{div}\,A|_- = 0 \quad \text{on} \quad S^o \quad .$$

Thus φ := div A turns out as a solution of the capacitor problem for $\overline{D}$, i.e. φ is continuous in $\overline{D}$ and hamronic in D with constant boundary values on S where especially for bounded or unbounded D, φ vanishes on S^o or in the infinity, respectively. Now with regard to any C^2-regular closed surface $S^*\subset D$, the flow vanishes because of STOKES theorem,

$$\iint_{S^*} \frac{\partial\varphi}{\partial n}\,ds = \iint_{S^*} (n,\text{grad}\,\text{div}\,A)\,ds = \iint_{S^*} (n,\text{rot}\,\text{rot}\,A)\,ds = 0 \quad ,$$

and so by means of the corresponding capacity tensor it follows that [3]

$$\varphi = 0 \quad \text{on} \quad S^1,\ldots,S^p \quad .$$

Finally the maximum principle for $\overline{D}$ yields $\varphi = 0$ in D and thus

$$\text{rot}\,v = \text{rot}\,\text{rot}\,A = \text{grad}\,\text{div}\,A = \text{grad}\,\varphi = 0 \quad \text{in} \quad D \quad .$$

References

[1] D. COLTON and R. KRESS, Integral equation methods in scattering theory, Wiley, New York 1983.

[2] O.D. KELLOGG, Foundations of Potential Theory, Die Grundlehren der mathematischen Wissenschaften Bd.31, Springer, Berlin/Heidelberg/New York 1967.

[3] E. MARTENSEN, Potentialtheorie, Teubner, Stuttgart 1968.

[4] E. MARTENSEN, Eine Verallgemeinerung der Sprungbeziehungen der Potentialtheorie, Archiv der Mathematik XXVI (1975), 620-626.

ON A UNIFIED CHARACTERIZATION OF CAPACITY

George C. Hsiao and Ralph E. Kleinman

Department of Mathematical Sciences
University of Delaware
Newark, Delaware 19716 U.S.A.

INTRODUCTION

The notion of capacity as a fundamental quantity associated with a smooth surface was treated exhaustively by Polya and Szego in a series of papers over more than twenty years following earlier work by Fekete (1923) and many others who considered the problem of determining the charge accumulated on an electrostatic condenser maintained at a constant potential. See Polya and Szego (1951) and Payne (1967) for extensive references.

In three dimensions, the capacity, C, of a surface $\partial\Omega$ can be defined in terms of the conductor potential. That is, if Ω_c denotes the exterior of $\partial\Omega$, p is the position vector of a point and u is a function for which $\nabla^2 u = 0$ in Ω_c, $u = 1$ on $\partial\Omega$ and $u = 0(\frac{1}{|p|})$ then $C = -\frac{1}{4\pi}\int_{\partial\Omega} \frac{\partial u}{\partial n}\, ds$. Alternatively capacity may be defined in terms of the solution of the Helmholtz resonator problem, that is, if $B_R(0)$ is a ball of radius R with center at the origin of a coordinate system centered in Ω, $\partial\Omega$ is contained in $B_R(0)$, and u_1 is a function for which $\nabla^2 u_1 = 0$ in $\Omega_c \cap B_R(0)$, $u_1 = 0$ on $\partial\Omega$ and $u_1 = 1$ on $\partial B_R(0)$ then $C = \lim_{R\to\infty} \frac{1}{4\pi}\int_{\partial B_R(0)} \frac{\partial u_1}{\partial n}\, ds$.

These characterizations of capacity are equivalent in $\mathbf{R}^3$ since in general the solutions of the conductor potential and the Helmholtz resonator problems are related by $u = 1 - \lim_{R\to\infty} u_1$. For example if $\partial\Omega$ is a sphere of radius a then the conductor potential is simply

$$u = \frac{a}{r} \text{ and } C = -\frac{1}{4\pi}\int_0^{2\pi} d\phi \int_0^{\pi} d\theta\, \frac{\partial}{\partial r}\left(\frac{a}{r}\right)_{r=a} a^2 \sin\theta = a.$$

In this case the solution of the Helmholtz resonator problem is

$$u_1 = \frac{\frac{1}{r} - \frac{1}{a}}{\frac{1}{R} - \frac{1}{a}} \text{ and } \frac{1}{4\pi}\int_{\partial B_R(0)} \frac{\partial u_1}{\partial n}\, ds = \frac{1}{\frac{1}{a} - \frac{1}{R}}. \text{ Hence } C = \lim_{R\to\infty} \frac{1}{\frac{1}{a} - \frac{1}{R}} = a$$

as before.

Unfortunately neither of these problems leads to nontrivial characterizations of capacity in $\mathbf{R}^2$ because of the behavior of regular solutions of Laplace's equation. That is, if we ascribe to regular solutions of Laplace's equation in exterior domains in $\mathbf{R}^2$ the same behavior at infinity as the fundamental solution then the conductor potential problem $\nabla^2 u = 0$ in Ω_c, $u = 1$ on $\partial\Omega$ and $u = 0(\ln |p|)$ does not always have a unique solution ($A \ln |p|$ is a solution for any A when $\partial\Omega$ is a circle of radius 1) and while the Helmholtz resonator problem is uniquely solvable, the quantity comparable to capacity in $\mathbf{R}^3$ always vanishes, i.e. $\lim_{R\to\infty} \int_{\partial B_R(0)} \frac{\partial u_1}{\partial n} ds = 0$. For example if $\partial\Omega$ is a circle of radius a then $u_1 = \frac{\ln r - \ln a}{\ln R - \ln a}$ and $\int_{\partial B_R(0)} \frac{\partial u_1}{\partial n} ds = \frac{2\pi}{\ln R - \ln a}$ which obviously vanishes when $R \to \infty$.

One might try to alter the condition at infinity in an effort to arrive at a definition of capacity consistent with those in $\mathbf{R}^3$. But the obvious changes are not helpful. The problem $\nabla^2 u = 0$ in Ω_c; $u = 1$ on $\partial\Omega$ and $u = 0(1)$ as $|p| \to \infty$ has the unique solution $u = 1$ in which case $C = \int_{\partial\Omega} \frac{\partial u}{\partial n} ds = 0$ while if we change the condition at infinity to $u = o(1)$, then the problem has no solution. In either case one cannot define capacity as in $\mathbf{R}^3$.

Polya and Szego were, of course, aware of this behavior and as early as 1931 defined two dimensional capacity in terms of the solution of the problem $\nabla^2 u = 0$ in Ω_c; $u = \ln \frac{1}{|p|} + 0(\frac{1}{|p|})$ and $u =$ constant on $\partial\Omega$. The constant boundary value could not be prescribed but was part of the solution and they called this constant $\ln \frac{1}{C}$ with C being the capacity constant. In 1945 they changed the definition slightly by formulating the potential problem $\nabla^2 u = 0$ in Ω_c, $u = 0$ on $\partial\Omega$ and $u = \ln \frac{1}{|p|} - \ln \frac{1}{C} + 0(\frac{1}{|p|})$ where as before C could not be prescribed but appeared as part of the solution. They termed C the outer radius or transfinite diameter (following Fekete) because of its interpretation as the diameter of the circle whose exterior is the image of Ω_c under a conformal transformation that preserves the point at infinity. However they refrained from referring to C as the capacity constant despite their earlier designation. This may have stemmed from the apparent inconsistency with the definition of capacity in $\mathbf{R}^3$. One may infer from their repeated and varied treatments of this capacity question that, in addition to the study of isoperimetric inequalities, they hoped to achieve a uniform characterization of capacity in terms of the solution of a potential problem in all dimensions and remove the annoying necessity of using one definition in two dimensions and another in three.

More recently, Fichera (1961), applied integral equations of the first kind to a variety of elliptic problems. This was further developed by Hsiao and MacCamy (1973) in some specific cases including Laplace's equation where they showed how logarithmic capacity (essentially the logarithm of the transfinite diameter) could be obtained as part of a solution of a system of first kind integral equations in $\mathbf{R}^2$. Symm (1967) and Jaswon and Symm (1977) also used first kind integral equations to characterize capacity in both $\mathbf{R}^2$ and $\mathbf{R}^3$ but not in a uniform manner.

In the present paper we show that the concept of outer radius, C, may be generalized to n dimensions through

$$\omega = -g_n(C)$$

where g_n is the fundamental solution of Laplace's equation (c.f. Eq. (2) below) and ω is a constant which appears as part of the solution of a uniquely solvable system of integral equations which coincide, for $n = 2$, to that studied by Hsiao and MacCamy

(1973). We also formulate a uniquely solvable exterior boundary value problem for the Laplacian in $\mathbf{R}^n$, which generalizes that defined by Polya and Szego (1931) for $n = 2$, and contains as part of the solution the constant ω.

NOTATION, DEFINITIONS AND REPRESENTATION FORMULA

Let Ω be a bounded simply connected domain in $\mathbf{R}^n$ with smooth (Lyapunoff) boundary $\partial\Omega$. Denote the complement of the closure by Ω_c and the unit exterior normal at each point $q \in \partial\Omega$ by $\hat{n}_q$. Let S_n denote the surface area of a unit ball in $\mathbf{R}^n$, i.e.,

$$S_n = \int_{|q|=1} ds_q = \frac{2\pi^{n/2}}{\Gamma(n/2)} \tag{1}$$

where $\Gamma(\cdot)$ is the usual gamma function and $|\cdot|$ is the Euclidian distance. Denote the fundamental solution of Laplace's equation for p and q in $\mathbf{R}^n$ by

$$g_n(|p-q|) = \begin{cases} -\dfrac{1}{S_2} \ln \dfrac{1}{|p-q|} , & n = 2 \\ \dfrac{-1}{(n-2)S_n} |p-q|^{2-n} , & n > 2 \end{cases} . \tag{2}$$

The normal derivative of g_n is seen to be

$$\frac{\partial g_n(|p-q|)}{\partial n_q} = \frac{\hat{n}_q \cdot (q-p)}{S_n \; |p-q|^n} . \tag{3}$$

Let $B_\in(p)$ denote a ball of radius $\in$ and center p and define

$$\alpha(p) = \lim_{\in\to o} \int_{\Omega\cap\partial B_\in(p)} \frac{\partial}{\partial n_q} g_n(|p-q|)ds_q = \begin{cases} 1 & , p \in \Omega \\ 1/2 & , p \in \partial\Omega \\ 0 & , p \in \Omega_c \end{cases} \tag{4}$$

in terms of which Gauss' integral may be written

$$\int_{\partial\Omega} \frac{\partial g_n(|p-q|)}{\partial n_q} ds_q = \alpha(p) . \tag{5}$$

It is convenient to have the following forms of Green's theorem for interior and exterior domains:

If $u \in C^2(\Omega) \cap C^1(\overline{\Omega})$ and $\nabla^2 u = 0$ in Ω then

$$\int_{\partial\Omega} \left\{ u(q) \frac{\partial}{\partial n_q} g_n(|p-q|) - g_n(|p-q|) \frac{\partial u}{\partial n_q}(q) \right\} ds_q = \alpha(p)u(p) ; \tag{6}$$

and

if $u \in C^2(\Omega_c) \cap C^1(\overline{\Omega}_c)$, $\nabla^2 u = 0$ in Ω_c , $u = Ag_n(r) + \omega + 0(1)$ and $\dfrac{\partial u}{\partial n} = A \dfrac{\partial g_n(r)}{\partial r} + o(\dfrac{1}{rlnr})$ as $r := |p| \to \infty$, where A and ω are constants but otherwise arbitrary, then

$$\int_{\partial\Omega} \left\{ g_n(|p-q|) \frac{\partial u(q)}{\partial n_q} - u(q) \frac{\partial}{\partial n_q} g_n(|p-q|) \right\} ds_q = [1-\alpha(p)]u(p) - \omega . \tag{7}$$

A function $u \in \Omega_c$ is said to be regular at infinity if

$$u = 0\,(\frac{1}{r^{n-2}}) \quad \text{and} \quad \frac{\partial u}{\partial r} = 0\,(\frac{1}{r^{n-1}}) \quad \text{as} \quad r \to \infty \ .$$

Thus when Green's theorem is applied to harmonic functions regular at infinity the constant ω will be absent in (7) for $n > 2$.

PREVIOUS DEFINITIONS OF CAPACITY AND RELATED QUANTITIES

In this section we present some of the existing definitions of capacity and related quantities.

If Ω_o is a bounded simply connected domain in $\mathbf{R}^3$ with smooth (Lyapunoff) boundary $\partial\Omega_o$ which contains $\overline{\Omega}$ in its interior then the capacity of $\partial\Omega$ with respect to $\partial\Omega_o$ is defined to be (see, e.g. Szego (1945) and Protter and Weinberger (1965))

$$C_o = -\frac{1}{4\pi} \int_{\partial\Omega} \frac{\partial u_o(q)}{\partial n_q} ds_q \tag{8}$$

where u_o is the capacitory potential of $\partial\Omega$ with respect to $\partial\Omega_o$ defined through

$$\begin{aligned} \nabla^2 u_o &= 0 \quad \text{in} \quad \Omega_o \setminus \overline{\Omega} \\ u_o &= 1 \quad \text{on} \quad \partial\Omega \\ u_o &= 0 \quad \text{on} \quad \partial\Omega_o \end{aligned} \ . \tag{9}$$

If $\Omega_o = B_\rho(0)$ then as $\rho \to \infty$, u_o becomes the solution of the conductor potential problem,

$$\begin{aligned} \nabla^2 u_1 &= 0 \quad \text{in} \quad \Omega_c \\ u_1 &= 1 \quad \text{on} \quad \partial\Omega \\ u_1 &= o(1) \quad \text{as} \quad r \to \infty \end{aligned} \tag{10}$$

and the capacity of $\partial\Omega$ is

$$C = \lim_{\rho\to\infty} C_o = -\frac{1}{4\pi} \int_{\partial\Omega} \frac{\partial u_1(q)}{\partial n_q} ds_q \tag{11}$$

In $\mathbf{R}^2$ one may define the capacity of $\partial\Omega$ with respect to $\partial\Omega_o$ analogously as

$$C_o = -\frac{1}{2\pi} \int_{\partial\Omega} \frac{\partial u_o(q)}{\partial n_q} ds_q \tag{12}$$

where u_o is as before with Ω and Ω_o in $\mathbf{R}^2$ rather than $\mathbf{R}^3$. However, in contrast with the three dimensional case it is not possible to define the capacity of $\partial\Omega$ simply by letting $\Omega_o \to \mathbf{R}^2$ as was pointed out in the introduction.

Alternatively the capacity of $\partial\Omega$ with respect to $\partial\Omega_o$ has been defined as Szego (1945), Polya and Szego (1945)

$$C_o = \frac{1}{S_n} \int_{\Omega_o \setminus \overline{\Omega}} |\nabla u_o|^2 \, dq \quad , \quad n = 2, 3 \ . \tag{13}$$

Moreover, it has been shown Kellogg (1953), Stakgold (1968), that u_o minimizes the Dirichlet integral over the class of admissable functions

$$U_{ad} := \left\{ u \;\middle|\; \begin{array}{l} u \in C^1(\Omega_o \setminus \overline{\Omega}) , \\ u = 1 \quad \text{on} \quad \partial\Omega , \\ u = 0 \quad \text{on} \quad \partial\Omega_o \end{array} \right\} \tag{14}$$

and C_o is the minimum value of the Dirichlet integral (13). As before the capacity of $\partial\Omega$ may be obtained as a limiting value of C_o as Ω_o grows to fill the entire space in $\mathbf{R}^3$ but not in $\mathbf{R}^2$.

Another characterization of capacity in $\mathbf{R}^3$ was given by Szego (1945), Riemann-Weber (1930), Stakgold (1968) where by C, also called the outer radius of $\partial\Omega$, is given by

$$\frac{1}{C} = \min_{\sigma \in V_{ad}} \int_{\partial\Omega} \int_{\partial\Omega} \frac{\sigma(p)\sigma(q)}{|p-q|} \, ds_p \, ds_q \tag{15}$$

where

$$V_{ad} = \left\{ \sigma \;\middle|\; \begin{array}{l} \sigma \in C^{\alpha}(\partial\Omega) , \; 0 < \alpha < 1 \\ \int_{\partial\Omega} \sigma(q) \, ds_q = 1 \end{array} \right\} . \tag{16}$$

Or, more appropriately in view of modern boundary element analyses Hsiao and Wendland (1977) and Nedelec (1977), the Holder continuous function space $C^{\alpha}(\partial\Omega)$ can be replaced by the Sobolev space $H^{-1/2}(\partial\Omega)$. We will subsequently show that this definition may be extended to $\mathbf{R}^n$ for any $n \geq 2$.

In $\mathbf{R}^2$, the outer radius or transfinite diameter has been characterized as the constant C occurring in the solution of the following problem (see Szego (1945))

$$\begin{aligned} \nabla^2 u &= 0 \quad \text{in} \quad \Omega_c \\ u &= 0 \quad \text{on} \quad \partial\Omega \\ u &= \log \frac{1}{r} - \log \frac{1}{C} + o(1) \quad \text{as} \quad r \to \infty . \end{aligned} \tag{17}$$

Alternatively, Symm (1967) introduced the Robin constant γ as a part of the solution of the coupled set of boundary integral equations

$$\begin{aligned} &\int_{\partial\Omega} \log |p-q| \, \sigma_1(q) \, ds_q + \gamma = -\log |p| \\ &\int_{\partial\Omega} \sigma_1(q) \, ds_q = 0 . \end{aligned} \tag{18}$$

The outer radius of transfinite diameter is given in terms of γ as

$$C = e^{-\gamma} . \tag{19}$$

The outer radius is also related Szego (1945), Symm (1967) the conformal mapping, f, of the domain Ω_c to exterior of the unit circle which preserves the point at infinity by

$$C = \lim_{z \to \infty} \frac{1}{|f'(z)|} . \tag{20}$$

Other characterizations of transfinite diameter have been given by Fekete (1923) and Polya and Szego (1931). Symm has extended the definition implicitly in (18) to $\mathbf{R}^3$ by introducing capacity through the solution of the integral equation

$$\int_{\partial\Omega} \frac{\sigma_1(q)}{|p-q|} \, ds_q = 1 \tag{21}$$

in which case

$$C = \int_{\partial\Omega} \sigma_1(q) \, ds_q \ . \tag{22}$$

In the next sections we will demonstrate how the characterization of outer radius or capacity in terms of the solution of a minimization problem in $\mathbf{R}^3$ (15), or a boundary value problem in $\mathbf{R}^2$ (17), or an integral equation in $\mathbf{R}^2$ or $\mathbf{R}^3$ (21) or (18) may be extended to $\mathbf{R}^n$ so that a fundamental constant associated with the surface emerges as the solution of problems which are formulated in exactly the same way for all dimensions.

A UNIFORM CHARACTERIZATION

We will generalize the characterization of outer radius in terms of the solution of a boundary value problem given by Szego (1945) in $\mathbf{R}^2$ to $\mathbf{R}^n$, then show that there is an equivalent characterization in terms of solutions of a system of boundary integral equations slightly different from those given by Symm (1967) and Jaswon and Symm (1977). In the next section we will show that the weak formulations of the boundary value problem and the integral equations give rise to the generalization to $\mathbf{R}^n$ of the characterization of outer radius as the minimum of a functional given by Szego (1945) and Riemann-Weber (1930).

We define the following classical boundary value problem for $u \in C^2(\Omega_c) \cap C^{1+\alpha}(\overline{\Omega}_c)$ and $\omega \in \mathbf{R}$

$$\begin{aligned} \nabla^2 u &= 0 \quad \text{in} \quad \Omega_c \\ u &= 0 \quad \text{on} \quad \partial\Omega \\ u &= g_n(r) + \omega + 0(r^{1-n}) \quad \text{as} \quad r \to \infty \end{aligned} \tag{23}$$

where g_n is defined in (2). The constant ω which appears as part of the solution pair (u,ω) is related to the outer radius C through

$$\omega = -\, g_n(C) \tag{24}$$

which constitutes an implicit definition of C, the capacity of $\partial\Omega$ for $n \geq 2$ (when $n = 2$ this is also called logarithmic capacity or transfinite diameter. The quantity ω is called the Robin constant in $\mathbf{R}^2$.) Explicitly for $n = 2$,

$$\omega = -\frac{1}{2\pi} \log C \ \Rightarrow \ C = e^{-2\pi\omega}$$

and for $n \geq 3$,

$$\omega = \frac{1}{(n-2)S_n \, C^{n-2}} \ \Rightarrow \ C = \{1/[(n-2)S_n \, \omega]\}^{1/n-2} \ .$$

The boundary value problem (23) will be shown to be equivalent to the system of boundary integral equations for $(\sigma,\omega) \in C^\alpha(\partial\Omega) \times \mathbf{R}$

$$\int_{\partial\Omega} \sigma(q) g_n(|p-q|\, ds_q = -\,\omega$$

$$\int_{\partial\Omega} \sigma(q)\, ds_q = 1 \ . \tag{25}$$

The unique solvability of (23) and (25) and the relation between the two formulations are established in the following theorems.

Theorem 1 – *The boundary value problem (23) has a unique solution pair* (u,ω) *for all* $n \geq 2$. *If* $n > 2$ *then* $\omega \neq 0$.

Theorem 2 – *The system of integral equations (25) has a unique solution pair* (σ,ω) *for all* $n \geq 2$. *If* $n > 2$ *then* $\omega \neq 0$.

Theorem 3 – *If* (u,ω_1) *is the solution of (23) and* (σ,ω_2) *is the solution of (25) then*

$$\sigma = \frac{\partial u}{\partial n} \quad \text{on} \quad \partial\Omega \ , \ \omega_1 = \omega_2 := \omega$$

$$\text{and} \quad u(p) = \int_{\partial\Omega} \sigma(q) g_n(|p-q|)\, ds_q + \omega \quad \text{in} \quad \overline{\Omega}_c \ .$$

Proof of Theorem 1

If $n > 2$ and $\omega = 0$ then the boundary value problem (23) admits only the trivial solution since the homogeneous exterior Dirichlet problem for functions regular at infinity has no nontrivial solution from, for example, the maximum principle. Hence we assume $\omega \neq 0$ and define

$$\mathrm{v} = 1 - \frac{u}{\omega} \tag{26}$$

in which case v solves the boundary value problem

$$\begin{aligned} \nabla^2 \mathrm{v} &= 0 \quad \text{in} \quad \Omega_c \\ \mathrm{v} &= 1 \quad \text{on} \quad \partial\Omega \\ \mathrm{v} &= -\frac{1}{\omega}\, g_n(r) + 0(\frac{1}{r^{n-1}}) = 0(\frac{1}{r^{n-2}}) \end{aligned} \quad . \tag{27}$$

But this has a unique solution $\mathrm{v} \in C^2(\Omega_c) \cap C^{1+\alpha}(\overline{\Omega}_c)$, since the boundary data is $C^{1+\alpha}(\partial\Omega)$. Now define

$$\omega := \lim_{r \to \infty} -\frac{g_n(r)}{\mathrm{v}(p)} \quad \text{and} \quad u(p) := \omega - \omega \mathrm{v}(p) \tag{28}$$

which establishes existence of at least one solution pair of (23) for $n > 2$.

For $n = 2$, we define

$$\mathrm{v} = u - g_2(|p|) \tag{29}$$

in which case v solves the boundary value problem

$$\begin{aligned} \nabla^2 v &= 0 && \text{in} \quad \Omega_c \\ v &= -g_2(|p|) && \text{on} \quad \partial\Omega \\ v &= \omega + 0(r^{-1}) = 0(1) && \text{as} \quad r \to \infty . \end{aligned} \tag{30}$$

This problem has a unique solution $v \in C^2(\Omega_c) \cap C^{1+\alpha}(\overline{\Omega}_c)$. In fact, following Mikhlin (1970), the solution can be constructed by taking it to be of the form

$$v(p) = \int_{\partial\Omega} \phi(q) \frac{\partial}{\partial n_q} g_2(|p-q|) ds_q + \int_{\partial\Omega} \phi(q) ds_q \quad , \quad p \in \Omega_c \ . \tag{31}$$

Here the density function ϕ satisfies the integral equation

$$-\frac{1}{2}\phi(p) + \int_{\partial\Omega} [\frac{\partial}{\partial n_q} g_2(|p-q|) + 1] \, \phi(q) ds_q = - g_2(|p|) \quad , \quad p \in \partial\Omega \tag{32}$$

which is always solvable (see Mikhlin (1970), p. 387).

Now define

$$\omega := \lim_{r \to \infty} v(p) = \int_{\partial\Omega} \phi(q) ds_q \ . \tag{33}$$

This together with (29) establishes existence of at least one solution pair of (23) for $n = 2$.

Now for the uniqueness, assume there exist two solution pairs (u_1, ω_1) and (u_2, ω_2) of (23). As usual, let $u = u_1 - u_2$ and $\omega = \omega_1 - \omega_2$. Then (u, ω) solves the boundary value problem

$$\begin{aligned} \nabla^2 u &= 0 && \text{in} \quad \Omega_c \\ u &= 0 && \text{on} \quad \partial\Omega \\ u &= \omega + 0(r^{1-n}) && \text{as} \quad r \to \infty . \end{aligned} \tag{34}$$

Since the Dirichlet integral

$$\int_{\Omega_c} |\nabla u|^2 dq = \lim_{R \to \infty} \int_{\partial B_R(0)} u \frac{\partial u}{\partial r} ds = 0 \ ,$$

it follows that $\nabla u = 0$ in Ω_c and hence u = constant in Ω_c. From the boundary condition $u = 0$ on $\partial\Omega$, this implies $u \equiv 0$ in Ω_c and consequently $\omega = 0$. This establishes uniqueness and completes the proof of Theorem 1.

Proof of Theorem 2

For $n > 2$ the exterior problem

$$\begin{aligned} \nabla^2 u^e &= 0 && \text{in} \quad \Omega_c \\ u^e &= 1 && \text{on} \quad \partial\Omega \\ u^e & && \text{regular at } \infty \end{aligned} \tag{35}$$

is uniquely solvable. From Green's Theorem (7) the boundary conditions and Gauss' integral (5), it follows that

$$\int_{\partial\Omega} g_n(|p-q|) \frac{\partial u^e}{\partial n_q} ds_q = 1 \quad \text{for} \quad p \in \partial\Omega \ . \tag{36}$$

Now there exists a solution $(\tilde{\sigma} = \frac{\partial u^e}{\partial n})$ of

$$\int_{\partial\Omega} \tilde{\sigma}(q) g_n(|p-q|)\, ds_q = 1 \; . \tag{37}$$

To show that (37) is uniquely solvable, assume two solutions $\tilde{\sigma}_1$ and $\tilde{\sigma}_2$ and let $\tilde{\sigma} = \tilde{\sigma}_1 - \tilde{\sigma}_2$ in which case

$$\int_{\partial\Omega} \tilde{\sigma}(q) g_n(|p-q|)\, ds_q = 0 \quad \text{on} \quad \partial\Omega \; . \tag{38}$$

Now define

$$\left.\begin{matrix} \mathrm{v}^e \\ \mathrm{v}^i \end{matrix}\right\} := \int_{\partial\Omega} \tilde{\sigma}(q) g_n(|p-q|) ds_q \quad \text{for} \quad \begin{cases} p \in \Omega_c \\ p \in \Omega \end{cases} \tag{39}$$

Then v^e is a solution of the homogeneous Dirichlet problem regular at infinity and v^i is a solution of the homogeneous interior Dirichlet problem. Hence both v^e and v^i vanish identically as do their normal derivatives in which case the usual jump conditions for the single layer imply that

$$\tilde{\sigma}(q) \equiv 0 \quad \text{on} \quad \partial\Omega$$

and therefore (37) has a unique solution. Next we show that if $\tilde{\sigma}$ solves (37) then

$$\int_{\partial\Omega} \tilde{\sigma}(q)\, ds_q \neq 0 \; . \tag{40}$$

Again define

$$\mathrm{v}^e(p) = \int_{\partial\Omega} \tilde{\sigma}(q) g_n(|p-q|)\, ds_q \;, p \in \overline{\Omega_c} \tag{41}$$

Then $\mathrm{v}^e(p)$ is regular at infinity and Green's theorem implies that

$$\begin{aligned} \int_{\partial\Omega} g_n(|p-q|) \frac{\partial \mathrm{v}^e}{\partial n_q} ds_q &= \frac{1}{2}\mathrm{v}^e(p) + \int_{\partial\Omega} \mathrm{v}^e(q) \frac{\partial g_n(|p-q|)}{\partial n_q} \quad \text{for } p \in \partial\Omega \\ &= 1 \end{aligned} \tag{42}$$

where the fact that $\tilde{\sigma}$ satisfies (37) and Gauss' integral (5) have been employed.

But since we have established the unique solvability of (37) it follows that

$$\tilde{\sigma}(p) = \frac{\partial \mathrm{v}^e}{\partial n_p} \tag{43}$$

Hence

$$\begin{aligned} \int_{\partial\Omega} \tilde{\sigma}(q)\, ds_q &= \int_{\partial\Omega} \frac{\partial \mathrm{v}^e}{\partial n_q}\, ds_q = \int_{\partial\Omega} \mathrm{v}^e(q) \frac{\partial \mathrm{v}^e}{\partial n_q}\, ds_q \\ &= -\int_{\Omega_c} |\nabla \mathrm{v}^e(q)|^2\, dq \; . \end{aligned} \tag{44}$$

Thus if $\int_{\partial\Omega} \tilde{\sigma}(q)\, ds_q = 0$ it follows that $\nabla \mathrm{v}^e \equiv 0$ in Ω_c which, with the boundary condition implies that $\mathrm{v}^e(p) \equiv 1$. However this contradicts the fact that v^e is regular at ∞. Therefore (40) is satisfied. We therefore define, for the uniquely defined solution of (37) satisfying (40)

$$\omega := -\frac{1}{\int_{\partial\Omega} \tilde{\sigma}(q)\, ds_q} \neq 0 \tag{45}$$

$$\sigma = \frac{\tilde{\sigma}(p)}{\int_{\partial\Omega} \tilde{\sigma}(q)\, ds_q} . \tag{46}$$

Clearly (σ , ω) satisfies the system (25). To establish uniqueness assume there exist two solution pairs (σ_1,ω_1) and (σ_2 , ω_2) and let

$$\sigma = \sigma_1 - \sigma_2$$

$$\omega = \omega_1 - \omega_2 .$$

Then,

$$\int_{\partial\Omega} \sigma(q) g_n(|p-q|) ds_q = - \omega \tag{47}$$

$$\int_{\partial\Omega} \sigma(q) ds_q = 0 . \tag{48}$$

If $\omega \neq 0$ then $\tilde{\sigma} := -\frac{\sigma}{\omega}$ satisfies (37) and since (40) is satisfied it follows that

$$\int_{\partial\Omega} \frac{\sigma(q)}{\omega}\, ds_q \neq 0$$

which contradicts (48). Thus we conclude that $\omega = 0$ or $\omega_1 = \omega_2$. Hence

$$\int_{\partial\Omega} \sigma(q) g_n(|p-q|) ds_q = 0 , \quad p \in \partial\Omega , \tag{49}$$

and using the same argument following (38) we find that $\sigma = 0$ or $\sigma_1 = \sigma_2$. This concludes the proof of Theorem 2 for $n > 2$.

For $n = 2$ the above analysis is not directly applicable because of the logarithmic growth of the fundamental solution. To establish the theorem in this case we follow the work of Hsiao and MacCamy (1973) based on results of Muskhelishvili (1953) and Fichera (1961). The desired result is contained in Theorem 3 of Hsiao and MacCamy (1973) and the main points of the proof specialized to the present case are included here for completeness.

The system which we wish to show uniquely solvable takes the form

$$\begin{aligned} &\int_{\partial\Omega} \sigma(q) \log |p-q|\, ds_q = - 2\pi\omega \\ &\int_{\partial\Omega} \sigma(q)\, ds_q = 1 . \end{aligned} \tag{50}$$

Every σ which is part of a solution pair (σ , ω) of (46) will satisfy

$$\frac{d}{ds_p} \int_{\partial\Omega} \sigma(q) \log |p-q|\, ds_q = 0 \tag{51}$$

where S_p denotes arc length in $\partial\Omega$. As shown in Muskhelishvili (1953), differentiation and integration may be interchanged yielding a singular integral equation with Cauchy kernel. Moreover the singular integral operator has index zero hence the operator and its adjoint have null spaces of the same dimension. Since it is easy to see that constants are solutions of the homogeneous adjoint equation, the dimension of the null space is at least one. To show that the dimension is exactly one, that is, that (50) has exactly one linearly independent solution we proceed as follows.

Assume that there exists two non-trivial solutions σ_i , $i = 1,2$, of (51). Then, integrating (51), we define two constants ω_i , $i = 1,2$, by

$$\omega_i := -\frac{1}{2\pi} \int_{\partial\Omega} \sigma_i(q) \log|p-q|\, ds_q \quad , \quad p \in \partial\Omega \quad , \quad i = 1,2 \ . \tag{52}$$

Furthermore define

$$A_i := \int_{\partial\Omega} \sigma_i(q)\, ds_q \quad , \quad i = 1,2 \tag{53}$$

and choose constants α_i, not both zero, such that

$$\alpha_1 A_1 + \alpha_2 A_2 = 0 \tag{54}$$

Next define

$$\sigma := \alpha_1 \sigma_1 + \alpha_2 \sigma_2 \tag{55}$$

and the single layer potential

$$u(p) = \int_{\partial\Omega} \sigma(q) \log|p-q|\, ds_q \quad , \quad p \in \mathbf{R}^2 \tag{56}$$

It is easy to see with (52)-(54) that u is a solution of the exterior Dirichlet problem

$$\begin{aligned} \nabla^2 u &= 0 \quad \text{in} \quad \Omega_c \\ u &= -2\pi(\alpha_1 \omega_1 + \alpha_2 \omega_2) \quad \text{on} \quad \partial\Omega \\ u &= 0(\frac{1}{r}) \quad \text{as} \quad r \to \infty \ . \end{aligned} \tag{57}$$

But this problem is solvable only if

$$\alpha_1 \omega_1 + \alpha_2 \omega_2 = 0$$

in which case u defined by (56) vanishes identically in $\overline{\Omega}_c$. In addition u vanishes in Ω since it is a solution of the homogeneous interior Dirichlet problem. Using standard arguments of potential theory, the vanishing of the normal derivatives and the jump conditions for derivatives of single layer potentials allow us to conclude that

$$\sigma = 0$$

in which case, with (55), σ_1 and σ_2 are seen to be linearly dependent. Moreover it has been shown Muskhelsihvili (1953) that every non-trivial solution σ_o of (51) has the property that

$$\int_{\partial\Omega} \sigma_o(q)\, ds_q \neq 0 \ .$$

Let σ be the unique solution of (51) such that

$$\int_{\partial\Omega} \sigma(q)\, ds_q = 1 \ ,$$

and define

$$\omega := -\frac{1}{2\pi} \int_{\partial\Omega} \sigma(q) \log|p-q|\, ds_q \ . \tag{58}$$

Then (σ , ω) is a solution pair of the system (50).

To prove uniqueness, assume the existence of two solution pairs, (σ_i , ω_i) $i = 1,2$. Set $\sigma = \sigma_1 - \sigma_2$ and $\omega = \omega_1 - \omega_2$ in which case

$$\int_{\partial\Omega} \sigma(q)\log|p-q|\, ds_q = -2\pi\omega$$

$$\int_{\partial\Omega} \sigma(q)\, ds_q = 0 \tag{59}$$

Then by the same argument as before (since differentiation of (59) with respect to arc length leads to (51)), we conclude that $\sigma = 0$ and $\omega = 0$ hence the system (50) is uniquely solvable. This completes the proof of Theorem 2.

Proof of Theorem 3

If (u , ω_1) is the solution of (23) then Green's Theorem (7) implies that

$$\int_{\partial\Omega} g_n(|p-q|)\frac{\partial u}{\partial n_q}\, ds_q = [1-\alpha(p)]\, u(p) - \omega_1 \ , \ p \ \varepsilon\ \overline{\Omega}_c \ . \tag{60}$$

For $p\ \varepsilon\ \partial\Omega$ we have

$$\int_{\partial\Omega} g_n(|p-q|)\frac{\partial u}{\partial n_q}\, ds_q = -\omega_1 \ . \tag{61}$$

The asymptotic form of g_n together with the behaviour of u as $r \to \infty$ and the representation (60) leads to

$$g_n(r) \int_{\partial\Omega} \frac{\partial u}{\partial n_q}\, ds_q = g_n(r) + 0(\frac{1}{r^{n-1}}) \ . \tag{62}$$

Hence

$$\int_{\partial\Omega} \frac{\partial u}{\partial n_q}\, ds_q = 1 \ . \tag{63}$$

Thus $(\frac{\partial u}{\partial n_q} , \omega_1)$ is a solution of the system (61), (63) which coincides with (25).

Conversely if (σ , ω_2) is the unique solution of (25) define

$$u(p) := \int_{\partial\Omega} \sigma(q) g_n(|p-q|)\, ds_q + \omega_2 \ \text{ in } \ \overline{\Omega}_c \ . \tag{64}$$

Then $(u(p) , \omega_2)$ is a solution of the boundary value problem (23) and because of uniqueness, is the only solution. Hence by the first part of the proof $(\frac{\partial u}{\partial n} , \omega_2)$ must be a solution of the system of integral equations (25). Because this system admits only one solution pair, it follows that

$$\sigma = \frac{\partial u}{\partial n} \ .$$

VARIATIONAL FORMULATION

In the previous section we showed how the Robin constant ω appeared as part of the solution of neither the boundary value problem (23) or the system of integral equations (25). In the case of the boundary value problem one may show that for $n \geq 3$ the weak formulation constitutes the necessary conditions for minimizing the functional

$$J[u] := \int_{\Omega_c} |\nabla u|^2 \, dq - 2\,\omega \tag{65}$$

over the space of functions

$$U_{ad} := \left\{ u \;\middle|\; \begin{array}{l} u \in C_1(\Omega_c) \\ u = g_n(r) + \omega + 0(\frac{1}{r^{n-1}}) \ \text{ as } \ r \to \infty \\ u = 0 \quad \text{on} \quad \partial\Omega \end{array} \right\} . \tag{66}$$

Moreover the functional assumes its minimum when u satisfies (23) and

$$\min_{u \,\varepsilon\, U_{ad}} J[u] = -\,\omega \ . \tag{67}$$

One may show that this is only a slight change of the standard variational treatment of the Dirichlet integral (see e.g. Kellogg (1929)).

Unfortunately there does not appear to be an analogue of this characterization for $n = 2$ because the Dirichlet integral does not exist for functions with logarithmic growth. However the variational formulation of the system of integral equations (25) is valid for $n = 2$ as well as $n \geq 3$ and these results are presented here.

Consider the following functional of $\sigma \,\varepsilon\, U_{ad}$ where

$$U_{ad} = \left[\sigma \mid \sigma \,\varepsilon\, H^{-1/2}(\partial\Omega) \ , \ \int_{\partial\Omega} \sigma(q) ds_q = 1 \right\} ; \tag{68}$$

$$J[\sigma] := -\int_{\partial\Omega}\int_{\partial\Omega} \sigma(p)\sigma(q) g_n(|p-q|) \, ds_q \, ds_p \ . \tag{69}$$

We will show that the Robin constant is intimately connected with the minimum of this functional for all $n \geq 2$. Further we show that the weak solution of the system of integral equations (25) minimizes this functional. To this end note that if σ and $\sigma + \varepsilon\eta$ are both in U_{ad} then

$$\int_{\partial\Omega} \eta(q) \, ds_q = 0 \tag{70}$$

and

$$\begin{aligned} J[\sigma + \varepsilon\eta] = J[\sigma] &- 2\varepsilon \int_{\partial\Omega}\int_{\partial\Omega} \eta(p)\, \sigma(q) g_n(|p-q|) ds_q \, ds_p \\ &- \varepsilon^2 \int_{\partial\Omega}\int_{\partial\Omega} \eta(p)\eta(q) g_n(|p-q|) ds_q \, ds_p \ . \end{aligned} \tag{71}$$

The necessary condition for $J[\sigma]$ to have an extreme value at $\sigma \varepsilon U_{ad}$ is

$$\int_{\partial\Omega}\int_{\partial\Omega} \eta(p)\,\sigma(q) g_n(|p-q|) ds_q\, ds_p = 0 \tag{72}$$

for all $\eta \varepsilon H^{-1/2}(\partial\Omega)$ satisfying (70). Furthermore if (72) is satisfied then $J[\sigma]$ is a minimum since

$$I[\eta] = -\int_{\partial\Omega}\int_{\partial\Omega} \eta(p)\,\eta(q)\, g_n(|p-q|)\, ds_q\, ds_p \ \geq\ 0 \quad \text{for all } \eta \in H^{-1/2}(\partial\Omega) \tag{73}$$

satisfying (70).

To show that (73) holds we follow the argument of Nedelec (1977) for $n \geq 3$ and Hsiao and Wendland (1977) for $n = 2$. For $\eta \varepsilon H^{-1/2}(\partial\Omega)$ satisfying (70) define

$$u(p) := \int_{\partial\Omega} \eta(q) g_n(|p-q|)\, ds_q \ , \quad p \ \varepsilon\ \mathbf{R}^n$$

The standard jump conditions for the single layer potential yield

$$\frac{\partial u}{\partial n^+} - \frac{\partial u}{\partial n^-} = \eta(q)\ .$$

Then

$$I(\eta) = -\int_{\partial\Omega} \eta(p) u(p)\, ds_q = -\int_{\partial\Omega} \left[\frac{\partial u}{\partial n^+} - \frac{\partial u}{\partial n^-}\right] u(p)\, ds_q$$

$$= \int_{\Omega^c \cap B_R(0)} |\nabla u|^2\, dp - \int_{\partial B_R(0)} \frac{\partial u}{\partial r} u\, ds_q + \int_{\Omega} |\nabla u|^2\, dp\ .$$

The growth of the Green's function is sufficient, for $n \geq 3$, to guarantee that

$$\int_{\partial B_R(0)} \frac{\partial u}{\partial r} u\, ds = 0(\frac{1}{R})\ .$$

However for $n = 2$ it is necessary to employ condition (70) to achieve this same result. Thus for $n \geq 2$, the Dirichlet integral always exists thus

$$I(\eta) = \int_{\mathbf{R}^n} |\nabla u|^2\, dp \geq 0$$

which guarantees that (73) holds.

Observe that (72) constitutes the weak formulation of the system of integral equations (25). Since we have shown that there exists a unique solution pair (σ, ω) of equation (25) it is easy to see that σ will satisfy (72). Moreover the minimum value of J is seen to be, with (25),

$$\begin{aligned} J[\sigma] &= -\int_{\partial\Omega}\int_{\partial\Omega} \sigma(p)\sigma(q) g_n(|p-q|) ds_q\, ds_p \\ &= \omega \int_{\partial\Omega} \sigma(p)\, ds_q = \omega\ , \end{aligned} \tag{74}$$

hence

$$\min_{\sigma \in U_{ad}} J[\sigma] = \omega \ .$$

This coincides with the characterization given Szego (1945) for $n = 3$.

CONCLUDING REMARKS

In summary we have characterized the capacity or outer radius implicitly by

$$\omega = - g_n(c) \tag{24}$$

where g_n is the fundamental solution of the n- dimensional Laplacian, (2), and ω emerges in the unique solution of the boundary value problem (23)

$$\begin{aligned} \nabla^2 u &= 0 \quad \text{in} \quad \Omega_c \\ u &= 0 \quad \text{on} \quad \partial\Omega \\ u &= g_n(r) + \omega + 0(\frac{1}{r^{n-1}}) \quad \text{as} \ r \to \infty \end{aligned} \tag{23}$$

or the system of boundary integral equations (25)

$$\begin{aligned} \int_{\partial\Omega} \sigma(q) g_n(|p-q|)\, ds_q \ + \ \omega = 0 \\ \int_{\partial\Omega} \sigma(q)\, ds_q = 1 \end{aligned} \tag{25}$$

or the optimization problem

$$\omega = \inf_{\sigma \in U_{ad}} \ - \int_{\partial\Omega}\int_{\partial\Omega} \sigma(p)\sigma(q) g_n(|p-q|)\, ds_q\, ds_p \tag{74}$$

where U_{ad} is defined in (68).

We conclude by showing how these characterizations encompass existing definitions summarized in section 3.

The definition of capacity (11) in terms of the solution of the conductor potential problem (10) for $n = 3$, u_1, is easily seen to coincide with ω, through (24), by substituting

$$u_1 = 1 - \frac{u}{\omega}$$

in equation (11).

We already noted at the beginning of section 5 that the characterization of capacity as the minimum of the Dirichlet integral when $n = 3$ is essentially equivalent to the weak form of (23). The generalization of (15) for $n \neq 3$ has been treated in section 5.

The relations between the system ot boundary integral equations (25) and those given by Symm are as follows. For $n = 2$, (σ_1 , γ) appearing in (18) are related to (σ , ω) of (25) by

$$\gamma = 2\, \pi\, \omega$$

$$\sigma_1 = \sigma + \frac{\partial u_o}{\partial n} - \frac{\partial g_2(r)}{\partial n}$$

where u_0 is the solution of the interior Dirichlet problem for the Laplacian with boundary values

$$u_0 = g_2(r) \quad \text{on} \quad \partial\Omega \ .$$

For $n = 3$ the relation between solutions of (21) and (25) is simply

$$\sigma_1 = \frac{\sigma}{4\pi\,\omega} \ .$$

Finally to establish the connection between the conformal mapping (c.f. Eq. (20)) and the boundary value problem (23) for $n = 2$, we let $f(z)$ be the conformal mapping of Ω_c to the exterior of the unit circle such that $|f(z)| = 1$ when $z \in \partial\Omega$ and $\frac{f(z)}{z} = 0(1)$ when $z \to \infty$. We use the complex representation z for the point p. Then

$$f(z) = z\, e^{\phi + i\psi}$$

where ϕ and ψ are harmonic conjugates regular at infinity (i.e. $\phi = 0(1)$). Then it is easy to see that ϕ is related to the solution of (23) by

$$u = \frac{1}{2\pi}\,\phi + \frac{1}{2\pi}\log r \ , \quad r = |z| \ .$$

Moreover, a simple computation shows that

$$\lim_{z\to\infty} |f'(z)| = \lim_{z\to\infty} e^{\phi} = e^{2\pi\omega}$$

from which equation (20) follows.

Acknowledgment – This work was partially supported by the National Science Foundation under Grant DMS-8507063.

References

Fekete, M., 1923, Uber die Verteilung der Wurzeln bei gewissen algebraischen Gleichungen mit ganzzahligen Koeffizienten, Mathematische Zeitschrift, Vol. 17, 228-249.

Fichera, G., 1961, Linear Elliptic Equations of Higher Order in Two Independent Variables and Singular Integral Equations *in* Partial Differential Equations and Continnum Mechanics, ed. R.E. Langer, The University of Wisconsin Press, 55-80.

Hsiao, G.C., and MacCamy, R.C., 1973, Solution of Boundary Value Problems by Integral Equations of the First Kind, SIAM Review, *15,* 687-745.

Hsiao, G.C., and Wendland, W., 1977, A Finite Element Method for some Integral Equations of the First Kind, J. Math. Anal. Appl. *58,* 449-481.

Jaswon, M.A., and Symm, G.T., 1977, "Integral Equation Methods *in* Potential Theory and Elastostatics", Academic Press, London,

Kellogg, O.D., 1953, "Foundations of Potential Theory, Dover Publications", New York.

Mikhlin, S.G., 1970, "Mathematical Physics, An Advanced Course", North- Holland Publishing Company, Amsterdam, London.

Muskhelishvili, N., 1953, "Singular Integral Equations", P. Noordhoff, Groningen, The Netherlands.

Nedelec, J.C., 1977, Approximation des Equations Integrals en Mecanique et en Phsique, Centre de Mathematiques Appliques, Ecole Polytechnique - Palaiseau.

Payne, L., 1967, Isoperimetric Inequalities, SIAM Review *9* (3), 453-488.
Polya, G., and Szego, G., 1931, Uber den transfiniten Durchmesser (Kapazitatskonstante) von ebenen und raumlichen Punktmengen, J. Reine Angew. Math. 165, 4-49.
Polya, G., and Szego, G., 1945, Inequalities for the Capacity of a Condenser, Amer. J. Math., 67, 1-32.
Polya, G., and Szego, G., 1951, "Isoperimetric Inequalities *in* Mathematical Physics", Princeton Univ. Press.
Protter, M.H., and Weinberger, H.F., 1965, On the Capacity of Composite Conducters, J. Math. and Phys. 44-45, 375-383.
Riemann-Weber (P. Frank and R.v. Mises), 1930, Die Differential und Integralgleichungen der Mechanik und Physik, Vol. 1, 8 ed. Braunschweig.
Stakgold, I., 1968, "Boundary Value Problems of Mathematical Physics", Vol. 2, Macmillan, New York.
Symm, G.T., 1967, Numerical Mapping of Exterior Domains Nu. Math. 10, 437-445.
Szego, G., 1945, On the Capacity of a Condenser, Bull. Amer. Math. Soc. 51, 325-350.

A NOTE ON THE COMPUTATION OF GREEN FUNCTIONS FOR SUBLAPLACIANS ON LIE ALGEBRAS OF TYPE H

H. Hueber

Fakultät für Mathematik
Universität Bielefeld
D-4800 Bielefeld

In [4] A. Kaplan introduced a class of step-2 nilpotent Lie groups for which the fundamental solution corresponding to the standard sublaplacian may be written down in an explicit elementary form. He also says that heuristic evidence suggests that these groups are the only ones which admit such nice explicit formulas. When I tried to convince myself of this evidence I found some modifications of Kaplan's proof, which give additional insight into the problem and may be - at least for people working in potential theory - regarded as a simplification.

Let us start with a more general situation. So let $\mathcal{N}$ be a real finite dimensional Lie algebra which has a direct sum decomposition $\mathcal{N} = V^1 \oplus \ldots \oplus V^r$ such that $[V^j, V^k] = V^{k+j} \neq \{0\}$ for $k + j \leq r$, and $[V^j, V^k] = \{0\}$ if $k + j > r$. We identify $\mathcal{N}$ with the simply connected Lie group N which belongs to $\mathcal{N}$ via the exponential mapping. Fix a basis $X_1, \ldots, X_n$ of V^1. We define vectorfields LX_i and RX_i, $i = 1, \ldots n$ on N by

$$LX_i f(x) = \frac{d}{dt}(f(x \exp(tX_i)))\Big|_{t=0}$$

$$RX_i f(x) = \frac{d}{dt}(f(\exp(tX_i)x))\Big|_{t=0} ,$$

in other words: LX_i is the left-invariant and RX_i is the right-invariant vectorfield with $LX_i f(0) = RX_i f(0) = \frac{df}{dx_i}(0)$. Let $L = LX_1^2 + \ldots + LX_n^2$,

$$R = RX_1^2 + \quad + RX_n^2 \quad , \text{ and}$$

$$S = L + R \quad .$$

It is well known that L, R and S are hypoelliptic differential operators, and furthermore each of these operators defines a nice harmonic space in a natural way (cf. [1], [2], [3]) .

<u>Proposition</u> – There exists a strictly positive smooth function $G : N \setminus \{0\} \to \mathbb{R}$ such that $Sg = 0$ and $\inf g = 0$. Any other function with this property is a positive multiple of g . For a certain constant c one has

$$L(cg) = R(cg) = \frac{1}{2} S(cg) = -\varepsilon_0 \quad .$$

<u>Proof</u> – Let G^L denote the Green function of L (cf. [3] Theorem 2.1), and let $g = G^L(\cdot,0)$. If $\tilde{f}(x) = f(x^{-1})$ then $Lf = \widetilde{R\tilde{f}}$ is easily checked, and as G^L is symmetric $g = \tilde{g}$ is clear. From this one derives $Rg = -\varepsilon_0$, and hence $\frac{1}{2} S(g) = -\varepsilon_0$. That g is unique up to positive multiples follows almost word by word as the first half of the proof of Theorem 2.1. in [3] (cf. also [2], Théorème 4). □

One nice feature is the following: Whenever one has a guess what g might be in a concrete situation the equation $Sg = 0$ is usually much easier to check than $Lg = 0$. Observing

$$2S = \sum_{i=1}^{n} (LX_i + RX_i)^2 + \sum_{i=1}^{n} (LX_i - RX_i)^2$$

it is easy to check that in the step-2 case

$$(*) \quad \begin{aligned} Sf(x) &= 2 \sum_{i=1}^{n} D^2 f(x)(X_i, X_i) \\ &+ \frac{1}{2} \sum_{i=1}^{n} D^2 f(x)([x,X_i],[x,X_i]) \quad , \end{aligned}$$

where as usual $D^2 f(x)$ is the bilinearform which is the second derivative of f at x .

Now we shall see that for groups of type H (cf. [4]) this formula may be improved. So let $r = 2$, and assume that N is equipped with an inner product $\langle \cdot , \cdot \rangle$ such that $V := V^1$ is orthogonal to $Z := V^2$. For each

$v \in V$ let as usual $ad_v(w) = [v,w]$ and denote the orthogonal complement in V of $\ker(ad_v)$ by V_v. Following Kaplan we say that N is of type H if ad_v maps V_v isometrically onto the center Z for each $v \in V$ with $|v| = 1$. Let Δ_V, Δ_Z denote the Laplacians on the subspaces V,Z and let p denote the projection onto V.

<u>Theorem</u> – The following statements are equivalent:

1. N is of type H.
2. There exists an orthonormal basis $X_1,\dots,X_n$ of V such that for all $z \in Z$ and all $x \in N$

   $$\sum_{i=1}^{n} \langle z,[x,X_i]\rangle^2 = |(x)|^2 |z|^2 .$$

3. For every orthonormal basis $X_1,\dots,X_n$ of V, all $z \in Z$ and all $x \in N$

   $$\sum_{i=1}^{n} \langle z,[x,X_i]\rangle^2 = |p(x)|^2 |z|^2 .$$

4. There exists an orthonormal basis $X_1,\dots,X_n$ of V such that

   $$S = 2\Delta_V + \frac{1}{2}|p|^2\Delta_Z ,$$

   where R,L and S are constructed starting with this basis.
5. For every orthonormal basis $X_1,\dots,X_n$ of V

   $$S = 2\Delta_V + \frac{1}{2}|p|^2\Delta_Z .$$

<u>Proof</u> – For every $v \in V$ and every $z \in Z$ let $A_v(z)$ denote the unique element in V for which

$$\langle A_v(z),y\rangle = \langle z,[v,y]\rangle$$

holds for all $y \in V$. Note that $A_v(Z) \subset V_v$.

$1 \Rightarrow 3$. Fix $v \in V$ with $|v| = 1$. Now 1) implies that for $z_1,z_2 \in Z$

$$\langle ad_v(A_v(z_1)),z_2\rangle = \langle [x,A_v(z_1)],z_2\rangle$$
$$= \langle A_v z_1, A_v z_2\rangle = \langle z_1,z_2\rangle ,$$

and this shows that A_v is the inverse of $ad_v\big|_{V_v}$. Let $X_1,\dots,X_n$ be an orthonormal basis of V. Then for every $z \in Z$

$$\sum_{i=1}^{n} \langle z,[v,X_i]\rangle^2 = \sum_{i=1}^{n} \langle A_v(z),X_i\rangle^2 = |A_v(z)|^2 = |z|^2 \quad ,$$

and this implies 3).

3 ⇒ 5. Polarizing 3) gives

$$\sum_{i=1}^{n} \langle z_1,[v,X_i]\rangle\langle z_2,[v,X_i]\rangle = |v|^2\langle z_1,z_2\rangle$$

for $v \in V$ and $z_1,z_2 \in Z$. Substituting this in (*) shows that 5 holds.

5 ⇒ 4. is obvious.

4 ⇒ 2. Fix an orthonormal base $X_1,\ldots,X_n$ of V such that 4) holds. Let $Z_1,\ldots,Z_m$ be an orthonormal base of Z . Then 4) together with (*) shows

$$\sum_{i=1}^{n} \langle [x,X_i],Z_k\rangle\langle [x,X_i],Z_l\rangle = \delta_{kl} \cdot |p(x)|^2$$

for all $x \in N$, and this implies 2).

2 ⇒ 1. Fix $v \in V$ with $|v| = 1$. By 2) we have for $z_1,z_2 \in Z$

$$\begin{aligned}\langle ad_v(A_v(z_1)),z_2\rangle &= \langle [v,A_v z_1],z_2\rangle \\ &= \langle A_v z_1,A_v z_2\rangle \\ &= \sum_{i=1}^{m} \langle A_v z_1,X_i\rangle\langle A_v z_2,X_i\rangle \\ &= \sum_{i=1}^{n} \langle z_1,[x,X_i]\rangle\langle z_2,[x,X_i]\rangle \\ &= \langle z_1,z_2\rangle \quad .\end{aligned}$$

This shows that $ad_v\big|_{V_v}$ is the inverse of $A_v : z \to V_v$, and that A_v is an isometry, hence also $ad_v\big|_{V_v}$. □

<u>Corollary</u> – Assume that N is of type H and let G^L denote the Green function of L . Then there exists $c \in]0,\infty[$ such that

$$G^L((v,z),0) = c(|v|^4+16|z|^2)^{\frac{2-Q}{4}} , \quad v \in V , \ z \in Z , \text{ where}$$

$Q = \dim V + 2\dim Z$.

Obviously the only thing we have to check is that the operator S annihilates the right hand side of the above equation, and this is a straight forward computation. The interested reader will notice that for some of our arguments the group structure is not really essential. We may for example consider the Grushin operator $L = \frac{\partial^2}{\partial x^2} + x^2 \frac{\partial^2}{\partial t^2}$ on $\mathbb{R}^2 = \{(x,t)\}$. Our arguments show that there exists a constant $c \in]0,\infty[$ such that $g(x,t) = c(t^2 + \frac{1}{4}x^4)^{-\frac{1}{4}}$ satisfies $Lg = -\varepsilon_0$, and that every positive continuous function h which satisfies $Lh = 0$ in the complement of the origin is of the form $h = \alpha + \beta g$ with $\alpha, \beta \in [0,\infty]$. What we really needed is the nice behaviour under canonical dilations, and the very special form of the operator, and these arguments cover more general situations.

So in fact we have proven a more general result: Let $\mathbb{R}^d = \mathbb{R}^n \times \mathbb{R}^m = \{(x,y)\}$, and let Δ_x denote the Laplacian with respect to the x variables and Δ_y denote the Laplacian with respect to the y variables. Let $L = 2\Delta_x + \frac{1}{2}|x|^2\Delta_y$, and $g(x,y) = (|x|^4 + 16|y|^2)^{\frac{2-(n+2m)}{4}}$.

<u>Theorem</u> – Every continuous $h : \mathbb{R}^d \setminus \{0\} \to [0,\infty[$ which satisfies $Lh = 0$ in $\mathbb{R}^d \setminus \{0\}$ in the sense of distributions is of the form $h = \alpha + \beta g$ with $\alpha, \beta \in [0,\infty[$. Especially there exists a constant $c < 0$ such that $Lg = c\varepsilon_o$.

The only thing which is left in order to show that this theorem holds, is to show that the sheaf $\mathcal{H}_L$ of continous solutions h of $Lh = 0$ defines a Brelot space on $\mathbb{R}^d$. Let $X_i = \sqrt{2}\frac{\partial}{\partial x_i}$, and let $X_{i,j} = \frac{1}{\sqrt{2}} x_i \frac{\partial}{\partial y_j}$.

Then

$$L = \sum_{i=1}^{n} X_i^2 + \sum_{j=1}^{m} \sum_{i=1}^{n} X_{ij}, \text{ and as } [X_i, X_{ij}] = \frac{\partial}{\partial y_j}$$

we may use Bony's result [1] which shows that $\mathcal{H}_L$ defines a Brelot space.

<u>References</u>

1. Bony, J.M.: Principe du maximum, inégalité de Harnack et unicité du problème de Cauchy pour les opérateurs elliptiques dégénérés. Ann. Inst. Fourier 19, (1) 277 - 304 (1969).

2. Hervé, R.-M.; Hervé, M.: Les fonctions surharmoniques dans l'axiomatique de M. Brelot associées à un opérateur elliptique dégénéré. Ann. Inst. Fourier 22, (2) 131 - 145 (1972).

3. Hueber, H.: Wiener's criterion in potential theory with applications to nilpotent Lie groups. Math. Z. 190, 527 - 542 (1985).

4. Kaplan, A.: Fundamental solutions for a class of hypoelliptic PDE generated by composition of quadratic forms. T.A.M.S. 258, 147 - 153 (1980).

ON THERMOELASTIC POTENTIAL

Hung Do Duc
Department of Mathematics
University of Pedagogics
Hanoi

Jerzy Gawinecki
Department of Mathematics
Military Technical Academy
Warsaw

Adam Piskorek
Department of Mathematics, Informatics and Mechanics, Warsaw University
Warsaw

S.Klainerman, in [1] proved that for suficiently small nonlinear perturbations of linear dissipative equations and systems like: the wave equation, heat equation, linear isotropic elasticity equations etc., the corresponding solution of the initial-value problem for perturbed equations and system behave asymptotically like the solutions of the initial-value problem for linear unperturbed equations and system.

This proof was based on a special version of the Nash-Moser-Hormander scheme and L^{∞}-L^{p} time decay, i.e. the behaviour as $t \to \infty$ of the solution of the initial-value problem for linear unperturbed equations and systems. THis time decay of solution one obtains with aid of fundamental solution (cf. [1], p.76).

We consider the linear isotropic thermoelasticity equations which have a following form:

(1) $\partial_t^2 u_j - \sum_k (c_1^2 - c_2^2)\partial_j \partial_k u_k - c_2^2 \partial_k^2 u_j + \gamma \partial_j T = 0$

(2) $\sum_k \gamma \partial_k \partial_t u_k + \beta \partial_t T - \varkappa \Delta T = 0$

where c_1, c_2 are the speeds of progations of elastic waves, are given as follows in terms of the Lame's constants λ, μ and the density ρ, i.e.

$$c_1^2 = (\lambda + 2\mu)/\rho \qquad c_2^2 = \mu/\rho$$

γ, β, μ are the physical positive constants. Without loss of generality we assume that $\beta = \varkappa = 1$.

This system of equations is dissipative in sense of S.Klainerman (cf.[1], p.74, def.(D),(D')) and we can apply the Klainerman's results in nonlinear thermoelasticity if we have the matrix of fundamental solutions of system (1),(2).

In order to costruct the matrix E(t,x) of fundamental solutions of system (1),(2) we proceed as follows. Using the Hilbert-Levy method (cf.[3],[2]) we obtain the matrix *E*(t,x) of fundamental solutions for the principal part of system (1),(2) i.e.

(3) $\partial_t^2 u_j - \sum_k (c_1^2 - c_2^2)\partial_j \partial_k u_k - c_2^2 \partial_k^2 u_j = 0$

(4) $\sum_k \gamma \partial_k \partial_t u_k + \partial_t T - \Delta T = 0$

The matrix *E*(t,x) has the following block form:

(5) $$E(t,x) = \begin{bmatrix} H(t,x) & 0 \\ \mathrm{div} H * \Gamma_t(t,x), & \Gamma(t,x) \end{bmatrix}$$

where H(t,x) is (3×3)-matrix of fundamental solution for linear isotropic elasticity equations (cf.[4]);

$(\mathrm{div} H * \partial_t \Gamma)(t,x)$ - (1×3)-matrix of following form:

(6) $(\mathrm{div} H * \partial_t \Gamma)(t,x) = ((\partial_k H_{jk}(\cdot,\cdot) * \partial_t \Gamma(t,x))_{j,k=1,2,3}$

and $\Gamma(t,x)$ is the fundamental solution of heat conduction operator $\partial_t - \Delta$; $*$ - denotes the four-dimensional convolution.

The components of matrix (6) can be written as

(7) $$(\partial_k H_{jk} * \partial_t \Gamma)(t,x) = \partial_j \left\{ \frac{e^{-|\cdot|}}{|\cdot|} *_3 \Gamma(t,\cdot) - (\partial_t + \Delta) \times \right.$$

$$\left. \times \left[\frac{1}{2|x|} (e^{-(t+|x|)} - e^{-|t-|x||}) \right] \right\}$$

and have a nice behaviour, as $t \to \infty$.

Now, we seek the matrix $E(t,x)$ of fundamental solutions for the system (1),(2) using the Hilbert-Levy method in the form:

(8) $E(t,x) = \mathcal{E}(t,x) + (\mathcal{E} * W)(t,x)$

where the (4×4)-matrix W is the solution of following integral equation:

(9) $W(t,x) + (A(\partial_x)\mathcal{E}) * W(t,x) = -A(\partial_x)\mathcal{E}$

where:

$$A(\partial_x) = \begin{bmatrix} 0 & , & \gamma\partial_{x_k} \\ 0 & , & 0 \end{bmatrix}$$

Using the matrix $E(t,x)$ of fundamental solutions for the system (1),(2) we obtain the solution of initial-value problem for this system, which we express by the convolutions:

(10) $$\begin{bmatrix} u \\ T \end{bmatrix}(t,x) = E(t,\cdot) *_3 \begin{bmatrix} u^1_0 \\ \vartheta^0 \end{bmatrix} + E_t'(t,\cdot) *_3 \begin{bmatrix} u^0 \\ 0 \end{bmatrix} + E *_4 \begin{bmatrix} X \\ Q \end{bmatrix}(t,x)$$

where u^0, ϑ^0, X, Q are given.

This convolutions are called the thermoelastic potential.

REFERENCES

[1] S.Klainerman, Long-time behaviour of solutions to nonlinear evolution equations. Arch.Rat.Mech.Anal. 78, 73-98, 1982.

[2] Z.Domański, A.Piskorek, Matrices of fundamental solutions for the system of quasi-static equations of thermoelasticity and the system of dynamic equations of thermal stresses. Arch.of Mechanics, 23, 265-270, 1971.

[3] E.E.Levy, Uspekh.Matem.Nauk, 18, 1940.

[4] A.Piskorek, Radon-Transformation und hyperbolische differentialgleichungen der mathematischen Physik. Methoden und Verfahren der Mathematischen Physik, Mannheim, Wien, Zürich, 1973.

THE DIRICHLET PROBLEM ON ENDS

Teruo Ikegami

Department of Mathematics
Osaka City University
Osaka

INTRODUCTION

In the recent investigations of the Dirichlet problem in an axiomatic framework we focus our interest on thefollowing two problems: (1) the boundary behavior of the normalized PWB-solutions, originally studied by O.Frostman[3] and generalized by Constantinescu-Cornea[2] and Lukeš-Malý[11] and (2) the unicity of Keldych operators formulated by J.Lukeš[10], given by Bliedtner-Hansen[1] its essential contribution. The normalized PWB-solutions $H_f^{U,X}$ on an open set U is formed by the Perron-Wiener-Brelot's method taking the closure of U in the one-point compactified space and putting f to be 0 at the infinity, and the effect of the ideal boundary is neglected. In this article we consider above problem under the full influence of the ideal boundary and in the conclusing result we reflect that our result is a generalized version of that obtained formerly.

1.PRELIMINARIES AND NOTATIONS

1. Let X be a P-harmonic space of Constantinescu-Cornea[2] with a countable base and U be an open subset of X. For a numerical function f on the relative boundary ∂U of U, i.e., $\partial U = \overline{U} \setminus U$, we define

$$\overline{H}_f^{U,X}(a) = \inf\left\{ v(a); \begin{array}{l} v \text{ is hyperharmonic on } U, \text{ lower bounded,} \\ v \geq 0 \text{ outside of a compact subset of } X, \\ \liminf v \geq f \text{ on } \partial U \end{array} \right\}$$

and

$$\underline{H}_f^{U,X}(a) = -\overline{H}_{(-f)}^{U,X}(a) .$$

If $\overline{H}_f^{U,X} = \underline{H}_f^{U,X}$ and if this function is harmonic we denote it by $H_f^{U,X}$, the <u>normalized PWB-solution</u>, and f is called <u>(U,X)-resolutive</u>. Every function f which is continuous on ∂U and $|f| \leq p$ by a potential p on X is (U,X)-resolutive, and

$$H_f^{U,X}(a) = \int f(x)\, d\varepsilon_a^{\complement U}(x) \quad \forall a \in U.$$

Let $x \in \partial U$ and

$$M_x = \{\mu \text{ ; a Radon measure on } \partial U, \exists\{a_n\} \text{ ; } a_n \to x, \varepsilon_{a_n}^{\complement U} \to \mu \text{ vaguely}\}.$$

Then, we have the following theorem on the structure of M_x.

Theorem A [Frostman-Constantinescu-Cornea-Lukeš-Malý]

There are four types of M_x, i.e.,

(i) $M_x = \{\varepsilon_x\}$, x is regular

(ii) $M_x = \{\varepsilon_x^{\complement U}\}$, $\varepsilon_x^{\complement U} \neq \varepsilon_x$, x is semi-regular

(iii) $M_x = \{\varepsilon_x, \varepsilon_x^{\complement U}\}$, $\varepsilon_x^{\complement U} \neq \varepsilon_x$, x is weakly irregular

(iv) $M_x = \{t\varepsilon_x + (1-t)\varepsilon_x^{\complement U} ; 0 \leq t \leq 1\}$, $\varepsilon_x^{\complement U} \neq \varepsilon_x$, x is strongly irregular.

Let, for $E \subset X$, $C_p(E) = \{f \in C(\overline{E}); \exists p\text{:a continuous potential on } X, |f| \leq p\}$ and

$S_p(U) = \{s \in C_p(U); \text{ superharmonic on } U\}$.

A linear and positive mapping $L: C_p(\partial U) \to H(U)$ (the space of harmonic functions on U) is called a K-operator if $Ls \leq s$ for every $s \in S_p(U)$. Then, we have

Theorem B [Lukeš-Bliedtner-Hansen]

There is a unique K-operator if and only if the set of irregular boundary points is negligible, i.e.,

$$\varepsilon_a^{\complement U}(\complement\{x \in \partial U; M_x = \{\varepsilon_x\}\}) = 0 \quad \forall a \in U.$$

This theorem is obtained by J.Lukeš[10] for relatively compact open set and is generalized for arbitrary U in [7].

2. Let X^* be a resolutive compactification of X and $\Delta = X^* \setminus X$. Then every finite continuous function f on the ideal boundary Δ is resolutive. We denote the PWB-solution of f by H_f. There corresponds for each $a \in X$ a Radon measure λ_a on Δ, called the harmonic measure at a, such that

$$H_f(a) = \int f(x) d\lambda_a(x).$$

For an open subset U of X, we denote the closure of U in X^* by U^* and $\Delta U = U^* \setminus U = (\Delta U \cap X) \cup (\Delta U \cap \Delta) = \partial U \cup \delta U$. If there is a superharmonic function on X which is continuously extendable and strictly positive on X^*, U^* is a resolutive compactification, and for the PWB-solution we have[9]

$$H_f^U = H_{(f-u)}^{U,X} + u \qquad \forall f \in C(\Delta U),$$

where $u = H_g$ and g is a continuous extension of $f|\delta U$ (the restriction of f on δU) onto Δ, and

$$H_f^U(a) = \int f(x) d\lambda_a^U(x).$$

For $x \in \Delta U$, we define

$$N_x^U = \{\mu \text{ ; a Radon measure on } \Delta U, \exists\{a_\iota\}, a_\iota \to x, \lambda_{a_\iota}^U \to \mu \text{ vaguely}\}.$$

A set $A \subset \Delta U$ is called negligible if $\lambda_a^U(A) = 0$ for every $a \in U$. In the following we use the notations:

$\Gamma = \overline{(\bigcup_{a \in X} \operatorname{supp} \lambda_a)}$; the harmonic boundary of X^*,

$\Gamma(U) = \overline{(\bigcup_{a \in X} \operatorname{supp} \lambda_a^U)}$; the harmonic boundary of U^*,

$(\Delta U)_r$: the set of regular boundary points of U^*.

2. THE PROBLEM AND AN EXAMPLE

We are going to investigate the problem: how can we generalize the theorems stated in 1 under the existence of the ideal boundary? The situation seems to be delicate and complicated at an ideal boundary point which is either a limit point of the relative boundary or the exterior of the harmonic boundary. To show the latter case we shall give an example.

Example

Let X be the unit disc of the complex plane punctured by two points: $X = \{z \in \mathbb{C}; |z| < 1\} \setminus \{1/2, -1/2\}$, and $k_0(z) = \mathrm{Re}(\frac{1+z}{1-z})$. We consider the k_0-harmonic structure, i.e., the quotient of harmonic functions by k_0. We compactify X by taking first the topological closure and then identifying $\{1,-1\}$ with e and $\{1/2,-1/2\}$ with e'. Thus we obtain a resolutive compactification X^*. It is even semi-regular, i.e., the PWB-solutions are constant and therefore all boundary points are regular or semi-regular. If we consider $G = X \setminus K$, where K is a compact subset of X, then $N_e^G = \{\varepsilon_e, \mu\}$, $\mu \neq \varepsilon_e$ and $N_{e'}^G = \{\mu_1, \mu_2\}$, $\mu_i \neq \varepsilon_{e'}$ $(i = 1,2)$, $\mu_1 \neq \mu_2$. Therefore the point e' is not of any type in Theorem A.

To avoid difficulties caused by situations cited above, we shall restrict ourselves to the case of an end G, i.e., G is an open subset of X, and $X \setminus G$ is non-empty and compact, and consider only on the harmonic boundary $\Gamma(G)$. Then $\delta G = \Delta$ and we have[9] $\lambda_a^G|\partial G = \varepsilon_a^{\complement G}$ for every $a \in G$.

Further we make the following assumptions:

(I) X^* is a metrizable and semi-regular compactification,

(II) there exists $s_0 \in C(X^*)$ such that $s_0 > 0$ and superharmonic on X,

(III) for each $x \in \Gamma$ there is a superharmonic function $v > 0$ on X such that $\lim_{a \to x} v(a) = 0$.

Remarks-1°. If X has a metrizable and resolutive compactification there always exists a metrizable and semi-regular compactification X^*[4], [5],[6]. In X^*, every point of Γ is regular.

2°. The assumption (III) is equivalent to

(III') there is a bounded continuous potential p on X such that $p > 0$ and $\lim_{a \to x} p(a) = 0$ at every $x \in \Gamma$.

3°. If X^* is metrizable and semi-regular, Γ contains at least two points and λ_a is non-zero and non-atmic then (III) is valid.

4°. Without the assumption (III) a regular point is not to be regular in general, if we remove compact subsets of the space as is seen in the above example.

In the sequel we fix an end G.

3. PROBLEM A

Let $f \in C(\Delta G)$, $u = H_{f|\Delta}$ and $g = (f - u)1_{\partial G}$. Then, since $H_g^G = H^{G,X}_{(g|\partial G)}$, we have $H_f^G = u + H_g^G$ and the following proposition is an immediate consequence of the assumption (III).

Proposition 1

For an end G, every point of Γ is regular for G*, i.e., regular boundary points of X* remain to be regular if we remove compact subsets of X.

By Prop.1 and by the fact that every regular point belongs to the harmonic boundary, we have

Proposition 2

$\Gamma = \Gamma(G) \cap \Delta$.

Now,we can obtain the result corresponding to the theorem of Frostman-Constantinescu-Cornea-Lukeš-Malý:

Theorem 1

(i) for $x \in \Gamma(G) \cap \Delta$, $N_x^G = \{\varepsilon_x\}$, i.e., x is regular for G*,

(ii) for $x \in \Gamma(G) \cap \partial G$, we have three cases

$$N_x^G = \begin{cases} \{\varepsilon_x\}, & x \text{ is regular} \\ \{\varepsilon_x, \lambda_x^G\}, \quad \lambda_x^G \neq \varepsilon_x, & x \text{ is weakly irregular} \\ \{t\varepsilon_x + (1-t)\lambda_x^G;\ 0 \leq t \leq 1\}, \quad \lambda_x^G \neq \varepsilon_x, & x \text{ is strongly irregular} \end{cases}$$

and for a sequence $\{a_n\}$ converging to x $\lambda_{a_n}^G \to \lambda_x^G$ if and only if $\varepsilon_{a_n}^{\complement G} \to \varepsilon_x^{\complement G}$.

$$\lambda_x^G(f) = \varepsilon_x^{\complement G}(f - H_{f|\Delta}) + H_{f|\Delta}(x) \qquad \forall f \in C(\Delta G),$$

x is regular for G* if and only if $\varepsilon_x^{\complement G} = \varepsilon_x$.

$$\lambda_x^G \big|_{\partial G} = \varepsilon_x^{\complement G} \neq \varepsilon_x \text{ if x is not regular for G*.}$$

In fact, (i) is trivially derived from Prop.1 and Prop.2. (ii) is proved in [9].

Remark

The part (ii) is proved for an arbitrary open set U[9] and every point $x \in \partial G \setminus \Gamma(G)$ is semi-regular.

4.PROBLEM B

We consider the theorem of Bliedtner-Hansen-Lukeš in our setting. Let $T = \partial G \cup \Gamma$ and

$S'(G) = \{s;$ bounded superharmonic on G, continuously extendable on $T\}$.

The inf-stable convex cone S'(G) will play an important role in the investigation. It is easily checked that S'(G) separates linearly points of T, i.e., $\forall x, y \in T$, $x \neq y$, $\forall \alpha \geq 0$ $\exists s \in S'(G)$: $s(x) \neq \alpha s(y)$. Thus the family of differences of restrictions of S'(G) on T is uniformly dense in C(T).

We define, for Radon measures μ, ν on G* with $\mu(\complement(G \cup T)) = \nu(\complement(G \cup T)) = 0$, $\mu \prec \nu$ if $\mu(s) \leq \nu(s)$ for every $s \in S'(G)$ and, for $x \in G \cup T$,

$\mathcal{M}_x(S'(G)) = \{\mu;$ a Radon measure on G*, $\mu(\complement(G \cup T)) = 0$, $\mu \prec \varepsilon_x\}$.

The following propositions are proved quite in the same way as [8]:

(1*) If f is a numerical function on T bounded above and μ is a Radon measure on G* with $\mu(\complement T) = 0$ then

$\mu(\hat{f}) := \mu(\inf\{s; s \in S'(G),\ s \geq f \text{ on } T\}) = \inf\{\mu(s); s \in S'(G),\ s \geq f \text{ on } T\}$.

(2*) Let $f \in C(T)$. For each Radon measure μ on G* with $\mu(\complement(G \cup T)) = 0$ there is a Radon measure ν on G* with $\nu(\complement T) = 0$, $\nu \prec \mu$ and

$\nu(f) = \mu(\hat{f}) = \sup\{\lambda(f);\ \lambda$ is a Radon measure on G*, $\lambda(\complement T) = 0, \lambda \prec \mu\}$.

(3*) Let μ be a Radon measure on G^* with $\mu(\complement T) = 0$. Then μ is minimal with respect to $\prec$ if and only if $\mu(\hat{t}) = \mu(t)$ for every $t \in -S'(G)$.

First we prove:

Proposition 3
$(G^*, S'(G))$ is geometrically simplicial, i.e.,
$\forall\, s, -t \in S'(G)$, $t \leq s$ on T $\exists\, h \in S'(G) \cap [-S'(G)]$: $t \leq h \leq s$ on $G \cup T$.

Proof – Let $f \in C(\Gamma)$ satisfying $t \leq f \leq s$ on Γ. By the assumption (III') there is a strictly positive potential p on X such that $\lim_x p = 0$ at every point $x \in \Gamma$. Then, we can find $c > 0$ with the following inequalities on ∂G:
$-cp < t - H_f \leq s - H_f < cp$.

Letting $v = \inf(s - H_f, cp)$ and $u = \sup(t - H_f, -cp)$ we have $-u, v \in S'(G)$ and $v - u \geq 0$ on T, thus $v - u \geq 0$ on $G \cup T$. On the other hand, by [1], there is a function w such that w is continuous on $G \cup \partial G$, harmonic on G and $u \leq w \leq v$ on $G \cup \partial G$. Therefore $h = w + H_f$ is the desired one.

From Prop. 3, we can derive as in [8]

Proposition 4
For every $x \in G \cup T$, $\mathcal{M}_x(S'(G))$ has the unique minimal measure μ_x.

To determine the unique minimal measure μ_x, we recall the convex cone in [1]
$S_p(G) = \{s \in C_p(G);\ s \text{ is superharmonic on } G\}$.
By virtue of [1] and our previous consideration it is not difficult to prove the following result:

Proposition 5
For $t \in -S'(G)$ and $x \in G \cup T$, we have

$$\mu_x(t) = \begin{cases} t(x) & x \in \Gamma \\ \varepsilon_x^Z(t - H_t) + H_t(x) & x \in G \cup \partial G, \end{cases}$$

where $Z = \complement \overline{G} \cup Ch_{S_p(G)} G$ and $Ch_{S_p(G)} G$ is the Choquet boundary in [1].

Remark
Putting $N_x^G = \{\lambda_x^G\}$ for $x \in G$, N_x^G has the unique minimal measure λ_x with respect to the relation $\prec$, where

$$\lambda_x = \begin{cases} \varepsilon_x & x \in \Gamma \\ \lambda_x^G & x \in G \cup \partial G \end{cases}$$

We define $Ch_{S'(G)} G^* = \{x \in G \cup T;\ \mathcal{M}_x(S'(G)) = \{\varepsilon_x\}\}$. Then,

Proposition 6
$Ch_{S'(G)} G^* = \Gamma \cup Ch_{S_p(G)} G$.

Now we are ready to discuss the problem B, that is the unicity of the Keldych operators.
We shall define that a linear and positive mapping $L: C(T) \to H(G)$ is to be a Keldych operator if $Ls \leq s$ for every $s \in S'(G)$. It is clear that the mapping $Df(a) = H_f^G(a)$ is a Keldych operator.

Theorem 2

The following statements are equivalent:

(1) $\complement(\Delta G)_r$ is negligible,

(2) $\mu_x = \lambda_x$ for every $x \in G \cup T$,

(3) if L is a Keldych operator then $Lf \doteq H_f^G$ for every $f \in C(T)$.

Proof – (1) $\Longrightarrow$ (2): (1) is equivalent to $\varepsilon_a^{\complement G}(\{x \in \partial G;\ \varepsilon_x^{\complement G} \neq \varepsilon_x\}) = 0$ for every $a \in G$, and the latter is equivalent to $\varepsilon_x^{\complement G} = \varepsilon_x^Z$ for every $x \in G \cup \partial G$ by [1]. This implies (2) since $\lambda_x = \mu_x = \varepsilon_x$ for every $x \in \Gamma$.

(2) $\Longrightarrow$ (3): a Keldych operator L defines a Radon measure ν_a on T for every $a \in G$, i.e., $Lf(a) = \nu_a(f)\ \forall f \in C(T)$. Then as in [7] we have $\mu_a \prec \nu_a \prec \lambda_a^G$.

(3) $\Longrightarrow$ (1): the mapping $Mf(a) := \mu_a(f)$ is a Keldych oprator. Since $\mu_a = \lambda_a^G$ for every $a \in G$, $\varepsilon_a^Z = \mu_a|\partial G = \lambda_a^G|\partial G = \varepsilon_a^{\complement G}$ for every $a \in G$, which implies, by [1], $\varepsilon_a^{\complement G}(\{x \in \partial G;\ \varepsilon_x^{\complement G} \neq \varepsilon_x\}) = 0$, q.e.d..

Remark

It is interesting to reflect two extremal measures at $x \in G \cup \partial G$,

$$\lambda_x(f) = \varepsilon_x^{\complement G}(f - H_{f|\Delta}) + H_{f|\Delta}(x), \quad \lambda_x|\partial G = \varepsilon_x^{\complement G},$$

$$\mu_x(f) = \varepsilon_x^Z(f - H_{f|\Delta}) + H_{f|\Delta}(x), \quad \mu_x|\partial G = \varepsilon_x^Z.$$

REFERENCES

[1]J.Bliedtner-W.Hansen: Simplicial cone in potential theory, Invent. Math. 29 (1975) 83-110.

[2]C.Constantinescu-A.Cornea: Potential theory in harmonic spaces,Grundlehr. der mat. Wiss. 158 Berlin-Heidelberg-New york 1972.

[3]O.Frostman: Les points irréguliers dans la théorie du potentiel et la citère de Wiener, Medd. Lunds Univ.Mat.Sem. 4 (1936) 1-10.

[4]T.Ikegami: A note on axiomatic Dirichlet problem, Osaka J. Math. 6 (1969) 39-47.

[5]T.Ikegami: On the regularity of boundary points in a resolutive compactification of a harmonic space, Osaka J. Math. 14 (1977) 271-289.

[6]T.Ikegami: Remarks on the reguality of boundary points in a resolutive compactification, Osaka J. Math. 17 (1980) 177-186.

[7]T.Ikegami: On a generalization of Lukeš' theorem, Osaka J. Math. 18 (1981) 699 -702.

[8]T.Ikegami: On the simplicial cone of superharmonic functions in a resolutive compactification of a harmonic space, Osaka J.Math. 20 (1983) 881-898.

[9]T.Ikegami: On the boundary behavior of the Dirichlet solutions at an irregular boundary point, Osaka J. Math. 21 (1984) 851-858.

[10]J.Lukeš: Théorème de Keldych dans la théorie axiomatique de Bauer des fonctions harmoniques, Czechoslovak Math. J. 24 (1974) 114-125.

[11]J.Lukeš-J.Malý: On the boundary behavior of the Perron generalized solution, Math Ann. 257 (1981) 355-366.

DIFFUSION KERNELS OF LOGARITHMIC TYPE

Masayuki Itô

Department of Mathematics, Nagoya University
Chikusa-ku, Nagoya 464, Japan

1. INTRODUCTION

Let X be a locally compact, non-compact Hausdorff space with countable basis. We denote by:

$C_K(X)$ the usual topological vector space of all finite continuous functions with compact support;

$C(X)$ the usual Fréchet space of all finite continuous functions on X;

$M_K(X)$ the usual topological vector space of all real Radon measures with compact support;

$M(X)$ the topological vector space of real Radon measures on X with the weak* topology.

A diffusion kernel V on X means a continuous linear operator from $C_K(X)$ into $C(X)$. The dual kernel V^* of V is the continuous linear operator from $M_K(X)$ into $M(X)$ defined by for any $\mu \in M_K(X)$ and $f \in C_K(X)$,

$$\int f dV^*\mu = \int Vf d\mu.$$

A diffusion kernel V on X is said to be of logarithmic type if for any $\mu \in M_K(X)$ with $\int d\mu = 0$, $V^*\mu$ is of form

$$V^*\mu = \lim_{a\to\infty} \int_0^a T_t^*\mu dt \quad (= \int_0^\infty T_t^*\mu dt),$$

where $(T_t)_{t\geq 0}$ is a sub-Markovian diffusion semi-group on X, namely T_t is a positive diffusion kernel on X, the mapping $t \to T_t f$ is continuous for every $f \in C_K(X)$, T_0 = the identity I, $(T_t)_{t\geq 0}$ satisfies the semi-group condition, namely for any $t,s \geqq 0$ and $f \in C_K(X)$, $T_t(T_s f) = T_{t+s}f$, and T_t is a contraction, namely for any $f \in C_K(X)$, $0 \leqq T_t f \leqq 1$ if $0 \leqq f \leqq 1$.

In this case, $(T_t)_{t \geqq 0}$ is uniquely determined and called the diffusion semi-group of V.

Evidently the logarithmic kernel on the 2-dimensional Euclidean space is a diffusion kernel of logarithmic type. It is of convolution type. For the non-convolution type, there is a remarkable example. That is the Neumann kernel, namely for a domain $X = \Omega$ in the Euclidean space R^n $(n \geqq 2)$ where the Neumann problem is resolutive, putting

$$Vf(x) = \int N(x, y) f(y) dy$$

for every $f \in C_K(X)$, we see that V is a diffusion kernel of logarithmic type, where $N(x, y)$ is the Neumann kernel on Ω.

Evidently Hunt diffusion kernels are always diffusion kernels of logarithmic type. The above two examples are not Hunt diffusion kernels. A diffusion kernel V of logarithmic type is said to be of purely logarithmic type if the diffusion semi-group of V is reccurent.

The purpose of this paper is to examine the potential theoretical properties of diffusion kernels of logarithmic type and to determine diffusion kernels of logarithmic type in terms of potential theoritical properties. In the case of convolution type, the discussion is rather simple and very clear [1,2,3], but in the case of non-convolution type, it is quite complicated.

2. POTENTIAL THEORETICAL PROPERTIES OF DIFFUSION KERNELS

We denote by $C_K^+(X)$, $C^+(X)$, $M_K^+(X)$ and by $M^+(X)$ the subsets of $C_K(X)$, $C(X)$, $M_K(X)$ and of $M(X)$ formed by all non-negative elements, respectively.

Let V be a diffusion kernel on X. We denote by $\mathcal{D}(V)$ the linear subspace of $C(X)$ formed by all $f \in C(X)$ satisfying $\int |f| d|V^* \varepsilon_x| < \infty$ for every $x \in X$ and $x \to \int f dV^* \varepsilon_x \in C(X)$, where ε_x denotes the Dirac measure at x. Define

$$Vf(x) = \int f dV^* \varepsilon_x$$

for every $f \in \mathcal{D}(V)$. Then V is a linear operator from $\mathcal{D}(V)$ into $C(X)$. Similarly we denote by $\mathcal{D}(V^*)$ the linear subspace of $M(X)$ formed by all $\mu \in M(X)$ satisfying $\int |Vf| d|\mu| < \infty$ for every $f \in C_K(X)$. Then $C_K(X) \ni f \to \int Vf d\mu$ defines a real Radon measure. We denote it by $V^*\mu$.

Then V^* is a linear operator from $\mathcal{D}(V^*)$ into $M(X)$. Put

$$\mathcal{D}^+(V) = \mathcal{D}(V) \cap C^+(X) \quad \text{and} \quad \mathcal{D}^+(V^*) = \mathcal{D}(V^*) \cap M^+(X).$$

We say that:

V is bounded if for any $f \in C_K(X)$, Vf is bounded on X.

V is semi-bounded if for any $0 \neq f$, $0 \neq g \in C_K^+(X)$, there exists a non-negative constant a such that $V(f - ag)$ is bounded.

We easily obtain the following two propositions.

Proposition 1 – If V is a diffusion kernel of logarithmic type, then V is semi-bounded.

Proposition 2 – If V is semi-bounded and unbounded, then there exists $\eta_0 \in M^+(X)$ such that for any $f \in C_K(X)$ with $\int f d\eta_0 = 0$, Vf is bounded, and η_0 is uniquely determined except for muliplications of positive constants.

In Proposition 2, we call η_0 the basic invariant measure for V. For a bounded diffusion kernel V, we always assume that the basic invariant measure η_0 for V is equal to 0.

We say that V satisfies the semi-complete maximum principle if for any $f, g \in C_K^+(X)$ and any non-negative constant a, $Vf - Vg \leqq a$ on X if $Vf - Vg \leqq a$ on $\text{supp}(f)$ and if $\limsup_{x \to \delta}(Vf(x) - Vg(x)) \leqq a$, where $\text{supp}(f)$ denotes the support of f and where δ denotes the Alexandroff point of X.

More generally, for two diffusion kernels V_1 and V_2 on X, we denote by $V_1 [V_2$ if for any $f, g \in C_K^+(X)$ and any non-negative constant a, $V_2 f - V_2 g \leqq a$ if $V_1 f - V_1 g \leqq a$ on $\text{supp}(f)$ and if $\limsup_{x \to \delta}(V_1 f(x) - V_2 g(x)) \leqq a$.

From the definitions it follows

Proposition 3 – Let V be a diffusion kernel on X and I the unit kernel, namely I is the identity mapping from $C_K(X)$ to itself. Then V satisfies the semi-complete maximum principle if and only if for any constant $c > 0$, $V + cI [V$.

Remark 1 – If V satisfies the usual complete maximum principle, then V satisfies the semi-complete maximum principle.

Proposition 4 – If V is a diffusion kernel of logarithmic type, then V satisfies the semi-complete maximum principle.

In fact, let $(V_p)_{p>0}$ be the resolvent defined by the diffusion semi-group of V, namely for the diffusion semi-group $(T_t)_{t \geqq 0}$ of V,

$$V_p = \int_0^\infty \exp(-pt) T_t dt.$$

Then the usual complete maximum principle for V_p gives the semi-complete maximum principle for V.

We shall define the semi-balayage principle and the semi-balayablity. For a diffusion kernel V on X and for an open set $\omega \neq \phi$ in X, we put

$$S_t(\omega; V) = \overline{\{V^* \mu \in M(X);\ \mu \in \mathcal{D}^+(V^*),\ \int d\mu = t,\ \mathrm{supp}(\mu) \subset \omega\}},$$

where t is a positive number and where the closure is in the sense of the weak* topology. Furthermore we write

$$H_t(V) = \bigcap_{K:\text{compact}} S_t(CK; V) \quad (t > 0) \quad \text{and} \quad H_0(V) = \bigcap_{t>0} H_t(V).$$

If V is semi-bounded, then $H_0(V) = \{a\eta_0;\ a \in R\}$, where η_0 is the basic invariant measure for V and where R denotes the set of all real numbers.

Let V_1 and V_2 be diffusion kernels on X. We denote by $V_1^* \prec V_2^*$ if for any $\mu \in M_K^+(X)$ and any relatively compact open set $\omega \neq \phi$ in X, there exist $\mu' \in \mathcal{D}^+(V_1^*)$ with $\int d\mu' \leqq \int d\mu$ and $\eta \in H_t(V_1)$ with $t = \int d\mu - \int d\mu'$ such that:

(B.1) $\mathrm{supp}(\mu') \subset \overline{\omega}$.

(B.2) $V_1^* \mu' + \eta = V_2^* \mu$ in ω.

(B.3) $V_1^* \mu' + \eta \leqq V_2^* \mu$ in X.

In this case, we call (μ', η) a semi-balayaged couple of μ to ω with respect to $V_1^* \prec V_2^*$ and for any $\mu \in \mathcal{D}^+(V_1^*)$, there exists a semi-balayaged couple of μ to ω with respect to $V_1^* \prec V_2^*$ if $\int d\mu < \infty$.

In particular, if $V^* \prec V^*$, we say that V^* satisfies the semi-balayage principle and that the above (μ', η) with respect to $V^* \prec V^*$ is a V^*-semi-balayaged couple of μ to ω.

If for any $\mu \in \mathcal{D}^+(V^*)$ with $\int d\mu < \infty$ and for any open set $\omega \neq \phi$ in X, there exists a V^*-semi-balayaged couple of μ to ω exists, we say that V^* is semi-balayable.

<u>Remark 2</u> — Let V be an unbounded and semi-bounded diffusion kernel on X and assume that V^* satisfies the semi-balayage principle. Then for any $\mu \in \mathcal{D}^+(V^*)$ with $\int d\mu < \infty$, any relatively compact open set $\omega \neq \phi$ in X and for any V^*-semi-balayaged couple (μ', η) of μ to ω, we have $\int d\mu' = \int d\mu$ and $\eta = a\eta_0$ with some $a \in R$, where η_0 is the basic invariant measure for V.

Similarly as in Proposition 4, we have the following:

Proposition 5 – If V is a diffusion kernel of logarithmic type, then V^* satisfies the semi-balayage principle.

The following theorem will show that the semi-complete maximum principle for V and the semi-balayage principle for V^* are mutually dual.

Theorem 1 – Let V_1 and V_2 be diffusion kernels on X. Assume that V_1 is semi-bounded. Then $V_1 \,[\, V_2$ and $V_1^* \prec V_2^*$ are equivalent.

The implication $V_1^* \prec V_2^* \to V_1 \,[\, V_2$ follows from the usual dual treatment. We shall give the sketch of the proof of the implication $V_1 \,[\, V_2 \to V_1^* \prec V_2^*$. We shall use the Choquet-Deny duality method[4].

Let $\omega \neq \phi$ be a relatively compact open set in X and $(\omega_n)_{n=1}^{\infty}$ an exhaustion of X satisfying $\overline{\omega} \subset \omega_1$. We set

$$A_n(\omega) = \left\{V_1^*\nu_1 + V_1^*\nu_2 + \lambda + a\eta_0;\ \begin{array}{l} \nu_j \in M_K^+(X)\ (j = 1, 2),\ \mathrm{supp}(\nu_1) \subset \overline{\omega} \\ \mathrm{supp}(\nu_2) \subset C\omega_n,\ \int d(\nu_1 + \nu_2) = 1 \\ \lambda \in M_K^+(X),\ \mathrm{supp}(\lambda) \subset C\omega \cap \overline{\omega_n},\ a \in R \end{array}\right\},$$

where η_0 is the basic invariant measure for V_1. Put

$$A(\omega) = \bigcap_{n=1}^{\infty} \overline{A_n(\omega)}.$$

Then the semi-boundedness of V_1 shows that $A(\omega)$ is a compact convex set in $M(X)$ and $V_1 \,[\, V_2$ gives $V_2^*\mu \in A(\omega)$ for every $\mu \in \mathcal{D}^+(V_2^*)$ with $\int d\mu = 1$. This shows that a semi-balayaged couple of μ to ω with respect to $V_1^* \prec V_2^*$ exists.

This theorem leads to the following corollaries.

Corollary 1 – Let V be a diffusion kernel on X. Assume that V is semi-bounded and that V satisfies the semi-complete maximum principle. Then for any $t > 0$, $\eta \in S_t(V) = S_t(X; V)$ and for any open set $\omega \neq \phi$ in X,

$$\min\,\{\xi \in S_t(\omega; V);\ \xi \geqq \eta \ \text{ in } \ \omega\}$$

exists, where the minimum is in the sense of measures.

We set

$$R_\omega^V \eta = \min\,\{\xi \in S_t(\omega; V);\ \xi \geqq \eta \ \text{ in } \ \omega\}$$

and call it the reduced measure of η to ω with respect to V^*. Put

$$R_\delta^V \eta = \inf_{K:\text{compact}} R_{CK}^V \eta ;$$

and we call it the reduced measure of η at the infinity δ with respect to V^*. Then $R_\delta^V \eta$ is in $M(X)$ or equal to $-\infty$, namely

$$\inf_{K:\text{compact}} \int f dR_{CK}^V \eta = -\infty$$

for every $0 \neq f \in C_K^+(X)$. The reduced measure $R_\delta^V \eta$ of η at δ gives the estimation of the behavior of η near δ.

We say that η diverges to $-\infty$ at δ in the sense of mean convergence if $R_\delta^V \eta = -\infty$.

For $t_1 \geqq 0$, $t_2 \geqq 0$ and $\eta_1 \in S_{t_1}(V)$, $\eta_2 \in S_{t_2}(V)$, we say that $\eta_1 - \eta_2$ vanishes at δ in the sense of mean convergence if there exists an exhaustion $(\omega_n)_{n=1}^\infty$ of X such that, putting $K_n = \overline{\omega_n}$, we have

$$w^*\text{-}\lim_{n\to\infty} (R_{CK_n}^V \eta_1 - R_{CK_n}^V \eta_2) = 0.$$

<u>Remark 3</u> - Let V be a semi-bounded diffusion kernel on X and assume that V satisfies the semi-complete maximum principle. Then V is unbouded if and only if $R_\delta^V V^* \varepsilon_x = -\infty$ for every $x \in X$.

<u>Corollary 2</u> - Let V be a semi-bounded diffusion kernel on X. Assume that V satisfies the semi-complete maximum principle. Then for any non-negative constant c, any relatively compact open set $\omega \neq \phi$ and any $\mu \in M_K^+(X)$, there exist $\mu'_{\omega,c} \in M_K^+(X)$ with $\int d\mu'_{\omega,c} \leqq \int d\mu$ and $\eta_{\omega,c} \in H_t(V)$ with $t = \int d\mu - \int d\mu'_{\omega,c}$ such that:

(1) $$\text{supp}(\mu'_{\omega,c}) \subset \overline{\omega}.$$

(2) $$V^* \mu'_{\omega,c} + c\mu'_{\omega,c} + \eta_{\omega,c} = V^* \mu \quad \text{in } \omega.$$

(3) $$V^* \mu'_{\omega,c} + c\mu'_{\omega,c} + \eta_{\omega,c}$$

$$= \min\left\{V^* \nu + c\nu + \eta ;\ \begin{matrix} \nu \in \mathcal{D}^+(V^*),\ \int d\nu + s = \int d\mu \\ V^* \nu + c\nu + \eta \geqq V^* \mu \ \text{ in } \omega \end{matrix}\right\}.$$

If $c > 0$, the above couple $(\mu'_{\omega,c}, \eta_{\omega,c})$ is uniquely determined. For a relatively compact open set $\omega \neq \phi$ and for a positive number p, we define the operator

$$W_{\omega,p} : M_K(X) \ni \mu \to \frac{1}{p}\mu'_{\omega,1/p} \in M(X),$$

where $\mu_{\omega,1/p} = (\mu^+)'_{\omega,1/p} - (\mu^-)'_{\omega,1/p}$. Then $W_{\omega,p}$ is a positive linear operator from $M_K(X)$ into $M(X)$ and $(W_{\omega,p})_{p>0}$ is a sub-Markovian resolvent on ω, namely for any $p > 0$ and $q > 0$,

$$W_{\omega,p}\mu - W_{\omega,q}\mu = (q-p)W_{\omega,p}(W_{\omega,q}\mu) \quad \text{and} \quad p\int dW_{\omega,p}\mu \leqq \int d\mu$$

for every $\mu \in M_K^+(X)$.

For an exhaustion $(\omega_n)_{n=1}^{\infty}$ of X, we have $W_{\omega_n,p}\mu \geqq W_{\omega_{n+1},p}\mu$ in for every $p > 0$ and $\mu \in M_K^+(X)$, so that for any $\mu \in M_K(X)$, we see that $w^*\text{-}\lim_{n\to\infty} W_{\omega_n,p}\mu$ exists in $M(X)$. Define

$$W_p : M_K(X) \ni \mu \to w^*\text{-}\lim_{n\to\infty} W_{\omega_n,p}\mu \in M(X).$$

Then W_p is independent from the choice of $(\omega_n)_{n=1}^{\infty}$ and $(W_p)_{p>0}$ is also a sub-Markovian resolvent on X, which is called the resolvent of the regular part of V^*.

3. DIFFUSION SEMI-GROUPS

We shall give three definitions about diffusion semi-groups.

A diffusion semi-group $(T_t)_{t\geqq 0}$ is transient if for any $\mu \in M_K^+(X)$, the integral $\int_0^\infty T_t^*\mu dt$ is in $M^+(X)$ and if the mapping $X \ni x \to \int_0^\infty T_t^*\varepsilon_x dt \in M(X)$ is locally bounded.

Remark 4 — Let $(T_t)_{t\geqq 0}$ be a diffusion semi-group on X. Then it is transient if and only if it is a diffusion semi-group of a certain Hunt diffusion kernel.

A diffusion semi-group $(T_t)_{t\geqq 0}$ is recurrent if there exists $\mu \in M_K^+(X)$ such that the integral $\int_0^\infty T_t^*\mu dt$ diverges.

A diffusion semi-group $(T_t)_{t\geqq 0}$ is semi-transient if for any $\mu \in M_K(X)$ with $\int d\mu = 0$, $(\int_0^t T_s^*\mu ds)_{t>0}$ is bounded in $M(X)$ and if for a fixed $x_0 \in X$, the family of the mappings

$$(K \ni x \to \int_0^t T_s^*(\varepsilon_x - \varepsilon_{x_0})ds)_{t>0}$$

is uniformly bounded for every compact set K in X.

Theorem 2 — Let $(T_t)_{t\geqq 0}$ be a sub-Markovian diffusion semi-group on X. Then $(T_t)_{t\geqq 0}$ is a diffusion semi-group of a certain diffusion kernel of logarithmic type if and only if it is semi-transient.

We see easily that the "only if" part holds, so that we shall give an outline of the proof of the "if" part. Since $(T_t)_{t\geq 0}$ is sub-Markovian, the resolvent $(V_p)_{p>0}$ defined by $(T_t)_{t\geq 0}$ exists. By the semi-boundedness of $(T_t)_{t\geq 0}$, $(V_p^* \mu)_{p>0}$ is relatively compact in $M(X)$ for every $p > 0$ and $\mu \in M_K(X)$ with $\int d\mu = 0$. Similarly as in the case of convolution type[1,2,3], we obtain the following results.

(I) For any increasing sequence $s_t = (t_n)_{n=1}^{\infty}$ of positive numbers with $\lim_{n\to\infty} t_n = \infty$, there exist a subsequence $s_t' = (t_{n_k})_{k=1}^{\infty}$ of s_t and a diffusion kernel $V_{s_t'}$ on X such that for any $\mu \in M_K(X)$ with $\int d\mu = 0$,

$$V_{s_t'}^* \mu = w^*\text{-}\lim_{n\to\infty} \int_0^{t_{n_k}} T_t^* \mu dt.$$

(II) For any decreasing sequence $s_p = (p_n)_{n=1}^{\infty}$ of positive numbres with $\lim_{n\to\infty} p_n = 0$, there exists a subsequence $s_p' = (p_{n_k})_{k=1}^{\infty}$ of s_p and a diffusion kernel $V_{s_p'}$ on X such that for any $\mu \in M_K(X)$ with $\int d\mu = 0$,

$$V_{s_p'}^* \mu = w^*\text{-}\lim_{n\to\infty} V_{p_{n_k}}^* \mu.$$

(III) For any s_t, s_p, $V_{s_t'}$ and $V_{s_p'}$ satisfying the above conditions, we have

$$V_{s_p'} \sqsubset V_{s_t'} \sqsubset V_{s_p'}.$$

By using $V_{s_p'}^* \nless V_{s_t'}^* \nless V_{s_p'}^*$ (see Theorem 1), we obtain $V_{s_p'} = V_{s_t'}$. This implies that for any $\mu \in M_K(X)$ with $\int d\mu = 0$,

$$w^*\text{-}\lim_{t\to\infty} \int_0^t T_s^* \mu ds \quad \text{and} \quad w^*\text{-}\lim_{p\to 0} V_p^* \mu$$

exist in $M(X)$ and that these are equal. Thus there exists a diffusion kernel of logarithmic type whose diffusion semi-group is equal to $(T_t)_{t\geq 0}$.

Theorem 2 is one of Tauberian type theorems. In the case that we discuss diffusion kernels of logarithmic type, semi-transient and sub-Markovian diffusion semi-groups play essential role.

4. MAIN THEOREMS

Finally we shall show our main theorem. In the case of convolution type, the analogous result is already known[1,2].

Theorem 3 – Let V be a diffusion kernel on X. Then V is a diffusion kernel of purely logarithmic type if and only if V satisfies the following five conditions:

(a) V is semi-bounded.

(b) V is non-degenerate, namely for any $x, y \in X$ with $x \neq y$, $V^*\varepsilon_x$ and $V^*\varepsilon_y$ are not proportional.

(c) V satisfies the semi-complete maximum principle.

(d) For any $x \in X$, $V^*\varepsilon_x$ diverges to $-\infty$ at δ in the sense of mean convergence.

(e) For any $x, y \in X$, $V^*\varepsilon_x - V^*\varepsilon_y$ vanishes at δ in the sense of mean convergence.

To show this theorem, we shall use the following known lemma.

Lemma[5] - Let V be a positive diffusion kernel on X and assume that V satisfies the usual complete maximum principle and that V is non-degenerate. If for any $f \in C_K^+(X)$ and for any compact set K in X,

$$\text{(R)} \qquad \inf_{F:\text{compact}} \sup_{x \in K} \int f dR_{CF}^V V^* \varepsilon_x = 0,$$

then V is a Hunt diffusion kernel.

The sketch of the proof of Theorem 3. The proof of the "only if" is not so complicated, so that we shall give only the proof of the "if" part. Let $(W_p)_{p>0}$ be the resolvent of the regular part of V^*. The condition (d) shows that $(W_p)_{p>0}$ is markovian, namely for any $\mu \in M_K^+(X)$, $p\int dW_p\mu = \int d\mu$. For any exhaustion $(\omega_n)_{n=1}^{\infty}$ of X and for any $f \in C_K(X)$, the conditions (d) and (e) show that for any $p > 0$, the sequence $(\int f dW_{\omega_n,p}\varepsilon_x)_{n=1}^{\infty}$ of functions of x converges uniformly to $\int f dW_p \varepsilon_x$ on any compact set as $n \to \infty$, where $(W_{\omega_n,p})_{p>0}$ is the resolvent on ω_n obtained in Corollary 2. Hence the mapping $X \ni x \to W_p\varepsilon_x \in M(X)$ is continuous for every $p > 0$. Define

$$V_p f(x) = \int f dW_p \varepsilon_x$$

for every $f \in C_K(X)$. Then V_p is a diffusion kernel on X. The condition (b) shows that V_p is also non-degenerate for every $p > 0$. Thus $(V_p)_{p>0}$ is a Markovian resolvent of non-generate diffusion kernels, so that for any $p > 0$, V_p satisfies the usual complete maximum principle. Furthermore the conditions (d) and (e) show that V_p satisfies the condition (R) in Lemma. Therefore there exists a uniquely determined Markovian diffusion semi-group $(T_t)_{t \geq 0}$ on X such that for any $p > 0$,

$$V_p = \int_0^{\infty} T_t \exp(-pt)\,dt.$$

The condition (a) and (d) show that the basic invariant measure η_0 for V does not vanish. Since $\eta_0 \in H_0(V)$ and $(T_t)_{t \geqq 0}$ is not transient, η_0 is $(T_t)_{t \geqq 0}$-invariant, namely $\eta_0 = T_t^* \eta_0$ for every $t \geqq 0$. By N. Suzuki's result[6], η_0 is a unique $(T_t)_{t \geqq 0}$-invariant measure ε on X except for the multiplications of positive constants. This shows that $(T_t)_{t\ 0}$ is semi-transient. By using Theorem 1 and the condition (e), we arrive at the conclusion that V is a diffusion kernel of purely logarithmic type.

Similarly as above, we have the following:

Theorem 4. Let V be a non-degenerate and semi-bounded diffusion kernel on X. If V is semi-balayable, then V is of form

$$V = V_0 + c\ _0 \quad \text{(i.e., } Vf = V_0 f + c\int f d\eta_0 \quad \text{for} \quad f \in C_K(X)),$$

where V_0 is a diffusion kernel of logarithmic type, η_0 is the basic invariant measure for V and where c is a constant.

It is a question if the above V_0 is of purely logarithmic type. In the case of convolution type, V_0 is always of purely logarithmic type[1,2,3], namely we know that Hunt convolution kernels are not semi-balayable.

REFERENCES

1. M. Itô, Les noyaux de convolution de type logarithmique, Théorie du potentiel, Proc. Orsay 1983, Lecture Notes in Math., Springer.
2. M. Itô, Une caractérisation des noyaux de convolution réels de type logarithmique, Nagoya Math. J., 97 (1985).
3. M. Itô, Le principe semi-complet du maximum pour les noyaux de convolution réels, Nagoya Math. J., 101 (1986).
4. G. Choquet and J. Deny, Aspects linéaires de la théorie du potentiel, Théorème de dualité et applications, C. R. Acad. Sc. Paris, 243 (1959).
5. M. Itô, On weakly regular Hunt diffusion kernels, Hokkaido Math. J., 10 (1981).
6. N. Suzuki, Invariant measures for uniformly recurrent diffusion kernels, Hiroshima Math. J., 13, 3 (1983).

A REMARK ON CAPACITARY MEASURES FOR DIFFUSION PROCESSES

Mamoru Kanda

Inst. of Math
Univ. of Tsukuba
Ibaraki 305

INTRODUCTION

In this note we shall give a characterization of capacitary measures of semipolar sets for diffusion processes, which is closely related to Dellacherie, Feyel and Mokobodzki[2]. In this study, the process $X = (X_t)$ is assumed to be a standard process on $(E,\mathcal{E})$, where E is a locally compact separable Hausdorff space and $\mathcal{E}$ is the Borel algebra on E. For X we suppose the following: i) X satisfies the strong duality hypotheses relative to a σ-finite measure $\xi(dx) = dx$; ii) $U(x,K)$ and $\hat{U}(x,K)$ are bounded on E for each compact subset K of E ; iii) the kernel density $u(x,y)$ of $U(x,dy)$ relative to ξ is continuous off the diagonal and $x \to u(x,y)$, $y \to u(x,y)$ are lower semicontinuous. The brief explanation of the notations above will be given in the subsequent section. But we assume that the reader is familiar with terminologies in Blumenthal and Getoor[1]. A process is called a diffusion process on E if it is a standard process on $(E,\mathcal{E})$ with continuous paths. For a subset B of E , T_B denotes the hitting time of B , that is, $T_B = \inf(t>0, X_t \in B)$.

A significant characterization of semipolar sets is given by Dellacherie et al.[2] : a set S is semipolar if and only if there exists a σ-finite measure m such that, for $A \subset S$, $m(A) = 0$ if and only if A is polar. We call the measure m the DFM-measure of S . In the following we shall construct a DFM-measure for a diffusion process with the conditions i), ii) and iii) without appealing DFM's theory. Although our process is much more limited than DFM's, our construction shows that, for a certain class of semipolar sets, the capacitary measure is a DFM-measure. This point is the crucial part of our result.

A similar result is also shown for a class of Markov processes (not necessarily diffusion processes) under a different assumption by different method[6].

CHARACTERIZATION OF CAPACITARY MEASURES

A pair of standard processes $X = (X_t)$ and $\hat{X} = (\hat{X}_t)$ on $(E,\mathcal{E})$ are said to be in strong duality with respect to $\xi(dx) = dx$, if the resolvents (U^λ), $(\hat{U}^\lambda)$ of X , $\hat{X}$ respectively, may be expressed in terms of potential kernel densities $u^\lambda(x,y)$ satisfying 1) $U^\lambda(x,dy) = u^\lambda(x,y)dy$, $\hat{U}^\lambda(x,dy) = u^\lambda(y,x)dy$ for every $x \in E$; 2) for every $y \in E$, $x \to u^\lambda(x,y)$ is λ-excessive relative to X and $x \to u^\lambda(y,x)$ is λ-excessive relative to $\hat{X}$;

3) $(x,y) \to u^\lambda(x,y)$ is in $\mathcal{E}^* \times \mathcal{E}^*$ for each $\lambda \geq 0$. Here $\mathcal{E}^*$ denotes the σ-algebra of universally measurable sets of E. We say that the process X on $(E,\mathcal{E})$ satisfies the strong duality hypotheses if there exists another standard process $\hat{X}$ on $(E,\mathcal{E})$ such that X and $\hat{X}$ are in strong duality. In what follows we assume the conditions i),ii) and iii) and write $U(x,dy)$, $u(x,y)$ in place of $U^0(x,dy)$, $u^0(x,y)$ respectively. As usual we add the symbol ^ on the top of the notation related to the process $\hat{X}$. If it holds $P_x(T_B < \infty) = \int u(x,y)\, \pi_B(dy)$ for some σ-finite measure π_B, we say that π_B is the capacitary measure of B for X. The total mass of π_B is called the capacity of B for X, denoted by $C(B)$. The important fact from the conditions i),ii) and iii) is that there exist bounded capacitary measures π_B, $\hat{\pi}_B$ of X, $\hat{X}$ respectively and $C(B) = \hat{C}(B)$, if B is a bounded Borel subset of E. See Getoor and Sharpe[5]. The switching identity:

$$E_x u(X_{T_B},y) = \hat{E}_y u(x,\hat{X}_{\hat{T}_B}), \quad B \text{ ; a Borel subset of } E \text{ ;} \tag{2.1}$$

is important in the sequal.

Theorem - Let S be a closed set with the property

$$S \subset S^\varepsilon = \{ x \text{ ; } P_x(T_S < \infty) < \varepsilon,\ \hat{P}_x(\hat{T}_S < \infty) < \varepsilon \} \tag{2.2}$$

for some ε, $0 < \varepsilon \leq 1$. Then there exists a σ-finite measure π such that 1) $\mathrm{supp}(\pi) \subset S$; 2) for each bounded Borel subset B of S, $\pi(B) = 0$ if and only if $C(B) = 0$; 3) for each bounded Borel subset B of S such that $B \subset S^\eta$ for some η with $\eta < 1$,

$$\pi_B(dy) = (1 - P_y(T_B < \infty))\, \pi|_B(dy), \tag{2.3}$$

where $\pi|_B$ is the restriction of π to the set B ; 4) in case $\varepsilon < 1$, if $P_x(T_S < \infty)$ is continuous off the set S and the capacitary measure π_S is defined, then (2.3) holds replacing B with S.

In the proof we shall use the notations below: A^r (resp. A^{co-r}) = $\{ x \text{ ; } P_x(T_A = 0) = 1$ (resp. $\hat{P}_x(\hat{T}_A = 0) = 1) \}$, A^{ir} (resp. A^{co-ir}) = $\{ x \text{ ; } P_x(T_A > 0) = 1$ (resp. $\hat{P}_x(\hat{T}_A > 0) = 1) \}$. First we prepare the following proposition.

Proposition - Let S be a closed set such that $S \subset S^\varepsilon$ for some ε, $\varepsilon < 1$ and such that the condition for S cited in 4) of the theorem holds. For each δ, $\varepsilon < \delta < 1$, define $G = \{ x \text{ ; } P_x(T_S < \infty) > \delta \}$ and set $D = G \cup S \cup (\bar{G})^c$. Writing $R = T_{D^c}$, we have

$$P_x(T_S < R) = 0 \quad \text{for} \quad x \in (\bar{G})^c \cup S, \tag{2.4}$$

and, for each Borel subset B of S, there exists a σ-finite measure μ_B on S such that

$$\mu_B \text{ does not carry any polar set and } \mathrm{supp}(\mu_B) \subset B \text{ ;} \tag{2.5}$$

$$P_x(T_B < R) = \int v(x,y)\, \mu_B(dy), \quad x \in D, \tag{2.6}$$

where $v(x,y)$ is defined through

$$u(x,y) = v(x,y) + E_x u(X_R,y). \tag{2.7}$$

Proof - Step 1 - We shall first show

$$G \text{ ; open,} \quad S \subset \bar{G}^{ir}, \quad S - \{\text{a polar set}\} \subset (G^c)^{co-ir}, \\ (\partial G - S) \subset \bar{G}^r \cap (G^c)^r \subset (\partial G - S)^r. \tag{2.8}$$

The hitting probability $P_x(T_S < \infty) = \int u(x,y)\pi_S(dy)$ is lower semi-continuous by the condition iii) and continuous off S by the assumption. So G is open and $(\partial G - S) \subset \{ x ; P_x(T_S < \infty) = \delta \}$. In general, for an excessive function f, $A^r \ni x \to \inf\{ f(x), x \in A \} \leq f(x) \leq \sup\{ f(x), x \in A \}$ by Blumenthal and Getoor[1]. We have $S \subset \{ G \cup (\partial G - S)\}^{ir}$ and so $S \subset \overline{G}^{\,ir}$, because $S \subset S^{ir}$. Suppose that there exists a non-polar compact subset H of S such that $H \subset (G^c)^{co-r}$. Choose a sequence of open sets $\{O_n\}$ such that $O_n \downarrow H$ with $C(O_n) \downarrow C(H)$. Set $H_n = O_n \cap G^c$. Then $C(H) = \int_H \hat{P}_x(\hat{T}_{H_n} < \infty)\pi_H(dx) = \iint u(y,x)\hat{\pi}_{H_n}(dy)\pi_H(dx) = \int P_y(T_H < \infty)\hat{\pi}_{H_n}(dy) \leq \int P_y(T_S < \infty)\hat{\pi}_{H_n}(dy) \leq \delta\, C(H_n)$. Hence $C(H) \leq \delta\, C(H)$, which is absurd. Now we shall prove the last statement of (2.8). Indeed, letting $U(x) = \delta - P_x(T_S < \infty)$ and $V(x) = P_x(T_S < \infty) - \delta$, then U (resp. V) is a barrier for G (resp. $(G)^c$) at every point of $\partial G - S$. So $(\partial G - S) \subset \overline{G}^{\,r} \cap (G^c)^r$ by Dynkin[3]. By the continuity of paths, $\overline{G}^{\,r} \cap (G^c)^r \subset (\partial G - S)^r$.

Step 2 – Writing $R = T_{D^c}$, $\hat{R} = \hat{T}_{D^c}$, we consider the processes (X,R), $(\hat{X},\hat{R})$ obtained by killing X, $\hat{X}$ at R, $\hat{R}$ respectively. Set $E(R) = \{ x ; P_x(R = 0) = 0 \}$, $E(\hat{R}) = \{ x ; \hat{P}_x(\hat{R} = 0) = 0 \}$. Then it follows from (2.8) that $E(R) = D$ and $D - (E(R) \cap E(\hat{R}))$ is polar. Further $\xi(E(\hat{R}) - (E(\hat{R}) \cap D)) = 0$, because the set is included in S. The potential kernel density $v(x,y)$ for (X,R) is given by (2.7). Furthermore the restriction of (X,R) and $(\hat{X},\hat{R})$ to $E(R) \cap E(\hat{R})$ are in strong duality with respect to ξ, if $v^\lambda(x,y)$ is defined analogously as is $v(x,y)$. (However the restriction of (X,R), $(\hat{X},\hat{R})$ to $E(R) \cap E(\hat{R})$ are not always standard.) The important fact is that the switching identity also holds:

$$E_x(v(X_{T_A},y), T_A < R) = \hat{E}_y(v(x,\hat{X}_{\hat{T}_A}), \hat{T}_A < \hat{R}) \qquad (2.9)$$

for every Borel set A. For details on the statement above see Getoor[4]. In subsequent paragraphs we shall show (2.4). First we prove

$$P_x(T_G < R) = 0 \quad \text{for} \quad x \in (\overline{G})^c \cup S . \qquad (2.10)$$

Noting that $S \subset \overline{G}^{\,ir}$ by (2.8) and $X_{T_G} \in G \cup G^r$ almost surely on $\{ T_G < \infty \}$, we see that $X_{T_G} \in \partial G - S$ almost surely P_x on $\{ T_G < \infty \}$ if $x \in (\overline{G})^c \cup S$ by the continuity of paths. Hence $P_x(T_G \geq R) = 1$ for $x \in (\overline{G})^c \cup S$. Next we show

$$v(x,y) = 0 \quad \text{for} \quad x \in (\overline{G})^c \cup S,\ y \in G \cup \{ S - (\text{a polar set}) \}. \qquad (2.11)$$

Setting $T = \max(T_G,R)$, we have $u(x,y) = v(x,y) + E_x u(X_R,y) \geq v(x,y) + E_x u(X_T,y)$, because $x \to u(x,y)$ is excessive relative to X. By (2.10) $E_x u(X_T,y) = E_x u(X_{T_G},y)$ for $x \in (\overline{G})^c \cup S$. Using the switching identity (2.1), we have $u(x,y) \geq v(x,y) + \hat{E}_y u(x,\hat{X}_{\hat{T}_G})$ for $x \in (\overline{G})^c \cup S$. Since $G \subset G^{co-r}$ and $S - \{\text{a polar set}\} \subset G^{co-r}$ by (2.8), we have $\hat{E}_y u(x,\hat{X}_{\hat{T}_G}) = u(x,y)$ for $x \in (\overline{G})^c \cup S$ and $y \in G \cup \{ S - (\text{a polar set}) \}$. Hence (2.11) follows. Now we show (2.4). By the condition ii), $V(x,K)$ is bounded for all compact set K. So, combining (5.11) with (5.13) in III [1], we see that there exists a sequence $\{g_n\}$ of nonnegative bounded nearly Borel functions such that $Vg_n \uparrow 1$. Then, by the switching identity (2.9), $E_x(Vg_n(X_{T_S}). T_S < R) = \int g_n(y) \hat{E}_y(v(x,\hat{X}_{\hat{T}_S}), \hat{T}_S < \hat{R})\, dy$. But $\hat{X}_{\hat{T}_S} \in S$ on $\{ \hat{T}_S < \infty \}$ and therefore $v(x,\hat{X}_{\hat{T}_S}) = 0$ almost surely for $x \in (\overline{G})^c \cup S$ by (2.11). Hence $E_x(Vg_n(X_{T_S}), T_S < R) = 0$. The proof of (2.4) is finished.

<u>Step 3</u> — In this step we show that the process (X,R) restricted to $E(R) = D$ is standard and the measure ξ is an excessive measure for (X,R). Since the process (X,R) is continuous, the only point to check the standardness is that $x \to P_x(X_t \in B\,,\ t < R)$ is $\mathcal{E}(D)$-measurable for each $B \in \mathcal{E}(D)$, where $\mathcal{E}(D)$ is the Borel algebra on D. Set

$$T = T_{\partial G} \cdot 1_{\{T_{\partial G} < T_S\}} + (T_S + T_{\partial G} \circ \theta_{T_S}) \cdot 1_{\{T_S = T_{\partial G}\}} ,$$

where θ is the shift operator. Since X is continuous and ∂G, S are closed sets, $T_{\partial G}$, T_S and $T_S + T_{\partial G}\,\theta_{T_S}$ are $\sigma(X_t, t\geq 0)$-measurable. Further $P_x(T = R) = 1$ for $x \in E$. Indeed, if $x \in (\bar{G})^c \cup S$, $P_x(R = T_{\partial G}) = 1$ and $P_x(T = T_{\partial G}) = 1$ by (2.4). If $x \in G$, then $R = T_{\partial G}$ if $T_{\partial G} < T_S$ and $R = T_S + T_{\partial G}\circ\theta_{T_S}$ if $T_S = T_{\partial G}$, P_x-almost surely. For $x \in \partial G - S$, $P_x(T_{\partial G} = 0\,,\ T_S > 0) = 1$ by the last statement of (2.8). So $P_x(R = T) = 1$. Hence $P_x(X_t \in B\,,\ t < R)$ is $\mathcal{E}(D)$-measurable. It is obvious that ξ is an excessive measure for (X,R), because ξ is an excessive measure for X.

<u>Step 4</u> — Here we shall prove (2.5) and (2.6). Since $S \subset \bar{G}^{ir}$ by (2.8) and since (2.4) holds for S, we see that, for every Borel subset B of S and for any increasing sequence of stopping times $\{T_n\}$ with limit $T \geq R$ P_x-almost surely, we have $P_x(T_n + T_B \circ \theta_{T_n} < R) \to 0$ as $n \to \infty$. In other words, $P_x(T_B < R)$ is a (natural) potential with respect to (X,R). Since (X,R) is a standard process on D, there exists a unique additive functional A of (X,R) on D whose potential is $P_x(T_B < R)$. See Getoor and Sharpe[5]. The corresponding additive functional is σ-integrable by Revez[7], because $P_x(T_B < R)$ is bounded. Hence the Revez measure μ_B corresponding to A does not carry any polar set,[7] and so A does not carry any polar set. On the other hand, $D - (E(\hat{R}) \cap D)$ is polar, because $D - (E(\hat{R}) \cap D) \subset S \cap G^{co\text{-}r}$ and $S \cap G^{co\text{-}r}$ is polar by (2.8). So the additive functional A is carried by $E(\hat{R})$. Then, applying Theorem 7.5 in Getoor[4], we obtain (2.6). The first assertion of (2.5) is proved just before. For the proof of the second assertion, we note that the same argument of the top of p287 in Blumenthal and Getoor[1] shows that μ_B is carried by $B \cup B^{co\text{-}r}$. But $S \subset S^{co\text{-}ir}$ now. We have finished the proof of the proposition.

<u>Proof of Theorem</u> — First consider the case that the set S is contained in S^{ε} for $\varepsilon < 1$ and satisfies the condition cited in 4) of the theorem. Let μ_S and μ_B $(B \subset S)$ be those obtained in the proposition. Then we have

$$\mu_S|_B = \mu_B \quad \text{for every Borel subset } B \text{ of } S , \tag{2.12}$$

where $\mu_S|_B$ is the restriction of μ_S to B, and

$$E_x(f(X_{T_B}), T_B < R) = \int v(x,y)\, f(y)\, \mu_B(dy) . \tag{2.13}$$

Indeed, it follows from (2.4) and $S \subset \bar{G}^{ir}$ that $P_x(T_B < R) + P_x(T_{S-B} < R) = P_x(T_S < R)$, $x \in D$. Since μ_B, μ_{S-B} and μ_S are carried in $E(\hat{R})$ as is shown in Step 4 of the proof of the proposition, the uniqueness of measures of potentials[4] assures that $\mu_B + \mu_{S-B} = \mu_S$. Hence (2.12) follows from (2.5). We also have $P_x(X_{T_B} \in A\,,\ T_B < R) = P_x(T_A < R)$ by (2.4) for every Borel subset A of B. Hence $E_x(1_A(X_{T_B}), T_B < R) = \int v(x,y)\, \mu_A(dy) = \int_A v(x,y)\, \mu_S(dy) = \int_A v(x,y)\, \mu_B(dy)$. The proof of (2.12) and (2.13) is now finished. In the next we show that, if a Borel subset B of S has a capacitary measure π_B of X, then

$$\pi_B(dy) = (1 - P_y(T_B < \infty)) \mu_S|_B(dy) . \tag{2.14}$$

(Hence (2.14) is valid for every bounded Borel subset of S .) Indeed we have $P_x(T_B < R) = P_x(T_B < \infty) - P_x(R < T_B < \infty) = \int u(x,y) \pi_B(dy) - E_x(P_{X_R}(T_B < \infty), R < T_B) = \int v(x,y) \pi_B(dy) + \int E_x u(X_R,y) \pi_B(dy) - E_x(P_{X_R}(T_B < \infty), R < T_B) = \int v(x,y) \pi_B(dy) + E_x(P_{X_R}(T_B < \infty), T_B < R < \infty) = \int v(x,y) \pi_B(dy) + E_x(E_{X_{T_B}}(P_{X_R}(T_B < \infty)), T_B < R)$. Here we have used $P_x(R = T_B) = 0$. By (2.4) we have $E_y(P_{X_R}(T_B < \infty)) = E_y(P_{X_R}(T_B < \infty), R < T_B)$ for $y \in S$. The right term equals to $P_y(R < T_B < \infty)$ and so equals to $P_y(T_B < \infty)$ for $y \in S$. Hence we have $E_x(E_{X_{T_B}}(P_{X_R}(T_B < \infty)), T_B < R) = E_x(P_{X_{T_B}}(T_B < \infty), T_B < R)$. Using (2.13) , the last term equals to $\int v(x,y) P_y(T_B < \infty) \mu_B(dy)$. As a consequence we get $\int v(x,y) (1 - P_y(T_B < \infty)) \mu_B(dy) = \int v(x,y) \pi_B(dy)$. The uniqueness of measures of potentials assures that $\pi_B(dy) = (1 - P_y(T_B < \infty)) \mu_B(dy)$. Then (2.14) follows from (2.12) . Now we see that a Borel subset B of S is polar if and only if $\mu_S(B) = 0$. Indeed, if B is polar, then $C(K) = \pi_K(K) = 0$ for all compact subset K of B . Hence $\mu_S(K) = 0$, and so $\mu_S(B) = 0$. Conversely, if $\mu_S(B) = 0$, then $\pi_K(K) = 0$ for all compact subset K of B , and then B is polar. In conclusion we have proved the theorem by choosing μ_S as the required measure π , in case $S \subset S^\varepsilon$ $(\varepsilon < 1)$ and S satisfies the condition cited in 4) of the theorem. Finally we shall study the case $S \subset S^1$. Choose a sequence of compact sets $\{K_n\}$ such that $K_n \uparrow E$ and set $S_n = S \cap K_n \cap \{ x ; P_x(T_S < \infty) < 1 - 1/n , \hat{P}_x(\hat{T}_S < \infty) < 1 - 1/n \}$. Then $S_n \uparrow S$ and S_n satisfies the condition cited in 4) of the theorem. Furthermore, by (2.12) , $\mu_{S_n}|_{S_m} = \mu_{S_m}$ if $n > m$. So we obtain a limit measure π such that $\pi|_{S_n} = \mu_{S_n}$ for every n . It is clear that the measure π satisfies 1) and 2) cited in the theorem. Choose a bounded Borel set B such that $B \subset S^\eta$ for some $\eta < 1$. Then it follows from (2.14) that $\pi_B(dy) = (1 - P_y(T_B < \infty)) \mu_{S_n}|_B(dy)$ for large n such that $B \subset S_n$. The right term equals to $(1 - P_y(T_B < \infty)) \pi|_B(dy)$ by the definition of the measure π . The proof of the theorem is complete.

Remark - If $S \subset S^\varepsilon$ for $\varepsilon < 1$ and S satisfies the condition cited in 4) of the theorem (for example, S is compact), it follows from the theorem that, for every Borel subset B of S ,

$$\pi_S(B) = 0 \text{ if and only if } B \text{ is polar.}$$

REFERENCES

1. R. M. Blumenthal and R. K. Getoor, "Markov Processes and Potential Theory" Academic Press, New York (1968).
2. C. Dellacherie, G. Feyel, and G. Mokobotzki, Integrals de capacites fortment sous-additives, Seminaire de Prob. XVI, 1980 - 1981, Lecture Notes in Math. 920, 8-28 (1982).
3. E. B. Dynkin, "Markov Processes" Springer-Verlag, Berlin (1965).
4. R. K. Getoor, Multiplicative functionals of dual processes, Ann. Inst. Fourier, XXI, 43-83 (1971).

5. R. K. Getoor and M. J. Sharpe, Last exit times and additive functionals, The Annals of Probability, 1.1 (4), 550-569 (1973).
6. M. Kanda, A note on capacitary measures of semipolar sets, to be submitted.
7. D. Revuz, Measures associees aux fonctionnelles additives de Markov I, Trans. Amer. Math. Soc. 148, 501-531 (1970).

ON THE CONVEXITY OF LEVEL SETS FOR ELLIPTIC AND PARABOLIC EXTERIOR BOUNDARY VALUE PROBLEMS

Bernhard Kawohl

SFB 123, Universität Heidelberg
Im Neuenheimer Feld 294
D 6900 Heidelberg

The talk is divided into two parts. First I study an elliptic free boundary problem and show that sometimes it is useful to work with curvilinear coordinates. Then I present some results on parabolic problems, which I obtained in cooperation with I. Diaz.

1. A FREE BOUNDARY VALUE PROBLEM WITH BERNOULLI BOUNDARY CONDITION

Let $\Omega \subset I\!R^2$ be a bounded convex domain with smooth boundary $\partial\Omega$ and consider the problem of finding a function u and a smooth closed arc Γ, being the boundary of a domain $D \supset \Omega$, and such that

$$\Delta u = 0 \qquad \text{in } G := D \setminus \bar{\Omega}, \tag{1.1}$$

$$u = 1 \qquad \text{on } \bar{\Omega}, \tag{1.2}$$

$$u = 0, \qquad \frac{\partial u}{\partial n} = const. \qquad \text{on } \Gamma. \tag{1.3}$$

This problem occurs in the modelling of potential flow. For its physical background one can consult e.g. [9,5,10,12]. If Γ is sufficiently smooth (which is known in two dimensions but which is an open problem in higher dimensions) then one can infer the uniqueness of (u, Γ) and the convexity of D, see [21,10]. Here I shall prove the convexity of D without using conformal maps. The initial hope to generalize the convexity result to the analogous problem in higher dimensions (D and Ω are subsets of $I\!R^n$ with $n \geq 3$) is dampened by several obstacles. First of all the existence question can (at present) only be answered by a variational approach which admits (and sometimes produces) singular free boundaries. This was shown by Alt and Caffarelli [4]. If the free boundary is smooth one can show its uniqueness as in the paper of Tepper [21] with a scaling trick which is attributed to Lindelöf in [1]. But even if the C^2-regularity of the free boundary were known, there are problems with extending our proof of the convexity of D.

Let us see why this is the case. In Remark 2 I shall sketch a totally different proof, which was kindly pointed out to me by A.Acker, and which works in arbitrary dimensions.

THEOREM 1

Let u be a classical solution to problem (1.1)(1.2)(1.3) with smooth free boundary Γ. If Ω is convex, then D is convex.

For the proof one has to proceed in two steps. We consider the function $v(x) := |\nabla u(x)|^2$ and derive the following lemmata.

LEMMA 2

Under the assumptions of Theorem 1 the function $v(x)$ is positive on the set G.

The shortest proof I know for this Lemma uses conformal maps, see [21,10]. The proof given here works also in higher dimensions. We know from Hopf's Lemma that v is positive on ∂G, since $0 < u < 1$ in G. So it remains to study v in G. As Alt and Caffarelli found out [4], the solution of problem (2.1) (2.2) (2.3) is a global minimizer of the functional

$$J(v) := \int_{\{v>0\}} \{|\nabla v|^2 + 1\} \quad dx \tag{1.4}$$

over the admissible set of functions

$$A := \{v \in W^{1,2}_{\text{loc}}(I\!R^n) | v = 1 \text{ on } \bar{\Omega}\}. \tag{1.5}$$

In more than two dimensions the functional J can have other critical points which also solve (1.1)(1.2)(1.3) in a weak sense. But the minimizing ones have to have nonvanishing gradient. If $y \in \Omega$, then the auxiliary function $w(x) := (y - x)\nabla u(x)$ is nonnegative due to the results in [11] on starshaped rearrangements. Moreover w is harmonic in G so that w must be positive in G. Consequently v is positive in G as well. This completes the proof of Lemma 2, but more can be said about v and u if $n = 2$.

The function v satisfies the differential inequality

$$\Delta v = 2\nabla(\Delta u)\nabla u + 2\sum_{i,j=1}^{n} u_{ij}^2 \geq 0 \qquad \text{in } G, \tag{1.6}$$

and, if $n = 2$, it satisfies even the differential equality

$$\Delta v - 2\left(\frac{\nabla v}{v}\right)\nabla v = 0 \qquad \text{in } G, \tag{1.7}$$

as can be seen e.g. from [18,p.69]. So for $n = 2$ the function v attains its (positive) minimum on ∂G. Finally, for $n = 2$, the free boundary is known to be sufficiently smooth [3] to prove uniqueness for problem (1.1)(1.2)(1.3). This was done in [21].

LEMMA 3

Let u satisfy the assumptions of Theorem 1. If $v(x)$ is positive on $\bar{G}$ and if $n = 2$, then D is convex provided Ω is convex.

Notice that the combination of Lemma 2 and Lemma 3 implies Theorem 1. To prove Lemma 3 we observe that due to (1.7) the function v must attain its minimum on $\partial G(=\partial\Omega \cup \Gamma)$. But where on ∂G does this occur? The minimum cannot be attained on $\partial\Omega$ since there we would have

$$-2u_n u_{nn} < 0 \qquad \text{on } \partial\Omega. \tag{1.8}$$

This would contradict the fact that u is harmonic up to ∂G:

$$\Delta u = u_{nn} + (n-1)H(x)u_n = 0 \qquad \text{on } \bar{G}. \tag{1.9}$$

Here the subscript $_n$ denotes differentiation in the direction $n(x)$ of the exterior normal to the level set $\{u \geq 1\}$. If $-n(x)$ is the normal vector on the oriented surface $\partial\Omega$, then $H(x)$ denotes its mean curvature.Consequently the minimum of v must be attained on the free boundary Γ. But on Γ the function v is constant. So v attains its minimum in any point on the free boundary. If this boundary is sufficiently smooth, which we know from [4] in two dimensions, then we can again use (1.9) and Hopf's Lemma,

$$2u_n u_{nn} < 0 \qquad \text{on } \Gamma, \tag{1.10}$$

to conclude that $H(x)$, the mean curvature of the free boundary, is positive. For $n = 2$ this means that D is strictly convex.

REMARK 1

As mentioned above, if the space dimension exceeds 2, a proof of Theorem 1 based on maximum principles seems to be hard. In this case (1.7) is no longer available, only (1.6), and this is not sufficient for my proof of Lemma 3. In fact for $n = 3$ one cannot (in general) expect such a minimum principle. The following function f is harmonic in a zero-neighborhood, but $h = |\nabla f|^2$ has a strict local minimum at zero, cf. [15].

$$f(x,y,z) = z + x^2 - y^2 + (\tfrac{2}{3}z^2 - x^2 - y^2)z$$

If, however, D is already known to be convex, and if u solves problem (1.1)(1.2)(1.3) in $n = 3$ dimensions, then the function $v(x) = |\nabla u(x)|^2$ attains its minimum on the boundary of D. This follows from the fact that all level sets of u are convex due to [13] and thus the level surfaces $\{x \in D | u(x) = t\}$ have positive Gaussian curvature $g(x)$. Therefore one can use the remarkable identity [22] $\Delta(\log|\nabla u(x)|) = -g(x)$ in G and the maximum principle to conclude that v is minimal on Γ.

The case that $n = 2$ and Ω is not convex has been thoroughly studied by Acker [3]. He derived more relations between the shapes of $\partial\Omega$ and Γ. For $n = 2$ there is another tricky way to prove convexity of level sets, and not just the Gabriel-Lewis method. One can show that the auxiliary function $\kappa(x)/|\nabla u|$ is harmonic, provided that u is harmonic and has no critical points. Here κ denotes the curvature of a level curve of u, see [20].

REMARK 2

There is a variational proof of the convexity of the free boundary, which is due to A.Acker [2]. He wrote it up for two space dimensions but the variational argument is independent of dimension. In §9 of his paper he shows that problem (1.1)(1.2)(1.3) has at least one solution with convex free boundary. To do this he minimizes the capacity of a doubly connected region with convex boundaries under a volume constraint on the region. One then has to show that this variational solution solves (1.1)(1.2)(1.3) as well. To derive (1.3) one can follow arguments of Caffarelli and Spruck in [7, pp.1374ff]. Those arguments also show that the free boundary, which was obtained by Acker's variational method, is smooth.

If any free boundary is smooth, then one can use the previously quoted scaling arguments to show the uniqueness of solutions.

Therefore Theorem 1 remains valid in higher dimensions.

2. A REACTION DIFFUSION PROBLEM

Let $\Omega \subset I\!R^n$ be a convex container with smooth boundary $\partial\Omega$, and suppose that a chemical is diffusing out of the container into the environment, where it reacts with other substances. If this reaction is sufficiently strong, it will dominate the diffusion. So the chemical cannot get very far and its concentration $u(t,x)$ has a support which is finite in space. Therefore let $D \subset I\!R^n$ be a large convex set containing the support of u for any time $t > 0$, denote $D \setminus \bar{\Omega}$ by G, and consider the problem

$$\begin{aligned} u_t - \Delta u + f(u) &= 0 && \text{in } (0,\infty) \times G, && (2.1)\\ u &\equiv 1 && \text{on } (0,\infty) \times \bar{\Omega}, && (2.2)\\ u &= 0 && \text{on } (0,\infty) \times \partial D, && (2.3)\\ u(0,x) &= 0 && \text{in } G. && (2.4) \end{aligned}$$

A typical candidate for the nonlinearity f is $f(u) = (u^+)^\alpha$ with $0 < \alpha < 1$. I am interested in the convexity of level sets of $u(t,x)$ in time and space. Those sets are convex iff

$$Q(t_1,t_2,x_1,x_2) := u((t_1+t_2)/2, (x_1+x_2)/2) - \min\{u(t_1,x_1), u(t_2,x_2)\}$$

is nonnegative for any pair $(t_1,x_1),(t_2,x_2)$ of points in $(0,\infty) \times D$. Notice that this property is preserved under locally uniform convergence of a sequence of functions u_n. Therefore we can regularize a Höldercontinuous nonlinearity f by Lipschitzcontinuous nonlinearities f_n and study convexity of level sets of the corresponding solutions u_n. So without loss of generality it suffices to prove Theorem 2 for those functions $f : [0,1] \to [0,\infty)$ which are strictly monotone increasing, positive and differentiable in $(0,1)$. For those the support of $u(t,x)$ is all of $[0,\infty) \times \bar{D}$.

THEOREM 2

Let $f(0) = 0$ and let f be nonnegative, monotone nondecreasing and Höldercontinuous on $[0,1]$. Let u solve problem (2.1)(2.2)(2.3)(2.4).

Then the level sets $\{(t,x) \in ((0,\infty) \times D | u(t,x) > const.\}$ are convex subsets of $I\!R^{1+n}$.

The details of the proof can be found in the paper [8], but its idea should be outlined here. One argues by contradiction and shows that if the above auxiliary function Q is not nonnegative, there must be a pair of points $(t_1,x_1) = z_1, (t_2,x_2) = z_2$ in which Q attains its negative minimum. Moreover z_1, z_2 and its convex combination $(z_1 + z_2)/2$ all lie in the domain in which the differential equation (2.1) holds. So one can play with first and second derivatives of u with respect to time and space and reach a contradiction. For elliptic problems this approach has been successful in many examples, see e.g. [11], and for parabolic problems with $f \equiv 0$ Theorem 2 was derived by C.Borell [6]. In order to show that z_1, z_2 and $(z_1 + z_2)/2$ are all in $(0, \infty) \times G$, one has to derive a Lemma which implies starshapedness of level sets with respect to $(\infty, 0)$ and $(0, 0) \in I\!R^{1+n}$.

LEMMA 4

Under the assumptions above Theorem 2 we have

$$u_t > 0, \qquad x \cdot \nabla u < 0 \qquad \text{and} \qquad t \cdot u_t + x \cdot \nabla u \leq 0 \qquad \textit{in} \quad (0, \infty) \times G.$$

Now one can show, if Q has a negative minimum in (z_1, z_2), that

$$u(z_1) = u(z_2) > u(z_{12}), \tag{2.5}$$

where $z_{12} = (z_1 + z_2)/2$. So one has to study derivatives of Q under the side constraint that $u(t_1, x_1) = u(t_2, x_2)$. Using the notation $Du = (u_t, \nabla u)$ for the gradient in $I\!R^{1+n}$ and $A = |Du((z_1 + z_2)/2)|, B = |Du(z_1)|, C = |Du(z_2)|$ one can derive the following equality

$$\frac{1}{A} = \frac{1}{2}(\frac{1}{B} + \frac{1}{C}), \tag{2.6}$$

and the inequality

$$\frac{1}{A^2}\Delta(u(z_{12})) \geq \frac{\mu}{B^2}\Delta(u(z_1)) + \frac{1-\mu}{C^2}\Delta(u(z_2)), \tag{2.7}$$

where $\mu = C/(B + C) \in (0, 1)$. The three relations (2.5)(2.6)(2.7) can be shown to contradict each other, if one uses the differential equation (2.1). In fact due to the monotonicity of f, (2.5) and (2.7) lead to

$$\frac{1}{A^2}u_t((z_1 + z_2)/2) > \frac{\mu}{B^2}u_t(z_1) + \frac{1-\mu}{C^2}u_t(z_2). \tag{2.8}$$

Because $u_t > 0$ and because of (2.6) we have

$$\frac{1}{u_t(z_{12})} = \frac{1}{2u_t(z_1)} + \frac{1}{2u_t(z_2)}. \tag{2.9}$$

Now (2.8), (2.9) and the definition of μ give

$$0 > \frac{\mu}{B^2}u_t(z_1) + \frac{1-\mu}{C^2}u_t(z_2) - \frac{1}{A^2}(\frac{1}{2u_t(z_1)} + \frac{1}{2u_t(z_2)})^{-1} =$$

$$A(\frac{1}{2u_t(z_1)} + \frac{1}{2u_t(z_2)})^{-1}\left[(\frac{1}{2B^3}u_t(z_1) + \frac{1}{2C^3}u_t(z_2))(\frac{1}{2u_t(z_1)} + \frac{1}{2u_t(z_2)}) - \frac{1}{A^3}\right].$$

To complete the proof of Theorem 2 we have to show that $[...] \geq 0$, but

$$(\frac{1}{2B^3}u_t(z_1) + \frac{1}{2C^3}u_t(z_2))(\frac{1}{2u_t(z_1)} + \frac{1}{2u_t(z_2)})$$
$$\geq (\frac{1}{2}B^{-3/2} + \frac{1}{2}C^{-3/2})^2 \geq (A^{-3/2})^2 = A^{-3},$$

which is the desired contradiction.

REMARK 3

Analogous results hold for degenerate parabolic equations. For instance one can replace the Laplace operator Δu in (2.1) by the pseudo-Laplace operator $\Delta_p u = div(|\nabla u|^{p-2}\nabla u)$ with $p > 1$ or by the fast diffusion operator $\Delta(u^m)$ with $0 < m < 1$. I refer to [8] for details.

3. AN OPEN PROBLEM

The following is a simple looking interior boundary value problem, whose solution could contribute a lot to the study of more complicated problems. Again let $\Omega \subset I\!R^n$ be convex and let u solve the problem

$$u_t - \Delta u = 0 \qquad \text{in } (0,\infty)\times\Omega, \tag{3.1}$$

$$u = 0 \qquad \text{on } (0,\infty)\times\partial\Omega, \tag{3.2}$$

$$u(0,x) = u_0(x) \qquad \text{in } \Omega. \tag{3.3}$$

Suppose that the function $u_0(x)$ has convex level sets $\{x \in \Omega | u_0(x) > const.\}$. Is it true that for every fixed but positive t the function $u(t,x)$ has level sets $\{x \in \Omega | u(t,x) > const.\}$, which are convex in space?

If the space dimension is $n = 1$, the answer to this problem is positive and follows from any of the papers [14,15,17,19]. Unfortunately the proof that works in one dimension fails in higher dimensions.

REFERENCES

[1] Acker.A. Heat flow inequalities with applications to heat flow and optimization problems. SIAM J. Math. Anal. **8** (1977) 604-618

[2] Acker,A. Area-preserving domain perturbation operators which increase torsional rigidity or decrease capacity with applications to free boundary problems. J. Appl. Math. Phys. (ZAMP) **32** (1981) 434-448

[3] Acker,A. On the geometric form of free boundaries satisfying a Bernoulli boundary condition. I and II. Math. Methods Appl. Sci. **6** (1984) 449-456 and **8** (1986) 387-404

[4] Alt,H.W., L.A.Caffarelli. Existence and regularity for a minimum problem with free boundary. J. Reine Angew. Math. **325** (1981) 105-144

[5] Beurling,A. On free-boundary problems for the Laplace equation. Seminar on analytic functions **1** Inst. for Advanced Study Princeton (1957) 248-263

[6] Borell,C. Brownian motion in a convex ring and quasiconcavity. Commun. Math. Phys. **86** (1982) 143--147

[7] Caffarelli,L.A., J.Spruck. Convexity properties of solutions to some classical variational problems. Commun. Partial Differ. Eqs. **7** (1982) 1337-1379

[8] Diaz,J.I., B.Kawohl. Convexity and starshapedness of level sets in some nonlinear parabolic problems. Preprint 393, SFB 123, Heidelberg 1986.

[9] Friedrichs,K.O. Über ein Minimumproblem für Potentialströmungen mit freiem Rande. Math. Annalen **109** (1934) 60-82

[10] Hamilton,R.S. The inverse function theorem of Nash and Moser. Bull. Amer. Math. Soc. **7** (1982) 65-222

[11] Kawohl,B. *Rearrangements and convexity of level sets in PDE.* Springer Lecture Notes in Math. **1150** (1985)

[12] Lacey.A.A., M.Shillor. Electrochemical machining with a threshold current. manuscript 1987, to appear in IMA J. Appl. Math.

[13] Lewis,J. Capacitary functions in convex rings. Arch. Ration. Mech. Anal. **66** (1977) 201-224

[14] Matano,H. Nonincrease of the lap number of a solution for a one-dimensional semilinear parabolic equation. J. Fac. Sci. Univ. Tokyo, Sect. IA, **29** (1982) 401--441

[15] Nickel,K. Gestaltaussagen über Lösungen parabolischer Differentialgleichungen. J. Reine Angew. Math. **211** (1962) 78--94

[16] Polya,G. Liegt die Stelle der größten Beanspruchung an der Oberfläche? Zeitschr. Angew. Math. Mech. **10** (1930) 353-360

[17] Polya,G. Qualitatives über den Wärmeausgleich. Zeitschr. Angew. Math. Mech. **13** (1933) 125-128

[18] Sperb,R. *Maximum principles and their applications.* Acad. Press, New York, 1981

[19] Sturm,C. Sur une classe d'equations á differences partielles. J. Math. Pures Appl. **1** (1836) 373-444

[20] Talenti,G. On functions whose lines of steepest descent bend proportionally to level lines. Ann. Sc. Norm. Sup. Pisa Ser.4, **10** (1983) 587-605

[21] Tepper,D.F. Free boundary problem. SIAM J. Math. Anal. **5** (1974) 841-846

[22] Weatherburn,C.E. On families of surfaces. Math. Annalen **99** (1928) 473-478

HARNACK INEQUALITY AND SOME CHARACTERISTICS OF NON-REGULAR POINTS OF SIMPLEXES

D.G. Keselman

Astrachanskij techniceskij Institut
Astrachan, U.S.S.R.

INTRODUCTION

Let S be a compact convex set, E=E(S) - the set of all extreme points of S, $Sh_S=\overline{E}\setminus E$ - the set of all non-regular points of S; $C(\overline{E})$ - the vector space of all real continuous functions on $\overline{E}$. A(S) - the space of all real affine continuous functions on S; supp ν - the closed support of the probability measure ν on S. For a given $x \in S$ we introduce further notations:

$\mathcal{M}_x$ - the set of all lower semicontinuous affine functions $s:S\to]-\infty;+\infty]$ with $s(x) < +\infty$; face(x) - the smallest face of S containing x;

cl.face(x) - the smallest closed face of S containing x;

E_x - the set of all extreme points of cl.face(x);

$M_x^+(G)$ - the set of all probability measures , which represent the point x and concentrated on the Borel subset $G \subseteq S$;

$Sh_x=\overline{E}_x\setminus E_x$ - the set of all non-regular points of cl.face(x);

$C(\overline{E}_x)$ - the vector space of all real continuous functions on $\overline{E}_x$.

In [1] we have described the points $x \in S$, for which an analogue of the Harnack inequality is satisfied on face(x):

Theorem 1 - [1] Assume that the point $x \in S$ has the face(x) of the second category, then for any compact $K \subset$ face(x) there exists such a number $\alpha_K \geq 1$ that $\sup_K f \leqslant \alpha_K < f(x)$ for every continuous affine function f: face(x) $\to [0; +\infty[$.

Corollary 1.1 - [1] Assume that the point $x \in S$ has face(x) of the second category. Consider the sequence $\{h_n\}$ of continuous affine functions on face(x) with the property $h_n \leqslant h_{n+1}$ for all $n \in N$ and the function $h = \sup_n h_n$. So if $h(x) < +\infty$, then the sequence h_n uniformly converges to h on every compact $K \subset$ face(x). In particular if a set face(x) is metrizable or locally compact, then h is a continuous function on face(x).

Corollary 1.2 - [1] Let $f \in \mathcal{M}_x$. Assume that the point $x \in S$ has a locally compact face(x), then the restriction $f|_{\text{face}(x)}$ is continuous.

Let S be now metrizable Choquet simplex, μ_x - the (unique) maximal measure representing the point $x \in S$.

By the solution of the Dirichlet problem for a boundary function f on E we understand the affine function U_f defined on

$$D_f = \{x \in S : f \in L_1(\mu_x)\}$$

by

$$U_f(x) = \mu_x(f).$$

Notice that D_f is always a face.

In [2] we have described the points x of the simplex S for which an analogue of the Harnack inequality is satisfied on face(x).

If for such points x the face(x) possesses th additional properties: E_x has no isolated points and supp $\mu_x = \overline{E}_x$, we shall prove that every measure $\lambda \in M_x^+(\overline{E})$ has no atomic components in the set $\overline{E}_x$.

Besides that we shall obtain that if the point $x \in Sh_S$ and supp $\mu_x = \overline{E}$, then the face(x) is of the first category.

1. HARNACK INEQUALITY AND QUASI-INTERIOR IN CHOQUET SIMPLEX

Theorem 2 – [2] Let x be a point of S and $K \subset \text{face}(x)$ - a compact set. Then the following assertions are equivalent:

(1) (a Harnack type inequality) there is a constant $\alpha_K \in [1; +\infty[$ such that for every continuous affine function $a: S \to [0; +\infty[$ we have

$$\sup_K a < \alpha_K a(x) ,$$

(1*) there is a constant $\alpha_K \in [1; +\infty[$ such that for every concave function $s: \text{face}(x) \to [0; +\infty[$ we have

$$\sup_K s \leqslant \alpha_K s(x) ,$$

(2) (a Harnack type monotone convergence theorem) every increasing sequance $\{a_n\}$ of real affine continuous functions on face(x) with $\sup_n a_n(x) < +\infty$ converges uniformly on K;

(3) every increasing sequence $\{a_n\}$ of continuous real affine functions on S with $\sup_n a_n(x) < +\infty$ converges uniformly on K;

(4) for every function $s \in \mathscr{M}_x$ the restriction s to K is continuous;

(5) for every function $s \in \mathscr{M}_x$ the restriction s to K is bounded;

(6) for every function $f \in L^1(\mu_x)$ the restriction U_f to K is continuous.

Remark – The assertion (6) is equivalent to:

the map $h: y \to h_y = \dfrac{d\mu_y}{d\mu_x}$ defined on face(x) is a homeomorphism embedding K to the space $L^{1*}(\mu_x)$.

Proof – Let $y = \lim y_n$ on K, then $h_y = \text{w-lim}_n h_{y_n} \Longleftrightarrow (\forall f \in L^1(\mu_x))$ the equality $\lim_n \int f\, h_{y_n}\, d\mu_x = \int f\, h_y\, d\mu_x$ is valid, which is equivalent to

$$\lim_n \int f\, d\mu_{y_n} = \int f\, d\mu_y ,$$

that is $\lim_n U_f(y_n) = U_f(y)$.

Corollary 2.1 – [2] Let $x \in S$. Then the following assertions are equivalent:

(1) for every compact set $K \subset \text{face}(x)$ there is a constant $\alpha_K \in [1;+\infty[$ such that $\sup_K a \leqslant \alpha_K a(x)$

for any continuous affine function

$$a: \text{face}(x) \to [0;+\infty[\quad ,$$

(2) for every function $f \in L^1(\mu_x)$ the restriction U_f to face(x) is continuous.

Theorem 3 – [2] Let $x \in S$. Assume that one condition of the preceding corollary is satisfied and moreover $\overline{\text{face}(x)} = \text{cl.face}(x)$. Then U_f is continuous at all points of face(x) for every $f \in C(\overline{E}_x)|_{E_x}$.

Theorem 4 – [3] The face(x) is dense in cl.face(x) iff supp $\mu_x = \overline{E}_x$.

Corollary 4.1 – Let $x \in S$. Assume that one condition of the corollary 2.1. is satisfied, and moreover supp $\mu_x = \overline{E}_x$. Then U_f is continuous at all points of face(x) for every $f \in C(\overline{E}_x)|_{E_x}$.

Let's introduce quasi-interior A in Choquet simplex:

A – the set of continuity points of U_f for all $f \in C(\overline{E})|_E$. As the set cl.face(x) is also a simplex, then by analogy we shall introduce the quasi-interior A_x in cl.face(x):

A_x – the set of continuity points of U_f for all $f \in C(\overline{E}_x)|_{E_x}$.

Theorem 5 – [4] The set A has the following properties:

(1) $A = \{ x \in S : M_x^+(\overline{E}) = \{\mu_x\}\}$;

(2) there is a continuous function p convex on S such that

$$A = \{ x \in S : \ U_p|_{E(S)} \text{ is continuous at the point } x \} \quad ;$$

(3) A is of the type G_δ and is dense in S.

If $Sh_s \neq \emptyset$, then for the set $I = S\backslash A$ the following theorem holds:

Theorem 6 – [4]

(1) the set I is convex;

(2) the point $x \in I \iff \text{face}(x) \cap I \neq \emptyset$;

(3) for any two points $y \in S$ and $x \in I$ we have $]y;x] \subset I$.

Theorem 3* – Let $x \in S$. Assume that one condition of the corollary 2.1. is satisfied and moreover supp $\mu_x = \overline{E}_x$. Then $x \in A_x$.

2. APPLICATIONS – Some properties of the set $M_x^+(\overline{E})$.

Theorem 7 – Let the point $x \in A_x$, $\nu \in M_x^+(\overline{E})$ then $\nu(Sh_x) = 0$.

Proof - If $m = \nu(Sh_x) > 0$, then we shall consider two probability measures

$$\lambda_1 = \frac{1}{m}\nu\Big|_{Sh_x} \quad \text{and} \quad \lambda_2 = \frac{1}{1-m}\nu\Big|_{(S\ Sh_x)} .$$

Let y_i the barycenter of the measure λ_i, $i=1,2$.

As $\nu = m\lambda_1 + (1-m)\lambda_2$ and $A(S)$ separates the points of S we have:

$$x = my_1 + (1-m)y_2 .$$

Let's consider the measure $\lambda = m\lambda_1 + (1-m)\mu_{y_2}$.

As $y_2 \in \text{face}(x)$, then $\text{supp}\,\mu_y \subseteq \bar{E}_x$.

That's why $\text{supp}\,\lambda \subseteq \bar{E}_x$ and $\lambda \neq \mu_x$ which contradicts the equality:

$$M_x^+(\bar{E}_x) = \{\mu_x\}$$

and consequently it contradicts theorem 5.

Corollary 7.1 - If the point $x \in A_x$, $\nu \in M_x^+(\bar{E})$, $D \subset Sh_s$ and $\nu(D) > 0$, then $D \subset \text{cl.face}(x)\setminus\bar{E}_x$.

Example - Let's consider relatively compact regions $w_1, w_2 \subset R^2$;

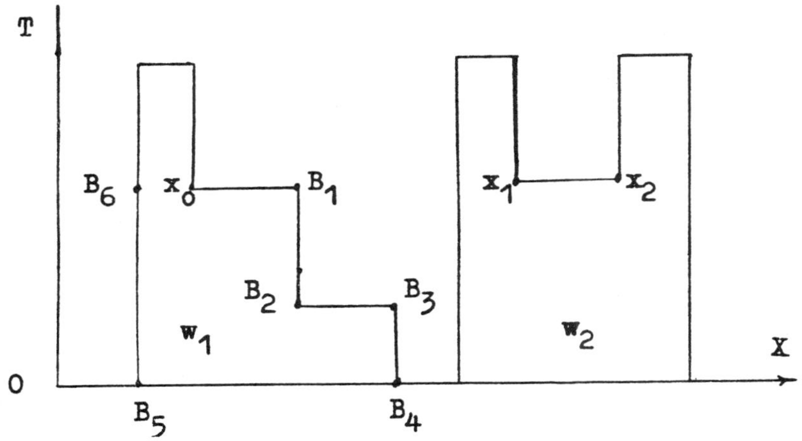

heat equation $\Delta U = 0$ in R^2, functional systems

$$H(\bar{w}_i) = \{f \in C(\bar{w}_i) : \Delta f = 0 \text{ in } w_i\} ,$$

and their state spaces

$$S_i = \{s \in H(\bar{w}_i) : s(\mathbb{1}) = 1 = \|s\|\} , \ i=1,2 ,$$

where $\mathbb{1}$ is the function equal to 1 on the whole of $\bar{w}_i$.

As it is known, S_1, S_2 are simplexes.

We shall consider every compact $\bar{w}_i$ as a closed subset of S_i, $i=1,2$.

The point x_o is a non-regular boundary point of w_1, but $x_o \in A_{x_o}$. In this case we have $\bar{E}_{x_o} = [B_1,B_2] \cup [B_3,B_4] \cup [B_4,B_5] \cup [B_5,B_6]$.

That's why by the theorem 7 we have $M^+_x(\overline{E}_{x_o}) = \{\mu_{x_o}\}$ and therefore $M^+_{x_o}(\overline{E}) = [\varepsilon_{x_o}, \mu_{x_o}]$.

Non-regular boundary points x_1, x_2 of w_2 belong to the same part of S_2, look [5, p.346] . Let

$$[x_1, y] = S_2 \cap \{x_1 + t(x_1 - x_2) : t \in [0; +\infty[\ \} ,$$

then $x_1 = tx_2 + (1-t)y$ for some $t \in]0; 1[$.

Therefore the measure $\nu = t\varepsilon_{x_2} + (1-t)\mu_y$ is the extreme point of $M^+_{x_1}(\overline{E(S_2)})$, i.e. the extreme boundary of the set $M^+_{x_1}(\overline{E(S_2)})$ consists of the three points $\{\varepsilon_{x_1}, \mu_{x_1}, \nu\}$.

By analogy with w_2 in R^2 we can make up examples of the regions for the heat equation where the extreme boundary of $M^+_{x_1}(\overline{E(S_2)})$ will consist of any finite or infinite sequence of measures. In contrast if the point z is a non-regular boundary point of a relatively compact region where we solve the Laplace equation, the set of probability measures with a barycenter in this point and concentrated on the Shilov's boundary is a segment $[\varepsilon_z, \mu_z]$.

Corollary 7.2 — Let the point x satisfy one of the conditions of the corollary 2.1., $\text{supp}\, \mu_x = \overline{E}_x$. Then for all measures $\nu \in M^+_x(\overline{E})$ the equality $\nu(Sh_x) = 0$ is valid.

Proof — By theorem 3* the point x belongs to A_x that's why the assertion of the corollary follows from theorem 7.

Theorem 8 — [6] Let $x \in S$. The following assertions are equivalent:

(1) $x \in E$;

(2) for every open neighbourhood $U(x)$ of a point x and for every number $\varepsilon > 0$ there exists such an open neighbourhood $V(x) \subseteq U(x)$ of a point x such that $\mu_y(S\ U(x)) < \varepsilon$, all $y \in V(x)$;

(3) every net x_α from S which converges to a point x is regular, i.e. $w\text{-}\lim_\alpha \mu_{x_\alpha} = \varepsilon_x$.

Theorem 9 — Let a point $x \in S$. Assume that one of the conditions of the corollary 2.1. is satisfied, the set E_x has no isolated points and supp $\mu_x = \overline{E}_x$, then every measure $\lambda \in M^+_x(\overline{E})$ has no atomic components in the set $\overline{E}_x$.

Proof-1. Let a point $k \in E_x$ and $m = \lambda(\{k\}) > 0$.

Consider the measure $\nu = (\lambda - m\varepsilon_k)/(1-m)$.

Let z be the barycenter of ν , then $\lambda = m\varepsilon_k + (1-m)\nu$ therefore

$x = mk + (1-m)z$, i.e. $k \in \text{face}(x)$.

Let's consider the function $f(y) = \begin{cases} 1 \text{ if } y \neq k \\ 0 \text{ if } y = k \end{cases}$ from $L^1(\mu_x)$.

By metrizability there is a decreasing sequence $\{V_n\}$ of open neighbourhoods of the point k in S which form a fondamental system ("a local base").

For all $n \in N$ let's consider a point $x_n \in V_n \cap (E_x \setminus \{k\})$. By the theorem 4 $\overline{\text{face}(x)} = \text{cl.face}(x)$ therefore by the theorem 8 for every point x_n there exists the neighbourhood $W_n = W_n(x_n) \subset V_n \setminus \{k\}$ and a point $y_n \in W_n \cap \text{face}(x)$ such that $\mu y_n(S \setminus W_n) < 1/n$.

Therefore $U_f(y_n) = \int f\, d\mu_{y_n} = \int_{W_n} f\, d\mu_{y_n} + \int_{S\setminus W_n} f\, d\mu y_n =$

$= \mu_{y_n}(W_n) + \int_{S\setminus W_n} f\, d\mu_{y_n} > 1 - 1/n + \int_{S\setminus W_n} f\, d\,\mu y_n$

and $\lim_n U_f(y_n) \neq 0 = U_f(k)$, i.e. U_f is discontinuous in the point k which contradicts the corollary 2.1.

2. If $k \in Sh_x$, then $\lambda(\{k\}) = 0$ by the corollary 7.2.

Remark 1 - If E_x has isolated points, then the theorem 9 may not be valid. Indeed let's consider the finite set $\{x_1,\ldots,x_n\} \subset E$, $n > 1$ and the point

$$x = (x_1 + \ldots + x_n)/n$$

It's evident that convex hull of the set $\{x_1,\ldots,x_n\}$ is equal to face(x) = cl.face(x), i.e. the set face(x) is locally compact in the induced topology. By the theorm 1 for the face(x) the assertion (1) of the corollary 2.1 is valid, but $\mu_x = (\varepsilon_{x_1} + \ldots + \varepsilon_{x_n})/n$.

Remark 2 - It is clearly seen that the proof of the 1-st part of the theorem 9 is still valid if for every compact $K \subset \text{face}(x)\setminus\{x\}$ there exists the number $\alpha_K \in [1; +\infty[$, such that $\sup_K a \leq \alpha_K\, a(x)$ (1) for every continuous affine function $a: \text{face}(x) \to [0; +\infty[$. In this case in the proof of theorem 9 we shall take the compact $\{k\} \cup \{y_n\}$ and use the theorem 2.

Corollary 9.1 – Let $x \in Sh_s$, $\text{supp}\ \mu_x = \overline{E}_x$, E_x has no isolated points and for every compact $K \subset \text{face}(x)\setminus\{x\}$ the inequality (1) is valid, then every measure $\lambda \in M_x^+(\overline{E})$ has no atomic components in the set E_x.

Theorem 10 – Let $x \in Sh_s$ and $\text{supp}\ \mu_x = \overline{E}$, then face(x) is the topological space of the first category.

Proof. Let's consider the contrary, then by the theorem 1 the Harnack inequality for every compact $K \subset \text{face}(x)$ is valid on the face(x). By the theorem 4 $\overline{\text{face}(x)} = S$ and from [6, theorem 3] there exists in face(x) such a sequence $x_n \to x$ that $\text{w-lim}_n\ \mu_{x_n} = \varepsilon_x$

Let's consider the function $f \in C(\overline{E})$ such that $f(x) = \varepsilon_x(f) \neq \mu_x(f) = U_f(x)$

Then $\lim_n U_f(x_n) = f(x) \neq U_f(x)$ that contradicts the corollary 2.1

REFERENCES

1.Keselman D.G.: Convex sets and Harnack inequality.
Comment. math. univ. Carolinae, 27, 2 (1986), 359-370

2.Keselman D.G.: Choquet simplexes and Harnack inequalities.
Proceedings of the 14-th Winter school on abstract analysis.
Srni (Bohemian weald), January, 1986, section of analysis (to appear).

3.Keselman D.G.: On some problems in Theory of Choquet simplexes connected with harmonic spaces. Sib.Mat.J., 1984,v.25, n°3,103-108 (in Russian)

4.Keselman D.G.: On some problems of Choquet theory connected with potential theory. Proceedings of the 12-th Winter school on abstract analysis. Srni, 15-29 January, 1984. Section of analysis. Supplemento ai Rendiconti del Circolo Matematico di Palermo, serie 11, N 5 (1984), 73-81.

5.Bear H.S.: Ordered Gleason parts. Pacific Journal of Math. 62 (1976) , 337-349.

6.Keselman D.G.: On Shilov boundary in Choquet simplex. The theory of functions and functional analysis and their applications, 1984, n°42, 62-67 (in Russian).

POLAR SETS IN A NONLINEAR POTENTIAL THEORY

T. Kilpeläinen

Department of Mathematics
University of Jyväskylä
SF-40100 Jyväskylä, Finland

1. INTRODUCTION

In this lecture we discuss nonlinear potential theory based on "A-superharmonic functions"; the theory can be viewed as a (nonlinear) extension of the classical study of superharmonic functions in $\mathbb{R}^n$.

To start with, we say that a function u is A-harmonic if $u \in C(G) \cap \text{loc}\, W_p^1(G)$, $1 < p < \infty$, (throughout this lecture G is an open set in $\mathbb{R}^n$, $n \geq 2$) and if

(1.1) $\nabla \cdot A(x, \nabla u) = 0$

in the weak sense, i.e.

(1.2) $\int_G A(x, \nabla u(x)) \cdot \nabla \phi(x)\, dx = 0$

for each $\phi \in C_0^\infty(G)$. We assume that the operator $A\colon \mathbb{R}^n \times \mathbb{R}^n \to \mathbb{R}^n$ satisfies for some numbers $1 < p < \infty$ and $0 < \alpha \leq \beta < \infty$:

(1.3) The function $x \mapsto A(x, h)$ is measurable for each $h \in \mathbb{R}^n$ and the function $h \mapsto A(x, h)$ is continuous for a.e. $x \in \mathbb{R}^n$.

For a.e. $x \in \mathbb{R}^n$ and for all $h \in \mathbb{R}^n$

(1.4) $A(x, h) \cdot h \geq \alpha\, |h|^p$,

(1.5) $|A(x, h)| \leq \beta\, |h|^{p-1}$,

(1.6) $(A(x, h_1) - A(x, h_2)) \cdot (h_1 - h_2) > 0$

whenever $h_1 \neq h_2$, and

(1.7) $A(x, \lambda h) = |\lambda|^{p-2} \lambda A(x, h)$

for all $\lambda \in \mathbb{R}$, $\lambda \neq 0$.

A prototype of the operators considered is the so-called p-Laplacian, $A(x, h) = |h|^{p-2} h$, in which case A-harmonic functions are customarily called p-harmonic, cf. e.g. [L]; if $p = 2$ the theory reduces to the classical theory in $\mathbb{R}^n$.

A lower semicontinuous (lsc) function $u: G \to \mathbb{R} \cup \{\infty\}$ is A-superharmonic if for each domain $D \subset\subset G$ and $h \in C(\overline{D})$, h is A-harmonic in D , the condition $h \leq u$ in ∂D implies that $h \leq u$ in D .

As mentioned above, A-superharmonic functions play a fundamental role in our theory. Thus the connections to the theory of nonlinear potentials studied by D. R. Adams, L. I. Hedberg, V. P. Khavin, V. G. Maz'ya, N. G. Meyers and others, cf. [Me], [MK], [HW], are not immediate. Our approach has applications in the theory of quasiregular mappings. As a matter of fact, in 1960's Yu. G. Reshetnyak [R] proved that if u is n-harmonic in G' and if $f: G \to f(G) = G'$ is quasiregular, then $u \circ f$ is A-harmonic for some A of the type $p = n$. In 1980 S. Granlund, P. Lindqvist and O. Martio initiated a study of conformally invariant variational integrals and quasiregular mappings; they introduced A-superharmonic functions if A is the gradient of a convex variational kernel; see [GLM 1-4], [LM 1-2]. This variational principle, the invariance of A-harmonic functions under quasiregular change of variables, is one of the basic tools in the theory of quasiregular mappings. For further applications of A-superharmonic functions in conformal geometry see e.g. [GLM 4].

The conformally invariant borderline case $p = n$ is somewhat different than the case $p < n$ (let it be remarked here that if $p > n$, then A-superharmonic functions are always continuous and locally in the Sobolev space W^1_p).

P. Lindqvist then published a study of p-superharmonic functions (i.e. the case of p-Laplacian) [L] where especially regularity properties were considered.

J. Heinonen and the present author took up the study of general A-superharmonic functions in [HK 1]; in particular, we worked out the connection between A-superharmonic functions and supersolutions of (1.1). The A-polar sets were analyzed in [HK 2] and applications of nonlinear potential theory to quasilinear obstacle problems were given in [HK 3].

All the results in this lecture are fruits of the joint work with J. Heinonen. For a more detailed discussion of the subject see [HK 1-3].

2. RESULTS

A set E in $\mathbb{R}^n$ is A-*polar* if there is a neighborhood U of E and an A-superharmonic u in U such that $u = \infty$ in E and that $u \not\equiv \infty$ in each component of U.

The p-*capacity* of a set E in G is defined as follows: The *inner* p-*capacity* of E in G is

$$_*\mathrm{cap}_p(E, G) = \sup_{C \subset E \text{ compact}} \inf_{u \in W(C,G)} \int_G |\nabla u|^p \, dx$$

where

$$W(C, G) = \{u \in C_0^\infty(G) : u = 1 \text{ in } C\} ;$$

the *outer* p-*capacity* of E in G is

$$^*\mathrm{cap}_p(E, G) = \inf_{U \supset E \text{ open}} {}_*\mathrm{cap}_p(U, G) .$$

A set E is of (*outer*) p-*capacity zero*, abbreviated $\mathrm{cap}_p(E) = 0$, if $^*\mathrm{cap}_p(E \cap G, G) = 0$ whenever G is an open set in $\mathbb{R}^n$. Note that for a bounded set E, $\mathrm{cap}_p(E) = 0$ as soon as $^*\mathrm{cap}_p(E, G) = 0$ for some bounded $G \supset E$, cf. [M 2], [R].

2.1. *Theorem* – *Let* E *be a set in* $\mathbb{R}^n$. *Then the following are equivalent*:

(i) E *is* A-*polar*.

(ii) E *is of* (*outer*) p-*capacity zero*.

(iii) *There is a nonnegative lsc function* w *in* $W_p^1(\mathbb{R}^n)$ *such that* $E \subset w^{-1}(\infty)$.

2.2. *Corollary* – *A countable union of* A-*polar sets is* A-*polar*.

Observe that Corollary 2.2 is not immediate since the sum of two A-superharmonic functions need not be A-superharmonic.

2.3. Theorem – Let $\mathcal{F}$ be a locally uniformly lower bounded family of A-superharmonic functions in G and let $w = \inf \mathcal{F}$. Then the lower regularization $\hat{w}$ of w, $\hat{w}(x) = \liminf_{y \to x} w(y)$, is A-superharmonic and $\hat{w} = w$ except on an A-polar set.

3. THE MAIN TOOLS

It goes without saying that in proving theorems the lack of linearity causes serious difficulties. The main tools are Harnack's inequality and principle, the comparison principle, the connection of A-superharmonic functions to supersolutions of (1.1) and last, but not least, the obstacle method. We also strongly employ the elementary fact that $\lambda u + \mu$ is A-superharmonic whenever u is and $\lambda, \mu \in \mathbb{R}$, $\lambda \geq 0$.

Recall that u in $\text{loc}\, W_p^1(G)$ is a supersolution of (1.1) if

$$\int_G A(x, \nabla u) \cdot \nabla \phi \, dx \geq 0$$

for each nonnegative $\phi \in C_0^\infty(G)$.

3.1. Proposition – If u is a supersolution of (1.1) then

$$v(x) = \operatorname{ess} \liminf_{y \to x} u(y)$$

defines an A-superharmonic function v with $v = u$ a.e.

Proof – Using the Moser iteration technique and standard estimates for supersolutions of (1.1) implies that v is lsc and that $v = u$ a.e. Then that v is A-superharmonic can be deduced from a comparison principle for supersolutions of (1.1); for details see [HK 1].

Note that

$$u(x) = \operatorname{ess} \liminf_{y \to x} u(y)$$

for each x if u is A-superharmonic [HK 1].

3.2. The obstacle problem – Let G be bounded and ψ a function in G. If $\theta \in W_p^1(G)$, $\theta \geq \psi$ a.e., then there is a unique solution u to the obstacle problem with the obstacle ψ and with boundary values θ, i.e., $u \geq \psi$ a.e., $u - \theta \in W_{p,0}^1(G)$ and

$$\int_G A(x, \nabla u) \cdot \nabla\phi \, dx \geq 0$$

for each $\phi \in W^1_{p,0}(G)$, $\phi \geq \psi - u$ a.e. Moreover, by Proposition 3.1, u can be assumed to be A-superharmonic.

If G is sufficiently regular (i.e. the complement of G is not Wiener thin at any point in ∂G , see [M 1]) and if $\theta \in C(\overline{G})$, then the solution $u \in C(\overline{G})$ and $u = \theta$ in ∂G . For this discussion see [HK 1]. A more careful analysis of obstacle problems can be found in [HK 3].

The efficient use of the obstacle problem has turned out to be the most powerful tool in this nonlinear theory. For example, it gives the following

3.3. _Proposition_ – _Let_ u _be_ A-_superharmonic_ _in_ G . _If_ u _is_ _locally_ _upper_ _bounded_, _then_ $u \in \text{loc } W^1_p(G)$. _If_ $u \in \text{loc } W^1_p(G)$, _then_ u _is_ _a_ _supersolution_ _of_ (1.1).

Proof – Let $D \subset\subset G$ be sufficiently regular and choose an increasing sequence $\phi_j \in C^\infty(\mathbb{R}^n)$ with $\lim \phi_j = u$ in $\overline{D}$. Let u_j be the solution to the obstacle problem in D with the obstacle and boundary values ϕ_j . Then $u = \lim u_j$ in D and it follows from a standard estimate [HK 1, 2.16] that u_j converges to u weakly in $W^1_p(D)$. Thus u is a supersolution of (1.1).

Finally, if $u \in W^1_p(D)$, then $w_k = \min(u, k)$ is bounded and A-superharmonic in D . Since $w_k \to u$ in $W^1_p(D)$, u is a supersolution of (1.1) by what was proved above.

Note that an A-superharmonic function u is not even locally in $W^1_p(G)$ in general.

4. PROOFS FOR THEOREMS

4.1. _Proof for Theorem 2.1_ – _(i) => (ii)_: Let B be a ball such that there is an A-superharmonic function u in B , $u \not\equiv \infty$ and $u = \infty$ in $F = E \cap \frac{1}{2}B$. It suffices to show that $^*\text{cap}_p(F, B) = 0$. Fix $\lambda > 1$ and pick an A-superharmonic v_λ in B , $0 \leq v_\lambda \leq \lambda$, such that $v_\lambda = \lambda$ in F and that $v_\lambda < 1$ in $\partial\frac{3}{4}B$. (Use a Poisson modification in a neighborhood of $\partial\frac{3}{4}B$). Then v_λ is a supersolution of (1.1), whence the estimate [HK 1, (2.25)] yields

$$\int_{\frac{3}{4}B} |\nabla \log v_\lambda|^p \, dx \le c$$

where c does not depend on λ. Now it can be easily seen that $^*\mathrm{cap}(F, B) = 0$, as desired.

(ii) => (iii): Let $B = B(0, R)$. We may clearly assume that $E \subset \frac{1}{2}B$. Choose open neighborhoods $U_i \subset \frac{1}{2}B$ of E such that

$$\sum_{i=1}^{\infty} {}^*\mathrm{cap}_p(U_i, B)^{\frac{1}{p}} < \infty .$$

Exhaust U_i by an increasing sequence of compact sets C_j (such that $B \setminus C_j$ is sufficiently regular) and let ϕ_i be the supremum of the capacity functions of C_j in B. Then define $\phi_i = 0$ in $\complement B$, whence ϕ_i is lsc, $\phi_i \in W_p^1(\mathbb{R}^n)$ and $\phi_i = 1$ in U_i. Moreover, by the Poincaré inequality,

$$\|\phi_i\|_{W_p^1(\mathbb{R}^n)} \le c\; {}^*\mathrm{cap}_p(U_i, B)^{\frac{1}{p}} ,$$

where c does not depend on i. Letting

$$w = \sum_{i=1}^{\infty} \phi_i$$

yields (iii).

(iii) => (i): We prove only the case $1 < p < n$. The case $p > n$ is trivial and the borderline case $p = n$ requires some special analysis, see [HK 2]. So let $1 < p < n$ and let v_i be the A-superharmonic solution to the obstacle problem in $B_i = B(0, i)$ with the obstacle and boundary values w. Since w is lsc the ess lim inf -property of v_i implies that $v_i(x) \ge w(x)$ for all $x \in B_i$. Since, by [HK 1, 2.8], the sequence v_i is increasing, $v = \lim v_i$ is A-superharmonic, and $v = \infty$ in E.

To complete the proof we show that $v \not\equiv \infty$. Using a Poincaré type inequality yields

$$\int_{B_i} \frac{|v_i(x) - w(x)|^p}{|x|^p} \, dx \le c \int_{B_i} |\nabla v_i(x) - \nabla w(x)|^p \, dx \le c \int_{\mathbb{R}^n} |\nabla w|^p \, dx$$

where c does not depend on i. (The last inequality follows since v_i is the solution to the obstacle problem in B_i.) Then letting $i \to \infty$ yields

$$\int_{\mathbb{R}^n} \frac{|v(x) - w(x)|^p}{|x|^p}\, dx \leq c \int_{\mathbb{R}^n} |\nabla w|^p\, dx < \infty$$

whence $v \not\equiv \infty$ as desired.

4.2. Proof for Theorem 2.3 — The function $\hat{w}$ is clearly A-superharmonic and we may assume that F consist of an decreasing sequence w_j such that $\lim w_j = w$. Thus the set

$$S_j = \{ x \in G : \hat{w}(x) + \frac{1}{j} < w(x) \}, \quad j = 1,2,\ldots,$$

is Borel, and by Corollary 2.2, it suffices to show that each compact C in S_j is A-polar. For that write $u = \hat{w} + \frac{1}{j}$ and let $D \subset\subset G$ be an open neighborhood of C. We may assume that $u > 0$ in D. Write

$$R_C^u = \inf \{ v \geq 0 : v \text{ A-superharmonic in } D \text{ and } v \geq u \text{ in } C \}$$

and

$$\hat{R}_C^u(x) = \liminf_{y \to x} R_C^u(y) .$$

Since, for each i, $w_i \geq u$ in C,

$$\hat{R}_C^u(x) \leq \hat{w}(x) < \hat{w}(x) + \frac{1}{j} = u(x)$$

for $x \in C$, and the claim follows from the next lemma.

4.3. Lemma — The set $S = \{ x \in D : R_C^u(x) > \hat{R}_C^u(x) \}$ is A-polar.

Proof — We may assume that $C = \partial C$. Let $\tilde{D} \subset\subset D$ be a regular neighborhood of C. Let $\phi_k \in C_0^\infty(\tilde{D})$ be an increasing sequence such that $\lim \phi_k = u$ on C. Let h_k be the A-harmonic function in $\tilde{D} \setminus C$ with $h_k - \phi_k \in W^1_{p,0}(\tilde{D} \setminus C)$. Then using comparison principles yields $R_C^u \geq h_k$ in $\tilde{D} \setminus C$. By the Kellogg property, see [HW], [M 1], there is an A-polar set $T \subset C$ with $\lim_{y \to x} h_k(y) = \phi_k(x)$ for all $x \in C \setminus T$. Thus it follows that

$$\hat{R}_C^u(x) \geq \phi_k(x)$$

whenever $x \in C \setminus T$. Letting $k \to \infty$ yields $S \subset T$ and the lemma is proved.

4.4. Remark – (a) As seen, many properties of superharmonic functions have their analogues in nonlinear theory, cf. [HK 1-3] and [K]. However, there are many open problems in this nonlinear situation. For example, let E be an A-polar set and $x_0 \in \mathbb{R}^n \setminus E$. Is there an A-superharmonic function u such that $u = \infty$ in E and that $u(x_0) < \infty$? It follows from results in [HK 3] that there is a bounded A-superharmonic v such that $\liminf_{x \to x_0} v(x) > v(x_0)$ provided $p > n - 1$.

(b) The annoying fact that the sum of two A-superharmonic functions is not A-superharmonic in general can be bypassed to some extent when one is dealing with functions in $W_p^1(G)$. It would be nice to find methods which do not single out A-superharmonic functions not in W_p^1.

REFERENCES

[GLM 1] S. Granlund, P. Lindqvist, and O. Martio, Conformally invariant variational integrals, Trans. Amer. Math. Soc. 277 (1983), 43-73.

[GLM 2] -"- F-harmonic measure in space, Ann. Acad. Sci. Fenn. Ser. A I Math. 7 (1982), 233-247.

[GLM 3] -"- Note on the PWB-method in the non-linear case, Pacific J. Math. 125 (1986), 381-395.

[GLM 4] -"- Phragmén-Lindelöf's and Lindelöf's theorems, Ark. Mat. 23 (1985), 103-128.

[HW] L. I. Hedberg and Th. H. Wolff, Thin sets in nonlinear potential theory, Ann. Inst. Fourier, Grenoble 33, 4 (1983), 161-187.

[HK 1] J. Heinonen and T. Kilpeläinen, A-superharmonic functions and supersolutions of degenerate elliptic equations, Ark. Mat. (to appear).

[HK 2] -"- Polar sets for supersolutions of degenerate elliptic equations (to appear).

[HK 3] -"- On the Wiener criterion and quasilinear obstacle problems, Preprint, Univ. Bonn (1987).

[K] T. Kilpeläinen, Convex increasing functions preserve the sub-F-extremality, Ann. Acad. Sci. Fenn. Ser. A I Math. 12 (1987), 55-60.

[L] P. Lindqvist, On the definition and properties of p-superharmonic functions, J. Reine Angew. Math. 365 (1986), 67-79.

[LM 1] P. Lindqvist and O. Martio, Two theorems of N. Wiener for solutions of quasilinear elliptic equations, Acta Math. 155 (1985), 153-171.

[LM 2] P. Lindqvist and O. Martio, Regularity and polar sets for supersolutions of certain degenerate elliptic equations (to appear).

[M 1] V. G. Maz'ya, On the continuity at a boundary point of solutions of quasi-linear elliptic equations, Vestnik Leningrad Univ. Mat. 13 (1970), 42-55, (Russian). Vestnik Leningrad Univ. Math. 3 (1976), 225-242, (English translation).

[M 2] -"- Sobolev spaces, Springer-Verlag, Berlin - Heidelberg - New York - Tokio (1985).

[MK] V. G. Maz'ya and V. P. Khavin, Non-linear potential theory, Russian Math. Surveys 27 (1972), 71-148.

[Me] N. G. Meyers, Continuity properties of potentials, Duke Math. J. 42 (1975), 157-166.

[R] Yu. G. Reshetnyak, Space mappings with bounded distortion, Izdat. Nauka, Novosibirsk (1982) (Russian).

A SHARPER FORM OF A THEOREM OF KOLMOGOROV

Boris Korenblum

Department of Mathematics
State University of New York
Albany, N.Y. 12222

1°. Let H^+ be the class of functions f(z) holomorphic in $\mathbb{D}=\{z\in\mathbb{C} : |z|<1\}$ and such that $\operatorname{Re} f(z)>0$. Functions in H^+ have the representation

$$f(z) = \int_{\partial\mathbb{D}} \frac{\zeta+z}{\zeta-z}\, d\mu(\zeta) + i\gamma \ , \tag{1}$$

where $\gamma\in\mathbb{R}$ and $d\mu>0$ is a Borel measure on $\partial\mathbb{D}$. It is also well know that f(z) have nontangential limit values almost everywhere:

$$f(e^{it}) = 2\pi\mu'(e^{it}) + i\tilde{\mu}(e^{it}) + i\gamma \ , \tag{2}$$

where $\mu'(e^{it})dt$ is the absolutely continuous part of $d\mu$ and

$$\tilde{\mu}(e^{it}) = \text{P.V.} \int_{-\pi}^{\pi} \cot\frac{s}{2}\, d\mu(e^{i(s+t)}) \, (\text{a.e.}) \tag{3}$$

(the sign for the principal value will be omitted in the sequel).

A classical theorem of Kolmogorov [1,p.111] asserts that $f(e^{it})$ is in the weak L^1 space, which can also be expressed in the following way. For a given $\lambda>0$ let f_λ denote the "λ-truncation" of f:

$$f_\lambda(e^{it}) = \begin{cases} \lambda \text{ if } |f(e^{it})|\geq\lambda \\ 0 \text{ if } |f(e^{it})|<\lambda \ . \end{cases}$$

Then $\|f_\lambda\|_1 = \frac{1}{2\pi}\int_{-\pi}^{\pi} f_\lambda(e^{it})dt \le C|f(0)|$ (here and in the sequel C denotes a numerical constant, not necessarily the same each time). If $(Uf_\lambda)(z)$ is the Poisson integral of f_λ, i.e. the harmonic function in $\mathbb{D}$ with boundary values $f_\lambda(e^{it})$, then $(Uf_\lambda)(0) \le C|f(0)|$. This immediately implies the conformally invariant statement

$$\sup_{\lambda>0}(Uf_\lambda)(z) \le C|f(z)| \quad (z\in\mathbb{D}) \ . \tag{4}$$

In this note it will be shown that there is a sharper*) form of (4) in which "level truncations" f_λ are replaced by "ψ-truncations" of f, where ψ's are essentially the kernels in (1). The existence of a sharper form of (4) was strongly suggested to the author by a theorem of A.V. Noell and T.H. Wolff [2, Proposition 0.6]. Note also that a similar result for the Hardy-Littlewood maximal function of a measure was established in [3].

The main motivation for this research was the need to know more about size properties of functions in H^+.

2°. To formulate our main result we first specify the notion of truncation. If f and g are functions on $\partial\mathbb{D}$ then the g-*truncation of* f is

$$f_g(e^{it}) = \begin{cases} |g(e^{it})| & \text{if } |f(e^{it})| \ge |g(e^{it})| \\ 0 & \text{if } |f(e^{it})| < |g(e^{it})| \ . \end{cases} \tag{5}$$

For a fixed $z\in\mathbb{D}$, Ψ_z will denote the set of functions of the form

$$\psi(w) = \psi(w;z,\zeta_1,\zeta_2,\lambda) = \lambda\,\frac{w-\zeta_1}{w-\zeta_2}\ , \tag{6}$$

where $\zeta_1, \zeta_2 \in\partial\mathbb{D}$ are the "end points" of a hyperbolic line through z, and $\lambda\in\mathbb{C}$; in particular,

$$\Psi_0 = \{\psi(w)=\lambda\,\frac{\zeta+w}{\zeta-w} : \zeta\in\partial\mathbb{D},\ \lambda\in\mathbb{C}\}\ .$$

*) See the example at the end of the paper.

Theorem – *If* $f\in H^+$, *then for all* $z\in\mathbb{D}$ *the following inequality holds:*

$$\sup_{\psi\in\Psi_z} \min\{|\psi(z)|, (Uf_\psi)(z)\}\le C|f(z)| . \tag{7}$$

3°. The statement (7) clearly is conformally invariant. It suffices therefore to prove it only for z=0:

$$\sup_{\psi\in\Psi_0} \min\{|\lambda|, (Uf_\lambda)(0)\}\le C|f(0)| . \tag{7'}$$

We list now some elementary properties of truncations (the proofs are left to the reader).

(i). $(g_1+g_2)_h \le 2[(g_1)_{h/2} + (g_2)_{h/2}]$.

(ii). If $0\le h\le h_1\le kh$ where $k>1$ is a constant, then $g_{h_1}\le kg_h$.

(iii). If $h=h_1$ on a set $E\subset\partial\mathbb{D}$, and $h=h_2$ on $\partial\mathbb{D}\setminus E$, where h,h_1,h_2 are nonnegative functions on $\partial\mathbb{D}$, then

$$g_h \le \max\{g_{h_1}, g_{h_2}\} .$$

Applied to (2) this gives

$$(Uf_\psi)(0)\le 2(U(2\pi\mu'+i\gamma)_{\psi/2})(0) + 2(U\tilde{\mu}_{\psi/2})(0)$$

$$\le 2\ \|2\pi\mu'+i\gamma\|_1 + 2(U\tilde{\mu}_{\psi/2})(0)\le C|f(0)|+2(U\tilde{\mu}_{\psi/2})(0).$$

It is therefore enough to prove:

$$\sup_{\lambda>0,\,s\in[-\pi,\pi)} \min\{\lambda, (U\tilde{\mu}_{\lambda\cot(t-s)/2})(0)\}\le C\|\mu\| , \tag{8}$$

where $\|\mu\|$ stands for the variation of the measure μ.

It is convenient now to replace $|\cot t/2|$ by a function $\tau(t)$ which is equal to $\tau_1(t)=\pi/(2|t|)$ for $|t|\le\pi/2$ and to $\tau_2(t)=2(\pi-|t|)/\pi$ for $\pi/2\le|t|\le\pi$; by the above properties of truncations it will then be sufficient to prove (8) with $\cot t/2$ replaced by either of the functions τ_1,τ_2. It is also convenient to change the setting to that of a real line (in this setting it is irrelevant whether μ is supported on $[-\pi,\pi]$, or on any other interval, or even spread over $\mathbb{R}$). We are thus led to the following two lemmas, which together give slightly more than what is required for the proof of our Theorem.

Lemma 1 - Let μ be a finite nonnegative Borel measure on $\mathbb{R}$ and

$$\tilde{\mu}(x) = \int_{-\infty}^{\infty} \frac{d\mu(t)}{t-x} \quad \text{(a.e.)} \quad .$$

For any function $\tau_1(x)=\lambda/|x-x_0|$ ($x_0 \in \mathbb{R}$, $\lambda \geq 4\|\mu\|$) the following estimate holds:

$$\|(\tilde{\mu})_{\tau_1}\|_1 \leq C\|\mu\| \quad . \tag{9}$$

Lemma 2 - Under the same assumptions on μ, for any function $\tau_2(x)=\lambda|x-x_0|$ ($x_0 \in \mathbb{R}$, $\lambda>0$) the following holds:

$$\|(\tilde{\mu})_{\tau_2}\|_1 \leq C\|\mu\| . \tag{10}$$

The ideas underlying the proofs of these two lemmas are similar: we try by a suitable substitution (change of variables) to reduce the lemmas to one of the forms of Kolmogorov's theorem. However, these substitutions are essentially different for Lemma 1 and Lemma 2.

Proof of Lemma 2 - We can assume that $x_0=0$. The L^1 norm in (10) is on $\mathbb{R}$, but if we can prove (10) with L^1 norm taken separately on $\mathbb{R}_+=(0,\infty)$ and $\mathbb{R}_-=(-\infty,0)$, then (10) will follow. Let $x>0$; we have

$$\tilde{\mu}(x) = \int_{-\infty}^{0} \frac{d\mu(t)}{t-x} + \int_{0}^{\infty} \frac{d\mu(t)}{t-x} = \tilde{\mu}_1(x) + \tilde{\mu}_2(x) \quad .$$

Since $|\mu_1(x)| \leq \|\mu_1\|/x$, a straightforward computation shows that $\|(\tilde{\mu}_1)_{\lambda x}\|_1 \leq \|\mu_1\|$ $(0<\lambda<\infty)$. To estimate $(\tilde{\mu}_2)_{\lambda x}$ we use the substitution $t=\sqrt{s}$, $x=\sqrt{y}$:

$$\tilde{\mu}_2(x) = \tilde{\mu}_2(\sqrt{y}) = \int_0^{\infty} \frac{d\mu(\sqrt{s})}{\sqrt{s}-\sqrt{y}} = 2\int_0^{\infty} \frac{\sqrt{y}\ d\mu(\sqrt{s})}{s-y} + \int_0^{\infty} \frac{d\mu(\sqrt{s})}{(\sqrt{s}+\sqrt{y})}$$

and

$$\frac{\tilde{\mu}_2(x)}{x} = \frac{\tilde{\mu}_2(\sqrt{y})}{\sqrt{y}} = 2\int_0^\infty \frac{d\mu(\sqrt{s})}{s-y} + \int_0^\infty \frac{d\mu(\sqrt{s})}{(\sqrt{s}+\sqrt{y})\sqrt{y}} . \tag{11}$$

On the other hand,

$$\int_0^\infty (\tilde{\mu}_2)_{\lambda x} dx = \int_{\{|\tilde{\mu}_2(x)|\geq \lambda x\}} \lambda x dx = \int_{\{|\tilde{\mu}_2(\sqrt{y})|\geq \lambda\sqrt{y}\}} \lambda\sqrt{y}\, d(\sqrt{y}) =$$

$$= \frac{\lambda}{2}\ |\{y\in\mathbb{R}_+ : |\tilde{\mu}_2(\sqrt{y})|/\sqrt{y} \geq \lambda\}| ,$$

which is $\leq$ weak-L^1 norm of $\tilde{\mu}_2(\sqrt{y})/\sqrt{y}$. To estimate the latter we use (11). The weak L^1 norm of the first term on the right side of (11) does not exceed $C\|\mu_1\|$ (by one of the forms of Kolmogorov's theorem). The second term is $\leq\|\mu_1\|/y$, and its weak-L^1 norm is $\leq C\|\mu_1\|$. Lemma 2 is proved.

<u>Proof of Lemma 1</u> - Again we take $x_0=0$. Using properties of truncations we get (on $\mathbb{R}_+$):

$$\tilde{\mu}_{\lambda/x} \leq 2(\tilde{\mu}_1)_{\lambda/2x} + 2(\tilde{\mu}_2)_{\lambda/2x} . \tag{12}$$

Since $\lambda\geq 2\|\mu\|$ and $|\tilde{\mu}_1(x)|\leq\|\mu_1\|/x$, the first term on the right side of (12) is 0. Therefore (9) is equivalent to

$$\int_0^\infty \left(\int_0^\infty \frac{d\mu(t)}{t-x}\right)_{\lambda/x} dx \leq \|\mu\|_{\mathbb{R}_+} \qquad (\lambda\geq 2\|\mu\|_{\mathbb{R}_+}) . \tag{13}$$

We now change the variables in (13): $x=e^\xi$, $t=e^\tau$; the resulting statement is

$$\int_{-\infty}^\infty e^\xi \left(\int_{-\infty}^\infty \frac{d\mu(e^\tau)}{e^\tau-e^\xi}\right)_{\lambda e^{-\xi}} d\xi \leq C\|\mu\| \qquad (\lambda\geq 2\|\mu\|) . \tag{14}$$

Clearly,

$$\left(\int_{-\infty}^\infty \frac{d\mu(e^\tau)}{e^\tau-e^\xi}\right)_{\lambda e^{-\xi}} = e^{-\xi}\left(\int_{-\infty}^\infty \frac{d\mu(e^\tau)}{e^{\tau-\xi}-1}\right) .$$

Therefore (14) becomes

$$\int_{-\infty}^\infty \left(\int_{-\infty}^\infty \frac{d\mu(e^\tau)}{e^{\tau-\xi}-1}\right)_\lambda d\xi \leq C\|\mu\| \qquad (\lambda\geq 2\|\mu\|) . \tag{15}$$

A simple computation shows that

$$\left|\frac{1}{e^{\tau}-1} - \frac{1}{\tau}\right| < 1 \qquad (-\infty<\tau<\infty)$$

and therefore

$$\left|\int_{-\infty}^{\infty}\frac{d\mu(e^{\tau})}{e^{\tau-\xi}-1}\right| \le \left|\int_{-\infty}^{\infty}\frac{d\mu(e^{\tau})}{\tau-\xi}\right| + \|\mu\| \quad .$$

Thus we obtain

$$\left(\int_{-\infty}^{\infty}\frac{d\mu(e^{\tau})}{e^{\tau-\xi}-1}\right)_{\lambda} \le 2\left(\int_{-\infty}^{\infty}\frac{d\mu(e^{\tau})}{\tau-\xi}\right)_{\lambda/2} \quad .$$

This shows that (15) is equivalent to

$$\int_{-\infty}^{\infty}\left(\int_{-\infty}^{\infty}\frac{d\mu(e^{\tau})}{\tau-\xi}\right)_{\lambda} d\xi \le C\|\mu\| \ ,$$

which is essentially the statement of Kolmogorov's theorem. The proof of Lemma 1 and of the Theorem is complete.

4°. In conclusion, we want to summarize some information on the size of $|f(z)|$ $(f\in H^{+})$ available from [3] and the above theorem, and then pose an open problem. Let $h(z)$ be a given positive subharmonic function in $\mathbb{D}$. If $h(z)=|f(z)|$ for some $f\in H^{+}$ then the following holds:

(i) The radial maximal function $h^{*}(e^{it}) = \sup_{0<r<1} h(re^{it})$ is in "balanced" $L^{1,\infty}$, which means that $M_w h^{*} \le Ch^{*}$ (see [3] for the definition of weak maximal function M_w).

(ii) The two function $h_1(z)$ and $h_2(z)$ $(z\in\mathbb{D})$ constructed from $h(e^{it})$ using, respectively, (4) and (7):

$$h_1(z) = \sup_{\lambda>0}(Uh_{\lambda})(z) \ , \tag{16}$$

$$h_2(z) = \sup_{\psi\in\Psi_2} \min\{\psi(z),(Uh_{\psi})(z)\} \tag{17}$$

are equivalent to $h(z)$.

We ask now this question: do these two properties guarantee that a given subharmonic $h(z)$ is dominated in $\mathbb{D}$ by the modulus of a function in H? Note that an example by

Noell and Wolff [2] implies that there are subharmonic functions enjoying all the above properties, except that $h_2(z)=+\infty$: essentially, this example is an $h_1(z)$ constructed by (16) from $h(e^{it})=[\mathrm{dist}(t,\{1,1/2,\ldots,1/n,\ldots\})]^{-1/2}$.

Acknowledgement - This paper was done while the author visited Institut d'Estudis Catalans, Centre de Recerca Matemàtica, Bellaterra (Barcelona). He wishes to thank this institution for its hospitality and support. He thanks the NSF for its financial support. He also wishes to thank Joaquim Bruna and Jośe Fernández for stimulating discussions.

References

1. J.B. Garnett, Bounded Analytic functions, Academic Press, 1981.

2. Alan V. Noell, Thomas H. Wolff, On Peak Sets for Lipα Classes (preprint).

3. Joaquim Bruna, Boris Korenblum, On Kolmogorov's theorem, the Hardy-Littlewood maximal function and the radial maximal function (to appear).

EXTREMAL PROBLEMS FOR THE DISTRIBUTED MOMENT PROBLEM

Ognyan Kounchev

Institute of Mathematics
Bulgarian Academy of Sciences
Sofia, Bulgaria

INTRODUCTION

The present work continues the investigation of the socalled distributed moment problem (DMP) (see [13] and references there). We use the last name instead of its alternative one, inverse potential problem [19], to stress the analogy with the one-dimensional moment problem [1,8,15].

First we prove a result about the uniqueness of the solution of the extremal problems for the inverse problem of gravimetry [3, 19,22]. After that we describe some geometric properties of the mass distributions for the same inverse problem. Then we prove that the element of best approximation by harmonic functions in the uniform norm is unique.

We chose thes items since they use similar technics based on a classical density result of Brelot [4] and Deny [5].

1. THE INVERSE POTENTIAL PROBLEM

Let $\overline{G}$ be a bounded domain in R^n with a regular boundary $F = \partial G$ and the operator

$$L(x,D) = \sum_{|\alpha|\leq 2m} a_\alpha(x) D^\alpha \tag{1.1}$$

be an elliptic differential operator of order 2m with sufficiently smooth coefficients in G [2,19], here we set as usually

$$D^\alpha = \partial^{|\alpha|}/(\partial x_1^{\alpha_1} \ldots \partial x_n^{\alpha_n}), \quad |\alpha| = \alpha_1 + \ldots + \alpha_n .$$

Following [19] we introduce the following notations

$$N(L) = \{h(x) : L(x,D)h(x) = 0, \quad x \in G, \quad h(x) \in C^{m-1}(\overline{G}) \cap C^{2m}(G)\}; \tag{1.2}$$

$$N_+(L) = \{h(x) \in N(L) : h(x) \geqq 0, \quad x \in \overline{G}\};$$

for the boundary operators

$$B_j(x,D) = \sum_{|\gamma| \leqq m_j} b_{j\gamma}(x) D^\gamma \qquad (j=1,\ldots,m) \tag{1.4}$$

We suppose that L, B_j, G satisfy the normality condition [2,19] which provides solubility of the boundary value problem

$$\begin{aligned} Lh(x) &= 0 \text{ for } x \in G, \\ B_j h(y) &= g_j(y) \text{ for } y \in F, \quad j = 1,\ldots,m, \end{aligned} \tag{1.5}$$

for sufficiently smooth boundary data g_j on F.

Through $M(\overline{G})$ and $M(F)$ we denote the set of all Radon measures on G and F respectively [10,18,19].

Further we suppose that L, B_j and G are such that a priori estimates for problem (1.5) can be established and a corresponding sweeping-out (balayage) operator can be defined by means of the following

<u>Proposition 1</u>- ([19]) Let $\mu \in M(\overline{G})$. Then there exists a vector-measure $\nu = (\nu^j)$, $1 \leqq j \leqq m$, $\nu^j \in M(F)$ such that for every $u \in N(L)$ the equality (1.6) holds:

$$\int_{\overline{G}} u(x)\,d\mu(x) = \sum_{j=1}^{m} \int_F B_j u(y)\,d\nu^j(y) \tag{1.6}$$

Instead of (1.6) we will often use the following abbreviations

$$\mu(u) = \nu(Bu) \text{ or } P\mu = \nu. \tag{1.7}$$

Now let $\nu \in M^m(F)$ be a given vector-measure. Consider the set

$$B(\nu) = \{\mu \in M(G) : P\mu = \nu, \quad \mu \geqq 0\}. \tag{1.8}$$

The inverse problem of potential theory consists in finding elements of the set $B(\nu)$ and more generally, in studying the structure of $B(\nu)$ [19].

Let w(x) be a continuous function defined in $\overline{G}$. We are interested in the following extremal problem

$$\int_{\overline{G}} w(x)\,d\mu(x) \to \max(\min), \quad \mu \in B(\nu) \tag{1.9}$$

for a given $\nu \in M^m(F)$.

For the case of the Laplace operator $L = \Delta = \sum_{i=1}^{n} \partial^2/\partial x_i^2$, in Theorem 1 below we prove that the minimization problem (1.9) has unique solution for subharmonic functions $w(x)$. In that case problem (1.9) may be interpreted as an extremal problem for the inverse problem of gravimetry [2,22]. For the proof we need the duality Proposition 3 [13]. In order to formulate it we introduce the sets

$$\underline{P}(w) = \{u \in N(L) : u(x) \leqq w(x),\ x \in \overline{G}\},$$

$$\overline{P}(w) = \{u \in N(L) : u(x) \geqq w(x),\ x \in \overline{G}\}.$$

They are not empty since all constants are in $N(L)$.

The measure $\nu \in M^m(F)$ is called positive if for every $h \in N_+(L)$ we have $\nu(Bh) \geqq 0$. It is called strictly positive if $\nu(Bh) > 0$ for every $h \in N_+(L)$, $h \not\equiv 0$, and singularly positive if it is positive but not strictly, i.e. there is an $h_0 \in N(L)$, $h_0 \not\equiv 0$ but $\nu(Bh) = 0$.

We have the following result:

Proposition 2- ([12]) For a given vector-measure $\nu \in M(F)$ the set $B(\nu)$ is not empty if and only if ν is a positive measure.

The following duality result is an analogue to a classical duality theorem [15, Ch. 4, Theorem 4.1]:

Proposition 3- Let the strictly positive vector-measure $\nu \in M^m(F)$ be given. Then

a) the equalities hold

$$\min\{\int_{\overline{G}} w(x)\,d\mu(x) : \mu \in B(\nu)\} = \max\{\nu(Bu) : u \in \underline{P}\}$$

$$\max\{\int_{\overline{G}} w(x)\,d\mu(x) : \mu \in B(\nu)\} = \min\{\nu(Bu) : u \in \overline{P}\};$$

b) a necessary and sufficient condition that the integral

$$\int_{\overline{G}} w(x)\,d\mu(x) \tag{1.10}$$

takes its maximal (minimal) value for $\mu = \mu_o \in B(\nu)$ is the existence of some $u_o \in \underline{P}(w)$ ($u_o \in \overline{P}(w)$) which coincides with $w(x)$ for $x \in \overline{G}$ where $d\mu(x) \neq 0$.

The proof of this proposition follows closely the idea in [15].

In the following paragraphs we will consider the case of the Laplace operator $L = \Delta$ and its powers defined by $\Delta^{i+1} = \Delta\,\Delta^i$ and $\Delta^o = 1$ for all positive integers i. We will impose no restrictions on the boundary operators B_j except the stated above. Other operators L for which the theorems may be proved are obtained through coordinate transformation of Δ and its powers. In the bidimensional case every strictly elliptic second order operator may be reduced to C.Δ [21] which extends our results to this generality.

2. UNIQUENESS FOR SOLUTIONS OF EXTREMAL PROBLEMS

For the sake of clarity, within the statement we formulate problem (1.9) for the Laplace operator $L = \Delta$ and the boundary operator $B_1 = 1$:

Theorem 1- Le ν be a given positive measure on F. Consider the set of positive measures

$$B(\nu) = \{\mu \in M(\overline{G}) : \mu \geqq 0, \int_{\overline{G}} h(x)\,d\mu(x) = \int_F h(y)\,d\nu(y) \tag{2.1}$$

for every function h(x), harmonic in G}.

If w(x) is a subharmonic function [7,16] then the extremal problem

$$\int_{\overline{G}} w(x)\,d\mu(x) \to \min \qquad (\mu \in B(\nu))$$

has unique solution.

Proof of Theorem 1- Since the set $B(\nu)$ is w*-compact in $M(\overline{G})$ [19] problem (2.2) has a solution. Let us call it μ_o. Applying Proposition 3 we obtain a harmonic in G function $h_o(x)$ satisfying

$$h_o(x) \leqq w(x) \qquad (x \in \overline{G}) \tag{2.3}$$

and such that $\mathrm{supp}(\mu_o) \subset M = \{x : h_o(x) = w(x)\}$.

If for some measure $\mu \in B(\nu)$ $\mathrm{supp}(\mu) \not\subset M$ then we get

$$0 = \int_{\overline{G}} (w(x) - h_o(x))\,d\mu_o(x) < \int_{\overline{G}} (w(x) - h_o(x))\,d\mu(x). \tag{2.4}$$

This obviously implies

$$\int_{\overline{G}} w(x)\,d\mu_o(x) < \int_{\overline{G}} w(x)\,d\mu(x)$$

since both μ, μ_o satisfy (2.1).

It follows that if $\mu_1 \in B(\nu)$ is another solution of problem (2.2) then $\mathrm{supp}(\mu_1) \subset M$.

On the other hand the set M is a set of zeros of the positive subharmonic function $w(x) - h_o(x)$ in $\overline{G}$. This implies that every point of M is regular for its complement CM [7,16]. We shall prove that the set CM is connected.

Let us suppose the opposite: the set CM has a compact component G_o. Then $w(x) - h_o(x) = 0$ on ∂G_o and a basic property of subharmonic functions [7] implies that $w(x) - h_o(x) < 0$ in G_o which contradicts (2.3).

Now a classical theorem [4,5] may be applied giving that the set of functions harmonic in G is dense in the set of continuous functions C(M) (in the uniform norm). Since we have $\mu_0(h) = \mu_1(h)$ for every $h \in N(\Delta)$ in G, the above implies $\mu_1 = \mu_0$. Q.E.D.

It is quite plausible that Theorem 1 holds for the case $L = \Delta^m$ with w(x) - subharmonic function of order m [17], i.e. a function satisfying $(-1)^m \Delta^m w(x) \leq 0$, $x \in \bar{G}$.

3. GEOMETRIC STRUCTURE OF THE MASS DISTRIBUTIONS

Here we prove a geometric property of the measures in $B(\nu)$ which have only regular points [7,16] in their support.

Theorem 2- Let $L = \Delta^m$ and the set $B(\nu)$, for a given vector-measure $\nu \in M^m(F)$, is not empty. If μ_1, μ_2 are two different measures in $B(\nu)$ which have supports regular sets and the distance between the two supports is not zero, i.e. $\mathrm{dist}(\mathrm{supp}(\mu_1), \mathrm{supp}(\mu_2)) > 0$, then one of the sets $\mathrm{supp}(\mu_1)$, $\mathrm{supp}(\mu_2)$ contains a closed surface Γ (the boundary of a simply connected domain in R^n) and the other one contains a point x_o which is inside Γ.

This result is especially interesting for the case of $L = \Delta$ since it justifies, in some sense, results of D. Zidarov [22] who studied methods for obtaining one mass distribution in $B(\nu)$ from another through a method of "concentration of masses" (see also [3]).

Proof of Theorem 2- We proceed in a way quite similar to the proof of the characterization theorem [6,11].

Consider the measure $d\mu = d(\mu_1 - \mu_2)$ which is not zero. We have that $\mu(h) = 0$ for every function h(x), harmonic in $\bar{G}$.

Now, let us suppose that $\mathrm{supp}(\mu)$ has a connected complement in R^n. Since $\mathrm{supp}(\mu)$ is a regular set we may apply a classical theorem [4,5] which states that the set of harmonic functions is dense in the set of continuous functions $C(\mathrm{supp}(\mu))$. According to a standard argument this contradicts $\mu(h) = 0$ for every $h \in N(\Delta)$ with a nonzero measure $d\mu$. It follows that $\mathrm{supp}(\mu)$ has nonconnected complement and we may find a compact simply connected component of $G \setminus \mathrm{supp}(\mu)$, let us call it G_o. Since the boundary of G_o is a connected set and $\mathrm{supp}(\mu_1)$, $\mathrm{supp}(\mu_2)$ are distant sets it is clear that $\Gamma = \partial G_o$ is contained either in $\mathrm{supp}(\mu_1)$ or in $\mathrm{supp}(\mu_2)$.

In order to prove the existence of a point x_o inside Γ, let us suppose that no such exists, i.e. every compact simply connected component of $G \setminus \mathrm{supp}(\mu)$ which has a boundary in $\mathrm{supp}(\mu_1)$ does not meet $\mathrm{supp}(\mu_2)$ and the same for $\mathrm{supp}(\mu_2)$. So we find a maximal open set G_1 which is simply connected (but in general not connected) with boundary in $\mathrm{supp}(\mu_1)$ and a similar G_2 for $\mathrm{supp}(\mu_2)$ such that G_1 and G_2 are distant and the set $G_1 \cup G_2 \cup \mathrm{supp}(\mu_1) \cup \mathrm{supp}(\mu_2)$ has a connected complement. The same theorem quoted above implies that there is a function $h_1(x)$, harmonic in $\overline{G}$ which is close to 1 on $G_1 \cup \mathrm{supp}(\mu_1)$ and close to -1 on $G_2 \cup \mathrm{supp}(\mu_2)$. The last contradicts $\mu(h_1) = 0$ or $\mu_1(h_1) = \mu_2(h_1)$, since $\mu_1(h_1)$ is a positive but $\mu_2(h_1)$ is a negative number. The proof is finished. Q.E.D.

Let us note that an interesting interpretation, from the point of view of elasticity theory, for the elements of the set $B(\nu)$ when $L = \Delta^2$ was recently considered in [9].

4. ON THE UNIFORM APPROXIMATION BY SOLUTIONS OF PDEs

For a given function $f(x) \in C(\overline{G})$ consider the approximation problem in the Chebyshev norm:

$$\|h-f\| = \max_{x \in \overline{G}} |h(x) - f(x)| \to \inf_{h \in N(L)} \quad (= \delta). \tag{4.1}$$

This problem has a solution for an appropriate extension of $N(L)$. For example, when $L = \Delta^m$, we consider the bounded in $\overline{G}$ solutions of $Lh(x) = 0$, $x \in G$ [17].

We define the important set where the norm is attained:

$$T(h) = \{x \in \overline{G}: |h(x)-f(x)| = \|h-f\|\} \tag{4.2}$$

For $L = \Delta$, when the element of best approximation in problem (4.1) is a continuous function, there is a characterization for it in the terms of the set $T(h)$ [6,11]. In the next theorem we show that this solution is unique:

<u>Theorem 3</u>. Let $L = \Delta$. Then the corresponding problem (4.1) has unique solution if its solution $\tilde{h}$ is a continuous function and the corresponding set $T(\tilde{h})$ in (4.2) has only regular points.

<u>Proof of Theorem 3</u>. According to [20] the function $\tilde{h} \in N(L)$ is an element solving (4.1) if and only if there exist Borel nonnegative measures μ_+, μ_- on $\overline{G}$ such that the measure $\mu = \mu_+ - \mu_-$ satisfies the following conditions:

$$\mathrm{Var}(\mu) = 1, \tag{4.3}$$

$$\int_{\overline{G}} h(x)\, d\mu(x) = 0 \text{ for every } h \in N(\Delta), \tag{4.4}$$

$$\operatorname{supp}(\mu_+) \subseteq T_+(\tilde{h}), \quad \operatorname{supp}(\mu_-) \subseteq T_-(\tilde{h}), \tag{4.5}$$

where the sets T_+ and T_- are defined as follows

$$T_\pm(\tilde{h}) = \{x \in \overline{G}: \tilde{h}(x) - f(x) = \pm\|\tilde{h}-f\| = \pm\delta\}.$$

Evidently, $T = T_+ \cup T_-$.

Let us suppose that problem (4.1) has two solutions h_1, h_2 with the properties stated in the theorem, and $\|h_1-f\| = \|h_1-f\| = \delta$. Then the function $\tilde{h} = (h_1+h_2)/2$ is a solution of (4.1) too. This may be seen from the relations

$$\|\tilde{h}-f\| = \|(h_1-f)/2 + (h_2-f)/2\| \leqq \delta/2 + \delta/2 = \delta.$$

The inclusions

$$T_\pm(h) \subseteq T_\pm(h_1) \cap T_\pm(h_2) \tag{4.6}$$

follow from the relations

$$\delta = |\tilde{h}(x) - f(x)| = |(h_1(x)-f(x))/2 + (h_2(x)-f(x))/2| \leqq$$
$$\leqq \delta/2 + \delta/2 = \delta.$$

This implies that the set $T(\tilde{h})$ has only regular points. Suppose further that the complement of the set $T(\tilde{h})$ is connected. According to the result in [4,5] we get that the set $N(L)$ is dense in the set of continuous functions $C(T(\tilde{h}))$ which contradicts (4.4) when applied to our function $\tilde{h}$, since (4.3) says that $d\mu$ is a nonzero measure. This contradiction shows that the complement of the set $T(\tilde{h})$ is not connected and we may find a compact component of $G \setminus T(\tilde{h})$ which is a simply connected set. Let us call it G_o. Since the sets $T_+(\tilde{h})$, $T_-(\tilde{h})$ are distant, and the boundary ∂G_o is a connected set, it follows that either $\partial G_o \subseteq T_+(\tilde{h})$ or $\partial G_o \subseteq T_-(\tilde{h})$. In both cases the inclusion (4.6) gives that $h_1(x) = h_2(x)$ for $x \in \partial G_o$. According to the maximum principle $h_1(x) = h_2(x)$ for $x \in G_o$ and the real analyticity of harmonic functions implies that $h_1(x) = h_2(x)$ everywhere in G. Q.E.D.

It is quite natural to conjecture that Theorem 3 holds for the polyharmonic operator $L = \Delta^m$.

REFERENCES

[1] Achieser, N.I.: The classical moment problem, Fizmatgiz, Moscow, 1961 (in Russ., there is Engl. transl.).

[2] Agmon, S., Douglis, A., Nirenberg, L.: Estimates near the boundary for the solutions of elliptic partial differential equations satisfying general boundary conditions. 1., Comm. Pure Appl. Math., 12 (1959), pp. 623-727.

[3] Anger, G.: Lectures on potential theory and inverse problems, in: Geodaetische und Geophysikalische Veroffentlichungen, Reihe 3, No. 45 (1980), pp. 15-95, published by the Nation. Committee for Geodesy and Geophysics, Acad. Sci. GDR, Berlin.

[4] Brelot, M.: Sur l'approximation et la convergence dans la theorie des functions harmoniques ou holomorphes, Bull. Soc. Math. France, 73 (1945), pp. 55-70.

[5] Deny, J.: Systemes totaux de functions harmoniques, Ann. Inst. Fourier (Grenoble), 1 (1949), pp. 103-113.

[6] Hayman, W.K., Kershaw, D., Lyons, T.J.: The best harmonic approximant to a continuous function, in: International Series of Numerical Mathematics, Vol. 65, Birkhauser Verlag, Basel, 1984.

[7] Helms, L.L.: Einfuehrung in die Potentialtheorie, De Gruyter, Berlin, 1973.

[8] Karlin, S., Studden, W.J.: Tchebycheff systems: with applications in analysis and statistics, Interscience Publishers, New York, 1970.

[9] Kleine, E.: An inverse problem for biharmonic potentials, Mathematische Nachrichten, 128 (1986), pp. 7-27.

[10] Kolmogorov, A.N., Fomin, S.V.: Elements of theory of functions and functional analysis, Nauka, Moscow, 1968 (in Russ., there is Engl. transl.).

[11] Kounchev, O.I.: On the harmonic function which deviates least from a given continuous one in the circle, DAN BSSR, Ser. fiz.-mat. nav., No. 4 (1985) (in Russ.).

[12] Kounchev, O.I.: Distributed moment problem and some related questions on approximation of functions of many variables, in: Mathemattics and education in mathematics, Publishing House of the Bulgarian Academy of Sciences, Sofia, 1985, pp. 454-458.

[13] Kounchev, O.I.: The distributed moment problem, in: Proc. Confer. Contr. and Identif. Distrib. Param. Sys., Vorau-Austria, 1986, Lect. Notes Inf. Contr. (to appear).

[14] Kounchev, O.I.: The Chebyshev alternance set and the theory of canonical representations of Markov, (in preparation).

[15] Krein, M.G., Nudel'man, A.A.: The problem of Markov and extremal problems, Nauka, Moscow, 1973 (in Russ., there is Engl. transl.).

[16] Landkof, N.S.: Foundations of modern potential theory, Nauka, Moskow, 1966 (in Russ., there is Engl. transl.).

[17] Nicolescu, M.: Opera matematica: Functii poliarmonice, Editura academiei Republicii Socialiste Romania, Bucuresti, 1980.

[18] Rudin, W.: Real and complex analysis, Mc Graw Hill Publ. Co., New York, 1976.

[19] Schulze, B.-W., Wildenhain, H.: Methoden der Potentialtheorie fuer elliptische Differentialgleichungen beliebiger Ordnung, Akademie Verlag, Berlin, 1977.

[20] Singer, I.: Best approximation in normed linear spaces by elements of linear subspaces, Springer, Berlin, 1970.

[21] Vekua, I.N.: Generalized analytic functions, GIFML, Moscow, 1959 (in Russ., there is Engl. transl.).
[22] Zidarov, D.: New approach to the solution of the integral equation of the first kind, in: Inverse and improperly posed problems in differential equations, Akademie Verlag, Berlin, 1979, pp. 271-288.

ON MATHEMATICAL MODELLING OF NON-NEWTONIAN FLOW BY PERTURBATION METHODS

R. Kreul and H. Kretzschmar

Department of Mathematics
Technical University of Karl-Marx-Stadt
GDR

INTRODUCTION

Isothermal incompressible flow in consideration of friction influences can generally be described by the non-linear system of partial differential equations

$$\begin{aligned} Re \frac{dv}{dt} &= -Re \cdot \text{grad } p + \text{DIV } T + f, \\ \text{div } v &= 0. \end{aligned} \tag{1}$$

There, basing on non-dimensional coordinates, v and p denote the velocity and pressure field, f - an external force influencing the flow system, Re - the REYNOLDS number and T - the stress deviator, for its concrete description system (1) has to be completed by a rheological equation of state such as $T = D$, $T = D + \varepsilon \cdot \frac{\mathcal{D} D}{\mathcal{D} t}$ etc. with $D = \frac{1}{2}(L + L^*)$, $W = \frac{1}{2}(L - L^*)$, $L = \text{grad } v$ and $\frac{\mathcal{D} \cdot}{\mathcal{D} t} = \frac{d \cdot}{dt} - (W \cdot - \cdot W)$ as the generalized JAUMANN derivative.
Already for the l i n e a r constitutive equation T = D, that means for Newtonian flow, solving of (1) is connected with a lot of difficulties. In studies of non-Newtonian flow, following the n o n l i n e a r character as well as the often complicated structure of the constitutive equation $T = T(v)$, some additional problems have to be solved, so that so far procedures for computing non-Newtonian flow have been developed to a relatively low degree[1,2].

That´s why in this paper an approach to the mathematical treatment of such classes of non-Newtonian flow, differing only insignificantly in some sense from an easier basic flow, is presented. It can be understood as a generalization of the concept of "primary and secondary flow", successfully applied in the literature to the approximative computation of flow problems of second order fluids[3], and especially enables the senseful use of known solution techniques for the (often Newtonian) basic flow for approximately solving the more complicated problem.

PERTURBATION METHODS FOR NON-NEWTONIAN FLOW PROBLEMS WITH A SMALL PARAMETER IN THE CONSTITUTIVE EQUATION

The concept of primary and secondary flow for second order fluids with *explicit* constitutive equation

$$T = D + \varepsilon \cdot \frac{\mathfrak{D} D}{\mathfrak{D} t}, \quad |\varepsilon| << 1, \tag{2}$$

described by Böhme[3], is based on the assumption, that the non-Newtonian properties of the second order fluid, in (2) represented by the additive term $\varepsilon \cdot \frac{\mathfrak{D} D}{\mathfrak{D} t}$, cause only insignificant differences from the Newtonian basic ("primary") flow. By substituting the expression (2) for T in the general mathematical model (1), namely, we get the following model, interpretable as NAVIER-STOKES equations, perturbed by a small parameter ε,

$$\left.\begin{aligned} &Re \frac{dv}{dt} = -Re \cdot \operatorname{grad} p + \Delta v + \varepsilon \cdot DIV \frac{\mathfrak{D} D}{\mathfrak{D} t} + f, \\ &\operatorname{div} v = 0, \end{aligned}\right\} \tag{3}$$

and this suggests the application of classical perturbation methods. With the asymptotic expansions

$$v = \sum_{i=1}^{\infty} \varepsilon^{i-1} v_i, \quad p = \sum_{i=1}^{\infty} \varepsilon^{i-1} p_i \tag{4}$$

the complicated *nonlinear* problem (3) for the flow of a second order fluid can be split up into a *sequence* of easier, generally *linear* problems of the type of Newtonian basic flow

$$\left.\begin{aligned} &Re \frac{dv_1}{dt} = -Re \cdot \operatorname{grad} p_1 + \Delta v_1 + f \\ &\operatorname{div} v_1 = 0 \end{aligned}\right\} \tag{5.1}$$

$$\left.\begin{aligned} &Re \left(\frac{\partial v_i}{\partial t} + v_1 \operatorname{grad} v_i + v_i \cdot \operatorname{grad} v_1 \right) = \\ &\qquad = -Re \cdot \operatorname{grad} p_i + \Delta v_i + F_i(v_1 \ldots v_{i-1}) \\ &\operatorname{div} v_i = 0 \\ &i = 2, 3, \ldots \end{aligned}\right\} \tag{5.2}$$

which differ only in the right sides F_i.

The successive analytical or numerical solution of these equations (5) then offers a possibility of the approximative computation of second order fluid flow[2,3,4].

Especially for non-Newtonian flow with the practically important equation of state

$$T + \varepsilon \frac{\mathfrak{D} T}{\mathfrak{D} t} = D \tag{6}$$

the approach presented above cannot be used because of the *implicit* character of the constitutive equation. However, the following considerations demonstrate that a generalization of the classical perturbation methods is possible in such a manner that non-Newtonian flow problems with the constitutive equation

$$T = T_{\text{basic flow}} + \varepsilon \cdot \mathcal{N}(D, T), \; 0 < |\varepsilon| << 1, \qquad (7)$$

can be treated approximately.
Taking into consideration that the mathematical model (1) can be universally applied to *all* isothermal incompressible flow problems with friction influences, whereas the *concrete* class of flow phenomena is characterized directly only by the constitutive equation $T = T(v)$, so (1) represents in general manner a coupled system of partial differential equations with *three* unknown quantities v, p and T. According to the basic ideas of classical perturbation methods, the special expression (7) for T then leads to the asymptotic expansions

$$v = \sum_{i=1}^{\infty} \varepsilon^{i-1} v_i, \; p = \sum_{i=1}^{\infty} \varepsilon^{i-1} p_i, \; T = \sum_{i=1}^{\infty} \varepsilon^{i-1} T_i, \qquad (8)$$

that means also to a perturbation series for T.
Now, with (8) the following sequence of reduced perturbation equations can be derived:

$$\left.\begin{aligned} &Re \frac{dv_1}{dt} = -Re \cdot \operatorname{grad} p_1 + DIV\, T_1 + f \\ &\operatorname{div} v_1 = 0 \\ &T_1 = T_1(v_1) = T_{\text{basic flow}}(v_1) \end{aligned}\right\} \qquad (9.1)$$

$$\left.\begin{aligned} &Re \left(\frac{\partial v_i}{\partial t} + v_1 \operatorname{grad} v_i + v_i \operatorname{grad} v \right) = \\ &\quad = -Re \cdot \operatorname{grad} p_i + DIV\, T_i + \tilde{F}_i(v_1 \ldots v_{i-1}) \\ &\operatorname{div} v_i = 0 \\ &T_i = T_i(v_1 \ldots v_i, T_1 \ldots T_i) \\ &i = 2, 3, \ldots \end{aligned}\right\} \qquad (9.2)$$

With (9) we again obtained a sequence of equations from the type of the basic flow, the successive solution of which represents an approach to the approximative computation of the more complicated flow problem (1), (7). There, the original perturbation method derived for second order fluids can be integrated as a special case in the more general procedure (9). With the asymptotic expansions (8) for v and T, the constitutive equation (2) namely can be split up into

$$T_1 = D(v_1),$$

$$T_i = D(v_i) + \frac{\partial D(v_{i-1})}{\partial t} + \sum_{k=1}^{i-1} (v_k \cdot \mathrm{GRAD}\, D(v_{i-k}) -$$

$$- D(v_k) \cdot W(v_{i-k}) + W(v_k) \cdot D(v_{i-k})),$$

$$i = 2, 3, \ldots ,$$

and so the sequences of equations (9) and (5) are identical.

REFERENCES

1. Crochet, M. J., Walters, K.
   Numerical methods in non-newtonian fluid mechanics
   Ann. Rev. Fluid Mech. 15(1983) 241 - 260

2. Kreul, R., Kretzschmar, H.
   Numerische Behandlung nicht-NEWTONscher Strömungen auf der Basis von Störungsmethoden
   TH Karl-Marx-Stadt, Preprint 10(1986)

3. Böhme, G.
   Strömungsmechanik nicht-newtonscher Fluide
   B. G. Teubner Stuttgart 1981

4. Kreul, R.
   Numerische Modellierung von Sekundärströmungen
   Dipl.-Arbeit, TH Karl-Marx-Stadt 1985

ON DIFFUSION SEMIGROUPS GENERATED BY SEMI-ELLIPTIC DIFFERENTIAL OPERATORS IN INFINITE DIMENSIONS

Gottlieb Leha

University of Passau
Passau, FRG

1. INTRODUCTION

We want to study some continuity properties of operator semigroups, generated by a semi-elliptic differential operator on a real separable Hilbert space $\mathbb{H}$. To this end, let us begin by writing the finite-dimensional semi-elliptic differential operator

$$(1) \quad \mathcal{L}u(x) = \frac{1}{2} \sum_{i,j=1}^{n} a_{ij}(x) \frac{\partial^2 u}{\partial x_i \partial x_j}(x) + \sum_{i=1}^{n} b_i(x) \frac{\partial u}{\partial x_i}(x)$$

on $\mathbb{H} = \mathbb{R}^n$ in coordinate-free form as

$$(1a) \quad \mathcal{L}u(x) = \frac{1}{2} \operatorname{tr} u''(x)(a(x)\cdot,\cdot) + u'(x)(b(x)),$$

where $\operatorname{tr} u''(x)(a(x)\cdot,\cdot)$ is the trace of the bilinear form $y, z \longrightarrow u''(x)(a(x)y, z)$, and where $a(x) = (a_{ij}(x))_{1\leq i,j\leq n}$ is the matrix of the diffusion coefficients, $b(x) = (b_i(x))_{1\leq i\leq n} \in \mathbb{H}$ the vector of the drift coefficients of the operator $\mathcal{L}$. Semi-ellipticity means, that the operator $y \longrightarrow a(x)\cdot y$ on $\mathbb{H}$ is nonnegative definite. Throughout the sequel it will be assumed that the coefficients a and b satisfy local Lipschitz conditions (cf. section 2). In order to make the first term on the right hand side of (1a) well-definded in infinite dimensions, we assume

$$(1b) \quad a(x) = \sigma(x)Q\sigma^*(x) \quad , x \in \mathbb{H}$$

where $\sigma(x) \in L(\mathbb{H},\mathbb{H})$ is a bounded linear operator on $\mathbb{H}$ for all $x \in \mathbb{H}$, $\sigma^*(x)$ its adjoint, and Q a positive definite, self-adjoint, nuclear operator on $\mathbb{H}$.

The continuity properties of the generated semigroup (P_t) we are interested in are as follows:

Give a topological vector space C such that

(2a) P_t operates on C, i.e. $P_t f \in C \;\; \forall f \in C \;\; \forall t \geq 0$

(2b) $f \longrightarrow P_t f$ is continuous $\forall t \geq 0$

(2c) $t \longrightarrow P_t f$ is continuous $\forall f \in C$,

where continuity means with respect to the topology on C.

We give a few remarks:

If C is a space of continuous real functions on $\mathbb{H}$ then condition (2a) means in particular that $x \longrightarrow P_t f(x)$, $x \in \mathbb{H}$, is continuous, i.e. that (P_t) is a Feller semigroup (cf. Ichikawa, 1984, for the Hilbert space context).

If C is again a space of functions and its topology is uniform convergence then, of course, condition (2b) follows from the positivity of (P_t).

Condition (2c) is usually called the C_o-property of (P_t) and in case (2c) holds, (P_t) a C_o-semigroup on C.

Perhaps the most important conclusions to be drawn from the continuity properties concern the infinitesimal generator G of the semigroup (P_t), i.e.,

(3a) $$Gf = \lim_{t \downarrow 0} \frac{P_t f - f}{t}, \quad \text{lim w.r.t. topology of } C, \text{ where}$$

(3b) $f \in D(G) = \{ f \in C, \text{ the limit in (3a) exists} \}$.

One such (immediate) conclusion would, e.g., be

(4a) If $f \in D(G)$ then $P_t f \in D(G)$ and $P_t Gf = GP_t f \quad \forall t \geq 0$

and therefore

$$(4b) \qquad P_t f = f + \int_0^t G\, P_t f \, dr \quad \forall\, t \geq 0,\ \forall\, f \in D(G).$$

The space C to be found should be such that $\mathcal{L}$ can be defined as an operator $\mathcal{L} : D(\mathcal{L}) \subset C \longrightarrow C$, where $\mathcal{L} \subset G$ and $D(\mathcal{L})$ is rich enough in some sense (e.g., separating measures on $\mathbb{H}$).

In finite dimensions the space C may be most often chosen to be a Banach space, as, e.g., the function space $C_o(\mathbb{H})$ of all continuous functions on $\mathbb{H}$ vanishing at infinity, endowed with uniform convergence (cf. Azencott, 1974, Dynkin, 1965), or as some L^p-space (cf. Pazy, 1983).

Since in infinite dimensions a natural reference measure is lacking, we concentrate on the two function spaces

$$(5a) \qquad C_b(\mathbb{H}) = \{\, f : \mathbb{H} \dashrightarrow \mathbb{R},\ f \text{ bounded and uniformly continuous on bounded sets of } \mathbb{H} \,\}$$

and

$$(5b) \qquad C_o(\mathbb{H}) = \{\, f \in C_b(\mathbb{H}),\ f \text{ small outside large balls}\}.$$

As usual, the second space $C_o(\mathbb{H})$ will be considered as a Banach space w.r.t. uniform convergence. The space $C_b(\mathbb{H})$ will be endowed with the topology of locally (i.e., on bounded sets) uniform convergence.

The semigroup generated by $\mathcal{L}$ will be defined via mathematical expectation of diffusion processes on $\mathbb{H}$, which are obtained as solutions of suitable stochastic differential equations on $\mathbb{H}$.

2. SOME PREREQUISITS FROM STOCHASTIC DIFFERENTIAL EQUATIONS ON $\mathbb{H}$

For the reader's convenience we give a brief account of some results on solutions of stochastic differential equations on $\mathbb{H}$. We consider the equation

$$(6) \qquad \begin{cases} d\, \xi(x,t) = \sigma(\xi(x,t))\, d\, \beta(t) + b(\xi(x,t))\, d\, t, \\ \xi(x,0) = x\ , \end{cases}$$

where $x \in \mathbb{H}$, $t \geq 0$ and $\sigma : \mathbb{H} \to L(\mathbb{H}, \mathbb{H})$, $b : \mathbb{H} \to \mathbb{H}$ satisfy Lipschitz conditions on bounded sets of $\mathbb{H}$ and where $\beta(t)$, $t \geq 0$, is centered Brownian motion on $\mathbb{H}$ with covariance operator Q. For more details and proofs we refer to Leha-Ritter (1984).

The theorems 3.2 and 4.5 of Leha-Ritter (1984) say that

a) for each $x \in \mathbb{H}$ there is exactly one continuous maximal (strong) solution $\xi(x,t)$, $0 \leq t < T$, of the equation (6), where $T = T(x) = \sup_n \inf \{ t \geq 0 : \| \xi(x,t) \| > n \}$ is life-time of $\xi(x,\cdot)$.

b) the process $\xi(x,\cdot)$ has for every $x \in \mathbb{H}$ infinite life-time, i.e., $T_x = +\infty$ a.s., if the coefficients a (cf. (1b)) and b fulfill the following growth conditions

$$\text{(I)} \qquad \operatorname{tr} a(x) \leq K(1+\|x\|^2) \qquad \forall x \in \mathbb{H}$$

$$\text{(II)} \qquad \langle x, b(x) \rangle \leq K(1+\|x\|^2) \qquad \forall x \in \mathbb{H}$$

where $a(x)$ is as in (1b), $\langle \cdot,\cdot \rangle$ is inner procuct, $\|\cdot\|$ is norm of $\mathbb{H}$, and K stands (also in the sequel) for a generic constant.

Under conditions (I) and (II) the semigroup (P_t) generated by $\mathcal{L}$ is then defined by

$$\text{(7)} \qquad P_t f(x) = E[f(\xi(x,t))], \; t \geq 0, \; f \in C_b(\mathbb{H}), \; x \in \mathbb{H}.$$

We remark that condition (II) is one-sided. It controls the outward drift which must not be too large if the process hasn't to go to infinity too fast.

3. THE CASE $C = C_o(\mathbb{H})$

Here we recall the following theorem from Leha-Ritter (1984), theorem 6.8:

Theorem 1: Suppose that the coefficients a and b satisfy in addition to the condition (I) from above the conditions

$$\text{(III)} \qquad \langle a(x)x, x \rangle \leq K\,(1+\|x\|^2) \qquad \forall x \in \mathbb{H}$$

$$\text{(IV)} \qquad |\langle x, b(x) \rangle| \leq K\,(1+\|x\|^2) \qquad \forall x \in \mathbb{H}.$$

Then the semigroup $(P_t)_{t\geq 0}$ operates on $C_o(\mathbb{H})$ and the properties (2a), (2b), (2c) hold for $C = C_o(\mathbb{H})$. If the operator $\mathcal{L}$ on $C = C_o(\mathbb{H})$ is defined by (1a,b) and

(1c) $D(\mathcal{L}) = \{ u \in C$: u twice continuously differentiable, $\mathcal{L}u \in C$, $u'': \mathbb{H} \to B(\mathbb{H},\mathbb{H})$ uniformly continuous $\}$, ($B(\mathbb{H},\mathbb{H})$ denotes the bilinear forms on $\mathbb{H}$)

then $\mathcal{L} \subset G$.

Note that conditions (I) and (II) give bounds on the size of the diffusion part, whereas condition (IV) (which includes (II)) controls the size of the drift towards infinity and towards the origin. These conditions are also heuristically convincing if P_t is to map $C_o(\mathbb{H})$ into itself.

4. THE CASE $C = C_b(\mathbb{H})$

If one wants to include examples where invariant measures for the semigroup exist, then often recurrence type behaviour of the process is necessary. In many cases such behaviour is provided by a strong drift towards the origin, which is excluded by condition (IV). Thus, we don't further claim this condition.

First we recall from Leha-Ritter (1984), prop. (5.5), the

Proposition 1: Suppose that a and b satisfy condition (I) and (II) from above, then the semigroup (P_t) operates on $C = C_b(\mathbb{H})$ (cf.(2a)).

Note that under conditions (I) and (II) we cannot expect C_o-property of (P_t), if we consider uniform convergence on $C_b(\mathbb{H})$. This is already shown by the simple example $\mathbb{H} = \mathbb{R}$, $\sigma \equiv 0$, $b(x) = -x^3$.

However, considering locally uniform convergence (i.e., on bounded sets of $\mathbb{H}$) in $C_b(\mathbb{H})$ we have the

Proposition 2: Suppose that a and b satisfy conditions (I) and (II), then property (2c) holds for $C = C_b(\mathbb{H})$, i.e., $t \to P_t f(x)$ is continuous for every $f \in C_b(\mathbb{H})$ and $x \in \mathbb{H}$, uniformly on bounded sets of $\mathbb{H}$.

Proof: Let $f \in C_b(\mathbb{H})$, $r > 0$, $\epsilon > 0$ be given. We show, that there exists $T > 0$ such that $| P_t f(x) - f(x) | < \epsilon$ for all $x \in \mathbb{H}$, $\| x \| \leq r$, and for all $t < T$.

a) Since σ and b satisfy local Lipschitz conditions, they are bounded on $B = B(r) = \{ x \in \mathbb{H}, \| x \| \leq r \}$. The argument in the first part of the proof of proposition (5.7) in Leha-Ritter (1984) shows that $E[\| \xi(x,t) - x \|^2] \leq K_1 t + K_2 t^2$ if $x \in B$. Using Tschebyschev's inequality we deduce for $\delta > 0$ and $C = C(x,t,\delta) = \{\|\xi(x,t)-x\|>\delta\}$:

$$P(C) \leq \frac{1}{\delta^2}(K_1 t + K_2 t^2), \quad \text{if } x \in B.$$

Here $P(C)$ is the probability of the event C.

b) On account of conditions (I) and (II) we infer from Leha-Ritter (1984), remark 4.6, that

$$E[\| \xi(x,t) \|^2] \leq g(x,t) = (\| x \|^2 + 1)e^{Kt} - 1.$$

Again we deduce by Tschebyschev's inequality for

$R > 0$ and $A = A(x,t,R) = \{ \| \xi(x,t) \| > R \}$:

$$P(A) \leq \frac{1}{R^2} K_3 \quad \text{if } x \in B,\ t \leq 1.$$

c) We now choose

$R \geq r$ large such that

$$2 \|f\|_\infty \frac{K_3}{R^2} < \epsilon/_3, \text{ where } \|f\|_\infty = \sup\{|f(x)|: x \in \mathbb{H}\},$$

and $\delta > 0$ small such that

$$|f(y)-f(z)| < \epsilon/_3 \quad \text{if } \|y\|, \|z\| \leq R,\ \|y-z\| < \delta,$$

and $0 < T \leq 1$ small such that

$$2 \|f\|_\infty \frac{1}{\delta^2}(K_1 t + K_2 t^2) < \epsilon/_3 \quad \text{if } 0 \leq t < T.$$

Then we have for $x \in B$, $0 \leq t < T$:

$$|P_t f(x)-f(x)| \leq E[|f(\xi(x,t) - f(x)|: A^c \cap C^c] + E[|\dots| : A] + E[|\dots| : C] < \epsilon/_3 + 2\|f\|_\infty P(A) + 2\|f\|_\infty P(C) < \epsilon.$$

We remark that in general we cannot expect property (2b) for (P_t) acting on $C_b(\mathbb{H})$, since in general $P(\{\|\xi(x,t)\|>r\}) > 0$ for all $x \in \mathbb{H}$, $r > 0$. Therefore, defining the infinitesimal generator of (P_t) as in (3a,b) for $C = C_b(\mathbb{H})$, we cannot immediately deduce conclusion (4a,b). Slightly modifying (3b) we define

$$\text{(3c)} \quad \tilde{D}(G) = \{f \in D(G) : \exists\, M = M(f) > 0 \text{ s.t. } |P_t f-f| \leq M\, t \text{ for small } t > 0\}.$$

Then we have

<u>Theorem 2</u>: Suppose that a and b satisfy conditions (I) and (II). Then the semigroup (P_t) operates on $C_b(\mathbb{H})$ and properties (2a), (2c) hold for $C = C_b(\mathbb{H})$.

Furthermore:

If $f \in \tilde{D}(G)$ then $P_t f \in \tilde{D}(G)$ and $P_t Gf = GP_t f \quad \forall\, t \geq 0$.

If the operator $\mathcal{L}$ is defined by (1a,b,c) for $C = C_b(\mathbb{H})$ then $\mathcal{L} \subset G$ and $D(\mathcal{L}) \subset \tilde{D}(G)$.

<u>Proof</u>: The first part comes from propositions 1 and 2. The second part is easily deduced from

(2b') $f \to P_t f$ is continuous on $C_b(\mathbb{H}) \cap \{ \|f\|_\infty \leq M \}$ for all $M > 0$.

The last part is obtained from the following characterization:

If $f \in C_b(\mathbb{H})$, then $f \in \tilde{D}(G)$ if and only if there exists $g \in C_b(\mathbb{H})$, in fact $g = Gf$, such that

$$f(\xi(x,t)) - f(x) - \int_0^t g(\xi(x,r))dr$$

is an L^2-martingale. For $f \in D(\mathcal{L})$ we may choose $g = \mathcal{L}f$ by Ito's formula. (For a similar result cf. Leha, 1984).

REFERENCES

Azencott, R. (1974), Behavior of diffusion semi-groups at infinity, Bull. Soc. Math. France 102, 193-240.

Dynkin, E. B. (1965), "Markov Processes", Vol. I, Springer, Berlin.

Ichikawa, A. (1984), Semilinear Stochastic Evolution Equations : Boundedness, Stability and Invariant measures, Stochastics 12, 1-39.

Leha, G. and Ritter, G. (1984), On Diffusion Processes and their Semigroups in Hilbert Spaces with an Application to Interacting Stochastic Systems, Ann. of Probab. 12 (4), 1077-1112.

Leha, G. (1984), Invarianz und Reversibilität für Diffusionsprozesse in Hilberträumen, Habilitationsschrift, Universität Erlangen.

Pazy, A. (1983), "Semigroups of Linear Operators and Applications to Partial Differential Equations", Appl. Math. Sc. Vol. 44, Springer, Berlin.

THE OBSTACLE PROBLEM IN A NON-LINEAR POTENTIAL THEORY

P. Lehtola

Department of Mathematics
University of Jyväskylä
SF-40100 Jyväskylä, Finland

INTRODUCTION

M. Brelot gave rise to the concept harmonic space when he extended classical potential theory on $\mathbb{R}^n$ to an axiomatic system on a locally compact space. I have recently constructed[1] a non-linear harmonic space by dropping the assumption that the sum of two harmonic functions is harmonic and considering some other axioms instead. This approach has its origin in the work of O. Martio, P. Lindqvist and S. Granlund[2,3,4], who have developed a non-linear potential theory on $\mathbb{R}^n$ connected with variational integrals of the type $\int F(x, \nabla u(x))\, dm(x)$, where $F(x, h) \approx |h|^p$.

In this lecture I will show how suitably this axiomatic system fits for the variational approach to potential theory, see also J. Heinonen and T. Kilpeläinen[5,6,7].

THE AXIOMS

Let X be a locally compact, locally connected, connected and non-compact Hausdorff space and let $\mathcal{H}$ be a sheaf of continuous real-valued functions on X. A function in $\mathcal{H}(U)$, $U \subset X$ open, is called an $\mathcal{H}$-function on U. A relatively compact open set G is said to be regular if

(i) for every $f \in C(\partial G)$ there exists a unique $H_f^G \in \mathcal{H}(G) \cap C(\bar{G})$ such that $H_f^G = f$ on ∂G, and

(ii) for every $f, g \in C(\partial G)$ the condition $f \leq g$ on ∂G implies that $H_f^G \leq H_g^G$ on G.

A component of a regular set is again regular. The following properties are taken as axioms:

A1. The sheaf $\mathcal{H}$ obeys Harnack's principle (i.e. if U is a domain and (h_n) is an increasing sequence in $\mathcal{H}(U)$, then either $\lim_{n\to\infty} h_n \in \mathcal{H}(U)$ or $\lim_{n\to\infty} h_n \equiv \infty$ on U).

A2. If $h \in \mathcal{H}(U)$ and $\lambda \in \mathbb{R}$, then λh and $h + \lambda \in \mathcal{H}(U)$, whenever U is an open set.

A3. For every open set $U \subset X$ and for every compact set $K \subset U$ there is a regular set G with $K \subset G \subset \overline{G} \subset U$.

Harnack's principle extends for upper directed families of $\mathcal{H}$-functions, by topological arguments. The axiom A3 implies that the regular domains form a base of X.

HYPER-$\mathcal{H}$-FUNCTIONS

Let U be an open set. A lower semi-continuous function $u: U \to \mathbb{R} \cup \{\infty\}$ is a <u>hyper-$\mathcal{H}$-function</u> on U, denoted by $u \in \mathcal{U}(U)$, if u satisfies the comparison condition: for every regular set $G \subset \overline{G} \subset U$ and for every $f \in C(\partial G)$ the condition $f \leq u$ on ∂G implies that $H_f^G \leq u$ on G. If u is not indentically ∞ in each component of U then u is called a <u>super-$\mathcal{H}$-function</u> on U, $u \in \mathcal{S}(U)$. Note that $\mathcal{U}$ need not be a sheaf.

<u>Proposition 1</u> - Let U be a relatively compact open set and $u \in \mathcal{U}(U)$, $v \in -\mathcal{U}(U)$. If

$$\varliminf_{x\to y} u(x) \geq \varlimsup_{x\to y} v(x)$$

for all $y \in \partial U$, then $u \geq v$ on U.

<u>Proof</u> - The set $K = \{x \in U : u(x) + \varepsilon \leq v(x)\}$ is compact, for every $\varepsilon > 0$. By A3, there is a regular set G with $K \subset G \subset \overline{G} \subset U$. Choose $f \in C(\partial G)$ such that $u + \varepsilon > f > v$ on ∂G. Then $u + \varepsilon \geq H_f^G \geq v$ on G and hence $u \geq v$ on U.

<u>Proposition 2</u> - A super-$\mathcal{H}$-function u on a domain U is finite on a dense set.

Proof – The proof is elementary and tedious and makes use of A1 and A2.

THE DIRICHLET PROBLEM

The Dirichlet problem in the non-linear case is solved by Perron's method, i.e. by using super-H-functions and their Poisson modifications.

Let U be a relatively compact open set and $f: \partial U \to \mathbb{R} \cup \{-\infty, \infty\}$ be a function. Let $\overline{H}_f$ be the infimum of the lower bounded hyper-H-functions u on U with

$$\underline{\lim}_{x\to y} u(x) \geq f(y)$$

for all $y \in \partial U$ and let $\underline{H}_f = -\overline{H}_{-f}$. Then we have

Theorem 1 – $\underline{H}_f \leq \overline{H}_f$, and

Theorem 2 – If U is a domain, then either $\overline{H}_f \in H(U)$ or $\overline{H}_f \equiv \infty$ or $\overline{H}_f \equiv -\infty$. The same alternatives hold for $\underline{H}_f$.

Proofs – Theorem 1 follows straightly from Proposition 1. M. Brelot[8] has proven Theorem 2 in a form which is applicable in the non-linear case.

THE OBSTACLE PROBLEM

We shall apply the above axiomatic system to solve an obstacle problem. Let G be a regular set and $f \in C(\overline{G})$. The problem is to find $u \in S(G) \cap C(\overline{G})$ such that

(i) $u \geq f$ on G

(ii) $u = f$ on ∂G

(iii) $u \in H(\{x \in G : u(x) > f(x)\})$.

To solve the problem we assume that ∂G contains at least two points and study the balayage function. Write

$$R_f = \inf \{v \in S(G) : v \geq f \text{ on } G\}$$

and

$$\hat{R}_f = \sup \{\phi \text{ l.s.c. on } G : \phi \leq R_f\}$$

Proposition 3 – $R_f = \hat{R}_f \in \mathcal{S}(G) \cap \mathcal{C}(G)$.

Proof – The proof follows the classical one.

Lemma 1 – Let $g \in \mathcal{C}(\bar{G})$, $y \in \partial G$ and let $K \subset G$ be a compact set. Then there is $h \in \mathcal{H}(G) \cap \mathcal{C}(\bar{G})$ such that $h \geq g$ on K , $h \geq g$ on ∂G and $h(y) = g(y)$.

Proof – Since ∂G contains two points, we may construct an increasing sequence of functions $h_n \in \mathcal{H}(G) \cap \mathcal{C}(\bar{G})$ with $h_n(y) = g(y)$ such that $h_n \nearrow \infty$ on G . This yields the claim.

Theorem 3 – $R_f \in \mathcal{C}(\bar{G})$ and $= f$ on ∂G .

Proof – It is enough to show that for every $y \in \partial G$ and $\varepsilon > 0$ there is $v \in \mathcal{S}(G)$ such that $v \geq f$ on G and that

$$\overline{\lim_{x \to y}}\, v(x) \leq f(y) + \varepsilon .$$

Let $K = \{x \in G : H_{f+\varepsilon}(x) \leq f(x)\}$ and let $h \in \mathcal{H}(G) \cap \mathcal{C}(\bar{G})$ be such that $h \geq f$ on K , $h \geq f + \varepsilon$ on ∂G and $h(y) = f(y) + \varepsilon$. Then $h \geq H_{f+\varepsilon}$ on G and so $h \geq f$ on G . Now $v = h$ is the required function.

Theorem 4 – $R_f \in \mathcal{H}(A)$, where $A = \{x \in G : R_f(x) > f(x)\}$.

Proof – The proof makes use of the Poisson modification of R_f in an obvious manner.

Remark – The solution is unique. In fact, let G be a regular set with more than one boundary point and let $f \in \mathcal{C}(\bar{G})$. If $u \in \mathcal{S}(G) \cap \mathcal{C}(\bar{G})$ satisfies the conditions (i) - (iii) then $u = R_f$. Moreover,

$$u = \inf \{v \in \mathcal{S}(G) \cap \mathcal{C}(\bar{G}) : v \geq F \text{ on } G ,\ v = F \text{ on } \partial G\} .$$

REFERENCES

1. P. Lehtola, An axiomatic approach to non-linear potential theory, Ann. Acad. Sci. Fenn. Ser. A I Math. Diss 62 (1986), 1-40.

2. S. Granlund, P. Lindqvist and O. Martio, Conformally invariant variational integrals, Trans. Amer. Math. Soc. 277 (1983), 43-73.

3. -"- Note on the PWB-method in the non-linear case, Pacific J. Math. 125 (1986), 381-395

4. P. Lindqvist and O. Martio, Two theorems of N. Wiener for solutions of quasilinear elliptic equations, Acta Math. 155 (1985), 153-171.

5. J. Heinonen and T. Kilpeläinen, A-superharmonic functions and supersolutions of degenerate elliptic equations, Ark. Mat. (to appear).

6. -"- Polar sets for supersolutions of degenerate elliptic equations (to appear).

7. -"- On the Wiener criterion and quasilinear obstacle problems, Preprint, Univ. Bonn (1987).

8. M. Brelot, Familles de Perron et problème de Dirichlet, Acta Sci. Math. (Szeged) 9 (1939), 133-153.

ON THE APPELL TRANSFORMATION

Heinz Leutwiler

Mathematisches Institut der Universität
Erlangen - Nürnberg
Bismarckstr. 1 1/2
D-8520 Erlangen, West Germany

INTRODUCTION

The main purpose of this paper is to show that the well known Appell transformation [1] is essentially - i.e. up to compositions with the obvious transformations of the form (1.2) mentioned below - the only transformation, mapping one solution of the n-dimensional heat equation $\Delta u = \frac{\partial u}{\partial t}$ into another (see Corollary 2.2 for a precise statement).

A corresponding result for the Laplace equation in $\mathbb{R}^3$ is mentioned in Kellogg's book [3] on page 235. It roughly says that - up to compositions with similarities - the Kelvin transformation is the only one which maps every harmonic function into another. For a proof of this result, which holds in an arbitrary n-dimensional space $\mathbb{R}^n$ $(n \geqq 3)$, as well as for its generalizations to Riemannian manifolds, we refer to the interesting article of B. Fuglege [2].

The method used to prove our main results may also be applied to other parabolic linear partial differential equations of second order, in order to find analogues of the classical Appell transformation.

1. THE APPELL TRANSFORMATION

We shall be concerned with the heat equation

$$Lu := \Delta u - \frac{\partial u}{\partial t} = o \tag{1.1}$$

in $\mathbb{R}^{n+1} = \mathbb{R}^n \times \mathbb{R}$ for $n = 1,2,\ldots$. The points of $\mathbb{R}^{n+1}$ will be denoted correspondingly by $(x,t) = (x_1,\ldots,x_n,t)$, where $x = (x_1,\ldots,x_n) \in \mathbb{R}^n$ and $t \in \mathbb{R}$. In particular, the Laplace operator refers to the first n coordinates:

$$\Delta = \sum_1^n \frac{\partial^2}{\partial x_i^2} \quad .$$

The scalar product of two vectors $x=(x_1,\ldots,x_n)$, resp. $y=(y_1,\ldots,y_n)$ of $\mathbb{R}^n$ will be denoted by $(x,y)=x_1y_1+\ldots+x_ny_n$, and the norm of x by $\|x\| = \sqrt{(x,x)}$. Solutions of (1.1), more precisely functions $u : U \to \mathbb{R}$ defined on an open subset U of $\mathbb{R}^{n+1}$ which are of class C^2 with respect to the space variable x and of class C^1 with respect to the time variable t and which satisfy (1.1), will be called <u>caloric functions</u> or <u>temperatures</u>.

Obviously, with every caloric function u, the function

$$(1.2) \qquad (Su)(x,t) := u(\lambda Rx + c, \lambda^2 t + \mu),$$

where $\lambda,\mu \in \mathbb{R}$, $\lambda > 0$, $R \in O(n)$ (the orthogonal group) and $c \in \mathbb{R}^n$, is again a caloric function. Furthermore, as observed by P. Appell [1] the function

$$(1.3) \qquad (Au)(x,t) := |t|^{-n/2} \exp(-\|x\|^2/4t)u(x/t,-1/t) \ ,$$

defined on $\mathbb{R}^{n+1}$ outside the hyperplane $t = 0$ is caloric, provided the function u is caloric on the same set.

With A and S at our disposal, we may of course consider compositions of A and S in order to get further examples of transformations, mapping caloric functions into caloric functions. All these transformations are of the form

$$(1.4) \qquad (Tu)(x,t) := \varphi(x,t)u(\phi(x,t)),$$

where ϕ maps some open set $U \subset \mathbb{R}^{n+1}$ bijectively onto some open set $V \subset \mathbb{R}^{n+1}$, φ is a (strictly) positive function on U, and u is a caloric function defined on V.

It is therefore natural to ask, if besides the examples mentioned so far, there are other transformations of type (1.4) associating to every caloric function u on V a caloric function Tu on U. In the next section we shall prove that this is not the case.

2. THE MAIN RESULT

Given the open sets U, V in $\mathbb{R}^{n+1}$, the bijective mapping $\phi : U \to V$ and the positive function $\varphi : U \to \mathbb{R}$, we shall say

DEFINITION – The transformation T, defined by (1.4), is a <u>caloric morphism</u>, provided the function Tu is caloric on U whenever u is caloric on V. (It would perhaps be better to call such a transformation T a caloric φ-morphism, but we shall not do so).

Examples of caloric morphisms are furnished by the transformations A and S, as well as their compositions. This, however, is already the whole group of caloric morphisms, as the following theorem affirms.

THEOREM 2.1 – <u>Let</u> $\alpha, \beta, \gamma, \delta, \kappa$ <u>be real numbers with</u> $\alpha\delta - \beta\gamma = 1, \kappa > o$, c,d <u>be arbitrary vectors in</u> $\mathbb{R}^n$ <u>and</u> $R \in O(n)$ <u>denote an arbitrary orthogonal matrix. Then</u>

$$(2.1) \qquad (Tu)(x,t) = \varphi(x,t)\, u\left(\frac{Rx + ct + d}{\gamma t + \delta}, \frac{\alpha t + \beta}{\gamma t + \delta}\right),$$

<u>where</u>

$$\varphi(x,t) = \begin{cases} \kappa|\gamma t+\delta|^{-n/2} \exp\left[-\dfrac{\|\gamma Rx+\gamma d-\delta c\|^2}{4\gamma(\gamma t+\delta)}\right], & \gamma \neq o \\ \kappa \cdot \exp\left[\dfrac{1}{4}\|c\|^2 t + \dfrac{1}{2}(c,Rx)\right], & \gamma = o, \end{cases}$$

<u>defines a caloric morphism on</u> $\mathbb{R}^{n+1}$ <u>outside the hyperplane</u> $\gamma t + \delta = o$. <u>Conversely, every caloric morphism is representable in this way</u>.

More precisely, the theorem says, that if the caloric morphism is defined on the open set $U \subset \mathbb{R}^{n+1}$, then there is a transformation of the form (2.1) whose domain of definition contains the set U and whose restriction to U represents the given morphism.

Observing that every transformation of the form (2.1) is representable as a composition of transformations of type S and A, defined in (1.2) and (1.3), we have

COROLLARY 2.2. <u>Every caloric morphism</u> T <u>is a composition of morphisms of type</u> A <u>and</u> S.

The proof of Theorem 2.1, given in the next section, may also be applied to some other parabolic linear partial differential equaltions of second order.

Choosing e.g., on $\mathbb{R}_+^4 \times \mathbb{R}$, where $\mathbb{R}_+^4$ denotes the upper half space of $\mathbb{R}^4$, the differential operator

$$(2.2) \qquad \tilde{L}u := \Delta u - \frac{2}{x_4}\frac{\partial u}{\partial x_4} - \frac{\partial u}{\partial t} ,$$

one finds that the transformation B defined by

$$(2.3) \qquad (Bu)(x,t) = \frac{1}{t}\exp(-\|x\|^2/4t)\; u(x/t,-1/t)$$

plays the same role for the solutions of the equation $\tilde{L}u = o$ as the Appell transformation does for the caloric functions.

3. PROOF OF THE MAIN THEOREM

It is based on the observation that the following polynomials are caloric on $\mathbb{R}^{n+1} = \{(x,t),\ x \in \mathbb{R}^n,\ t \in \mathbb{R}\}$:

$$(3.1) \qquad 1,\ x_i,\ x_i x_j \qquad (i \neq j) ,$$

$$(3.2) \qquad x_i^2 + 2t,$$

$$(3.3) \qquad x_i^3 + 6x_i t , \qquad\qquad (i,j = 1,\dots,n)$$

$$(3.4) \qquad x_i^4 + 12\, x_i^2 t + 12\, t^2 .$$

By hypothesis the transformation T of (1.4) maps solutions of $Lu = o$ into solutions. Denoting the mapping ϕ by $\phi = (f,\tau)$, where $f = (f_1,\dots,f_n)$ and τ are defined on U, it follows from (1.4) and (3.1) that φ and $\varphi \cdot f_i$ $(i = 1,\dots,n)$ must be caloric on U. On the other hand, for arbitrary functions u,v of class $C^2(U)$, we have

$$(3.5) \qquad L(uv) = u\,Lv + 2(\mathrm{grad}\ u,\ \mathrm{grad}\ v) + v\,Lu,$$

where $\mathrm{grad}\ u = \left(\frac{\partial u}{\partial x_1}, \dots, \frac{\partial u}{\partial x_n}\right)$ only refers to the space coordinates. Consequently, $L\varphi = o$ and $L(\varphi f_i) = o$ imply that

$$(3.6) \qquad \varphi L f_i + 2(\mathrm{grad}\ \varphi,\ \mathrm{grad}\ f_i) = o \qquad (i = 1,\dots,n).$$

From (3.2) we infer that $L[\varphi(f_i^2 + 2\tau)] = o$ and hence, by (3.5)

$$(3.7) \qquad \varphi L(f_i^2 + 2\tau) + 2(\mathrm{grad}\ \varphi,\ \mathrm{grad}(f_i^2 + 2\tau)) = o.$$

Using (3.5) again, we then conclude that, on account of (3.6), we must also have

$$(3.8) \qquad \varphi(\|\text{grad } f_i\|^2 + L\tau) + 2(\text{grad } \varphi, \text{ grad } \tau) = o.$$

Applied to $L[\varphi(f_i^3 + 6\, f_i\tau)] = o$, a consequence of (3.3), we infer from (3.5) and the fact that for arbitrary C^2 functions u

$$L(u^3) = 3\, u^2\, Lu + 6u \|\text{grad } u\|^2 ,$$

the relation

$$(3.9) \qquad \varphi(\text{grad } f_i, \text{ grad } \tau) = o.$$

We thereby made use of (3.6) and (3.8). In the same vain we finally conclude from

$$L[\varphi(f_i^4 + 12\, f_i^2\tau + 12\, \tau^2)] = o,$$

in conjunction with the relation

$$L(u^4) = 4\, u^3 Lu + 12\, u^2 \|\text{ grad } u\|^2 ,$$

valid for arbitrary C^2-functions u, that

$$(3.1o) \qquad \varphi\|\text{ grad } \tau\|^2 = o.$$

Since by hypothesis $\varphi > o$, this implies that $\text{grad}\,\tau = o$ and hence

$$(3.11) \qquad \tau = \tau(t).$$

Inserting this result in (3.8) we then obtain

$$(3.12) \qquad \frac{d\tau}{dt} = \|\text{grad } f_i\|^2 \qquad\qquad (i = 1,\ldots,n)$$

and, in particular,

$$(3.13) \qquad \|\text{ grad } f_i\| = \|\text{ grad } f_j\| \qquad\qquad (i,j = 1,\ldots,n).$$

We now claim that besides this relation we also have

$$(3.14) \qquad (\text{grad } f_i, \text{ grad } f_j) = o \qquad\qquad (i \neq j).$$

Indeed, by (3.1), $L[\varphi(f_if_j)] = o$ and hence by (3.5)

$$\varphi L(f_if_j) + 2(\text{grad } \varphi, \text{ grad}(f_if_j)) = o.$$

Applying (3.5) again, and making use of (3.6) twice, we find that (3.14) holds.
Equations (3.12) and (3.14) imply that, for each fixed $t \in \mathbb{R}$, the mapping $f = (f_1, \dots, f_n)$ is a similarity. This is easy if $n=2$. If $n \geq 3$ then according to a theorem of Liouville (3.13) and (3.14) imply that f is a Möbiustransformation and thus by (3.12) a similarity. Hence the mapping ϕ must have the form

$$(3.15) \qquad \phi(x,t) = (\lambda(t)R(t)x + k(t), \tau(t)),$$

where the positive scalar λ, the vector $k \in \mathbb{R}^n$ and possibly also the orthogonal matrix $R \in O(n)$ depend on t.

We shall verify next that in fact R is independent of t. Obviously, by (3.6) the components f_i of $f(x,t) = \lambda(t)R(t)x + k(t)$ satisfy the equation

$$(3.16) \qquad \varphi \frac{\partial f_i}{\partial t} = 2(\text{grad } \varphi, \text{ grad } f_i).$$

Introducing the function $\psi := \log \varphi$ we therefore have

$$(3.17) \qquad \frac{\partial f_i}{\partial t} = 2(\text{grad } f_i, \text{ grad } \psi) = 2\lambda(R \text{ grad } \psi)_i \quad (i=1,\dots,n).$$

Denoting the vector with the components $(\partial/\partial t)(f_i)$ by $\dot{f}$ and the transpose of the matrix R by R^t we conclude that

$$(3.18) \qquad \text{grad } \psi = \frac{1}{2\lambda} R^t \dot{f} = \frac{1}{2\lambda} R^t (\dot{\lambda}Rx + \lambda\dot{R}x + \dot{k})$$

$$= \frac{1}{2} (\frac{\dot{\lambda}}{\lambda} x + R^t \dot{R}x + \frac{1}{\lambda} R^t \dot{k}) ,$$

where we used the dot to denote the derivative with respect to t. Inserting (3.18) on both sides of

$$\frac{\partial}{\partial x_i} \left(\frac{\partial \psi}{\partial x_j}\right) = \frac{\partial}{\partial x_j} \left(\frac{\partial \psi}{\partial x_i}\right)$$

we find that $R^t\dot{R} = (\dot{R})^tR$. On the other hand $R^tR = \text{id}$ implies $(\dot{R})^tR + R^t\dot{R} = 0$, so that $\dot{R} = 0$.

Now that the orthogonal matrix R is independent of t we may assume w.l.o.g. that it is the identity matrix, i.e.

$$(3.19) \qquad \phi(x,t) = (\lambda(t)x + k(t), \tau(t)).$$

Otherwise simply replace ϕ by $\tilde{\phi}(x,t) := \phi(R^{-1}x,t)$.
From (3.16) we then get

$$(3.20)\qquad \varphi(\dot{\lambda}(t)x + \dot{k}(t)) = 2\lambda \operatorname{grad} \varphi.$$

Consequently

$$(3.21)\qquad \begin{aligned} 2\lambda \operatorname{div}(\operatorname{grad} \varphi) &= \operatorname{div}[\varphi(\dot{\lambda}x + \dot{k})] = \\ &= (\operatorname{grad} \varphi, (\dot{\lambda}x + \dot{k})) + n\dot{\lambda}\varphi \end{aligned},$$

and hence in view of (3.20)

$$(3.22)\qquad 4\lambda^2 \Delta\varphi = \varphi(\| \dot{\lambda}x + \dot{k} \|^2 + 2n\lambda\dot{\lambda}).$$

On the other hand, integration of (3.20) yields

$$(3.23)\qquad \log \varphi = \frac{\dot{\lambda}}{4\lambda} \| x \|^2 + \frac{1}{2\lambda} (x, \dot{k}) + \eta,$$

where the constant $\eta \in \mathbb{R}$ depends on t.
Differentiating now with respect to t, we obtain

$$(3.24)\qquad \frac{\partial\varphi}{\partial t} = \varphi\left[\frac{1}{4}\left(\frac{\dot{\lambda}}{\lambda}\right)^{\bullet} \| x \|^2 + \frac{1}{2}\left(x, \left(\frac{\dot{k}}{\lambda}\right)^{\bullet}\right) + \dot{\eta}\right].$$

But φ, being a coloric function, satisfies (1.1). Thus (3.22) and (3.24) imply that

$$(3.25)\qquad p\| x \|^2 + 2(q,x) + r = 0,$$

where

$$p := 2(\dot{\lambda})^2 - \lambda\ddot{\lambda}, \quad q := 2\dot{\lambda}\dot{k} - \lambda\ddot{k}, \quad r := \|\dot{k}\|^2 + 2n\lambda\dot{\lambda} - 4\lambda^2\dot{\eta}.$$

Since this equation holds for arbitrary $(x,t) \in U$, we conclude that $p = 0$, $q = 0$, and $r = 0$.
Integrating $p = 0$ yields

$$(3.26)\qquad \lambda(t) = \frac{1}{\gamma t + \delta},$$

γ, δ being arbitrary real numbers (not both equal to 0).
Subsequently $q = 0$ results in

$$(3.27)\qquad k = \frac{ct + d}{\gamma t + \delta},$$

where c and d are arbitrary vectors in $\mathbb{R}^n$.
In order to determine τ, we observe that on account of (3.12) $\dot{\tau} = \lambda^2$, wherefore

$$(3.28)\qquad \tau = \frac{\alpha t + \beta}{\gamma t + \delta},$$

with arbitrary $\alpha, \beta \in \mathbb{R}$ satisfying the condition $\alpha\delta - \beta\gamma = 1$.

The integration of $r = o$ must be done for $\gamma \neq o$ and $\gamma = o$ separately. Whereas in the first case

$$(3.29) \qquad \eta = -\frac{1}{4}\frac{\|\delta c - \gamma d\|^2}{\gamma(\gamma t + \delta)} - \frac{n}{2}\log|\gamma t + \delta| + \text{const},$$

we find in the second one

$$(3.3) \qquad \eta = \frac{1}{4}\|c\|^2 t + \text{const}.$$

Inserting in (3.23) we end up with (2.1), completing the proof of Theorem 2.1.

Finally let us remark that the proof only depends on the requirement that the polynomials (3.1) - (3.4) are mapped into caloric functions by T.

REFERENCES

1 Appell , P. Sur l'équation $\partial^2 z/\partial x^2 - \partial z/\partial y = o$ et la théorie du chaleur. J. Math. Pures. Appl. (4) 8(1892), 187 - 216.

2 Fuglede, B. Harmonic morphisms between Riemannian manifolds. Ann. Inst. Fourier 28, 2 (1978), 1o7-144.

3 Kellogg, O.D. Foundations of potential theory. Berlin, Springer 1929 (re-issued 1967).

PERTURBATION AND EXCESSIVE FUNCTIONS

Habib Maagli

Departement de Mathematiques
Faculte des Sciences de Tunis
Campus Universitaire 1060 Tunis - Tunisia

INTRODUCTION

Originally, N.Boboc and Gh.Bucur [2] studied perturbations of a resolvent or a semigroup on a measurable space, by bounded kernels. They gave a characterization of the perturbed excessive structures.

The aim of this paper is to give a similar characterization when perturbation is done by a measurable "negative" function. More precisely : we consider a resolvent $\mathbb{V} = (V_r)_{r \geq 0}$ on a measurable space $(X, \mathbb{B})$ possessing a proper kernel V_o. Let $\mathbb{V}_h = (V_{h+r})_{r \geq o}$ be the perturbed resolvent of $\mathbb{V}$ by a measurable positive function h(see [1]). We suppose that h is a V_o-exact function (see definition 1), then we characterize the supermedian (resp. excessive) function with respect to $\mathbb{V}_h$. For instance we show that a finite positive measurable function v on X is $\mathbb{V}_h$-excessive if and only if for any $r \geq o$:

$$rV_r v \leq v + V_r(hv) \text{ and } \lim_{r \to +\infty} rV_r v = v .$$

In the second part we consider a relatively bounded semigroup $\mathbb{P} = (P_t)_{t > o}$ on $(X, \mathbb{B})$. Let ${}^h\mathbb{P} = ({}^hP_t)_{t > o}$ be the perturbed semigroup by a measurable positive function h (see [3]). We suppose that h is a $\mathbb{P}$-exact function (see definition 2) and we characterize the supermedian (resp. excessive) functions with respect to ${}^h\mathbb{P}$. In particular we prove that a finite positive measurable function v is ${}^h\mathbb{P}$-excessive if and only if for any $t > o$:

$$P_t v \leq v + \int_0^t P_s(hv)ds \quad \text{and} \quad \lim_{t \to o} P_t v = v .$$

I-PERTURBATION OF RESOLVENT

On the measurable space $(X, \mathbb{B})$ we consider the set F (resp. F_b) of all measurable (resp. measurable bounded) functions on X. F^+(resp. F_b^+) denote the set of positive elements of F(resp. F_b). Let C be a cone in F^+ and let $u,v \in F^+$; we shall use the notation $u \leq_C v$ to express that there exists $w \in C$ such that $u+w = v$.

In the following let $\mathbb{V} = (V_r)_{r\geq 0}$ be a resolvent on $(X, \mathbb{B})$ possessing a proper potential kernel V_0. The set of all supermedian (resp. excessive) functions with resp. to $\mathbb{V}$ will be denoted by S_0 (resp. E_0). Similarly, the set of all supermedian (resp. excessive) functions with respect to $\mathbb{V}_h$ is denoted by S_h(resp. E_h).

<u>Definition 1</u> (cf. [1]) - A function $h \in F^+$ is called V_0 -exact, if there exists a unique kernel V_h such that:

$$V_0 = V_h + V_0(h\, V_h) = V_h + V_h(h\, V_0)$$

<u>Proposition 1</u> - For every V_0 -exact function h, the following holds:

a) If $v \in S_0$(resp.E_0) with $V_h(hv) < \infty$, then $v-V_h(hv) \in S_h$ (resp. E_h).

b) If $u \in S_h$(resp. E_h) with $V_0(hu) < \infty$, then $u+V_0(hu) \in S_0$ (resp. E_0).

<u>Proof</u> - a) Let $v \in S_0$ such that $V_h(hv) < \infty$ and let $\tilde{v}=\sup_{r>0} rV_r v$.

Then we have :

$$v-V_h(hv) \geq \tilde{v}-V_h(h\tilde{v}) = \lim_{n\to\infty} (V_0 f_n - V_h(hV_0 f_n)) = \lim_{n\to\infty} V_h\, f_n ,$$

where $f_n \in F_b^+$. We get $v-V_h(hv) \in F^+$. Moreover we have :

$$rV_{r+h}(v-V_h(hv)) = rV_{h+r}(\tilde{v}-V_h(h\tilde{v})) = rV_{r+h}(\lim_{n\to\infty} V_h\, f_n)$$

$$\leq \underline{\lim}\; r\, V_{r+h} V_h f_n \leq \underline{\lim}\; V_h f_n \leq v-V_h(hv).$$

This shows that : $v - v_h(hv) \in S_h$.

Now suppose that $v \in E_0$ and $V_h(hv) < \infty$. We get :

$$\lim_{r\to\infty} rV_{h+r}(hV_r v) \leq \lim_{r\to\infty} V_{r+h}(hv) = 0 \quad .$$

From the resolvent equation : $V_r v = V_{r+h} v + V_{h+r}(hV_r v)$, we obtain $\lim_{r\to\infty} rV_{r+h} v = \lim_{r\to\infty} rV_r v = v$.

Hence $\lim_{r\to\infty} rV_{r+h}(v-V_h(hv)) = v-V_h(hv)$.

b) Let $u \in S_h$, then :

$$rV_r(u+V_o(hv)) = rV_{r+h}\, u + rV_r(hV_{r+h}\, u) + rV_rV_o(hu)$$
$$\leq u+V_r(hu) + rV_r\, V_o(hu) = u + V_o(hu).$$

It follows that :

$$u + V_o(hu) \in S_o.$$

Since $V_o(hu) < \infty$, we have :

$$\lim_{r\to\infty} rV_r(hV_{r+h}u) \leq \lim_{r\to\infty} V_r(hu) = 0 .$$

Thus if $u \in E_h$, we have :

$$\lim_{r\to\infty} rV_r(u+V_o(hu)) = u+V_o(hu).$$

Corollary 1 – Let h be a V_o-exact function and $v \in F^+$ such that $V_o(hv) < \infty$. Then the following assertions are equivalent :

(1) $v \in S_o$(resp. E_o) ;

(2) $v - V_h(hv) \in S_h$(resp. E_h).

Corollary 2 – Let h be a V_o-exact function and $v \in F^+$ such that $V_o(hu) < \infty$. Then the following assertions are equivalent :

(1) $u \in S_h$(resp. E_h) ;

(2) $u + V_o(hu) \in S_o$(resp. E_o).

Lemma 1 – Let $W = (W_r)_{r \geq o}$ be a resolvent on a measurable space $(X, \mathbb{B})$ possessing a proper potential kernel W_o. Let $u,v,w \in F^+$ such that $W_ou+W_ow < \infty$. Then the following assertions are equivalent :

(1) $v + W_ow - W_ou$ is W-supermedian (resp.W-excessive) ;

(2) $\forall r \geq 0$; $rW_rv+W_ru \leq v+W_rw$(resp. $\forall r \geq o$; $rW_rv+W_ru \leq v+W_rw$ and $\lim_{r\to\infty} rW_rv = v$).

Theorem 1 – Let h be a V_o-exact function and $u,v \in F^+$ such that $V_ou + V_o(hv) < +\infty$. Then the following statements are equivalent :

(1) $V_ou \underset{S_o(\text{resp.}E_o)}{\preccurlyeq} v+V_o(hv)$

(2) $V_hu \underset{S_h(\text{resp.}E_h)}{\preccurlyeq} v$;

(3) $\forall r \geq o$; $rV_rv + V_ru \leq v+V_r(hv)$(resp.$\forall r \geq o$;$rV_rv+V_ru \leq v+V_r(hv)$ and $\lim_{r\to\infty} rV_rv = v$);

(4) $\forall r \geq o$; $rV_{r+h}v+V_{r+h}u \leq v$(resp.$\forall r \geq o$; $rV_{r+h}v+V_{r+h}u$ and $v \in E_h$).

Proof – (1)$\Longleftrightarrow$(3) : Using Lemma 1 with $w = hv$ and $(W_r)_{r\geq o} = (V_r)_{r\geq o}$.

(2) $\Longleftrightarrow$ (4) : Similarly using Lemma 1 with $w=o$ and $(W_r)_{r\geq o} = (V_{r+h})_{r\geq o}$

(1) $\Longrightarrow$ (2) : Let $w \in S_o$(resp.E_o) such that $w+Vu = v+V_o(hv)$.

Then we have :

$$w+V_hu + V_h(hV_ou)+V_h(hw) = v+V_o(hv)+V_h(hw).$$

This implies that :

$$w+V_hu + V_h(h(v+V_o(hv))) = v+V_o(hv)+V_h(hw).$$

Since $V_o(hv) < \infty$, we obtain :

$$w+V_hu = v+V_h(hw) \text{ and } V_h(hw) \leq V_o(hv) < \infty.$$

It follows that : $v = V_hu + (w-V_h(hw))$.

Now, by proposition 1;a) we get : $w-(v_h(hw)) \in S_h$(resp. E_h).
This means that :

$$V_ou \underset{S_h(\text{resp. } E_h)}{\preccurlyeq} v .$$

(2) $\Longrightarrow$ (1) : Let $g \in S_h$(resp. E_h) such that $g+V_hu = v$. Then we have :

$$V_o(hg) + V_o(hV_hu) = V_o(hv) .$$

It follows that :

$$v + V_o(hv) = h + V_o(hg) + V_ou .$$

Now $g \in S_h$(resp. E_h) with $V_o(hg) \leq V_o(hv) < \infty$; thus by proposition 1 b) we get $g+V_o(hg) \in S_o$(resp.E_o). Therefore

$$V_ou \underset{S_o(\text{resp. } E_o)}{\preccurlyeq} v+V_o(hv).$$

Theorem 2 – Let h be a V_o- exact function and $v \in F^+$ such that $v+V_o(hv)< \infty$. Let B a kernel on $(X, \mathbb{B})$. Then the following statements are equivalent :

(1) $V_o(Bv) \underset{E_o}{\preccurlyeq} v+V_o(hv)$;

(2) $V_h(Bv) \underset{E_h}{\preccurlyeq} v$;

(3) $\forall r \geq o$; $rV_rv + V_r(Bv) \leq v+V_r(hv)$ and $\lim_{r\to\infty} rV_rv=v$;

(4) $\forall r \geq o$; $rV_{r+h}v + V_{r+h}(Bv) \leq v$ and $v \in E_h$;

(5) $v \in E(\mathbb{V}_h(B))$.

Proof - Choose $u = Bv$ in Theorem 1. We obtain the equivalence between (1), (2), (3) and (4). Moreover the equivalence (2)<=>(5) holds by [2], Theorem 1.4, p. 164 .

II - PERTURBATION OF SEMIGROUPS

In this part we consider a relatively bounded semigroup $\mathbb{P} = (P_t)_{t>o}$ on the measurable space $(X, \mathbb{B})$. In [3], to every $h \in F_b$, there is associated a unique semigroup ${}^h\mathbb{P} = ({}^hP_t)_{t>o}$ defined by :

$$P_t = {}^hP_t + \int_o^t P_{t-s}(h\,{}^hP_s)ds = {}^hP_t + \int_o^t {}^hP_s(hP_{t-s})ds .$$

In the case of an unbounded positive function h, ${}^hP_t$ is defined by :

$${}^hP_t = \inf_n {}^{h_n}P_t ; \text{ where } h_n = \inf(h,n) .$$

The set of all supermedian (resp. excessive) functions with respect to a semi-group T will be denoted by $S(T)$ (resp. $E(T)$).

Proposition 2 - Let $h \in F_b$ and $v \in F^+$ such that for any $t > o$:

$\int_o^t P_s(hv)ds < \infty$. We consider the following assertions :

a) $P_tv \leq v + \int_o^t P_s(hv)ds \quad (\forall t > o)$;

b) ${}^hP_tv \leq v \quad (\forall t > o)$.

Then we have a) $\Longrightarrow$ b).

Proof - Given $r \geq o$ such that $|h| \leq r$ an let be $Q_t = e^{-rt}P_t$. We know by [3] , Theorem 1 , that ${}^hP_t = (Q(B))_t$ where $B = (r-h)id$. Thus it suffices to show (cf.[2] , Proposition 2.4, p.175) that :

$$Q_tv + \int_o^t Q_s(Bv)ds \leq v .$$

Indeed, the inequality : $P_tv \leq v + \int_o^t P_s(hv)ds$ implies that

$$\int_o^t Q_s(Bv)ds \leq (1-e^{-rt})v - e^{-rt} \int_o^t P_s(hv)ds.$$

Hence

$$Q_tv + \int_o^t Q_s(Bv)ds \leq \int_o^t Q_s(Bv)ds + e^{-rt}v + e^{-rt} \int_o^t P_s(hv)ds \leq v.$$

This completes the proof.

Corollary 3 - Let $h \in F_b^+$ and $v \in F^+$ such that for any $t > o$;

$\int_0^t P_s(hv)ds < \infty$. Then the following assertions are equivalent :

(1) $P_t v \leqq v + \int_0^t P_s(hv)ds \quad (\forall t > o)$;

(2) ${}^hP_t \leq v \quad (\forall t > o)$.

<u>Lemma 2</u> – Let $T = (T_t)_{t>o}$ be a semigroup on $(X, \mathbb{B})$ and let $(W_r)_{r \geqq o}$ be its associated resolvent. We suppose that $W_o = \int_0^\infty T_t dt$ is proper. Let $u,v,w \in F^+$ be such that $W_o u + W_o w < \infty$. Then the following assertions are equivalent :

(1) $T_t v + \int_0^t T_s u \, ds \leqq v + \int_0^t T_s w \, ds \quad (\forall t > o)$;

(2) $W_o u \underset{S(T)}{\preccurlyeq} v + W_o w$.

<u>Proposition 3</u> – Let $T = (T_t)_{t>o}$ be a semigroup as in Lemma 2. $u,v,w \in F^+$ such that $v+W_o(w) < \infty$. Then the following statements are equivalent :

(1) $\forall t > o$; $T_t v + \int_0^t T_s u \, ds \preccurlyeq v + \int_0^t T_s w \, ds$ and $\lim_{t \to o} T_t v = v$;

(2) $\forall r \geqq o$; $rW_r v + W_r u \leq v + W_r w$ and $\lim_{r \to \infty} rW_r v = v$;

(3) $W_o u \underset{E(T)}{\preccurlyeq} v + W_o w$.

<u>Proof</u> – (1)$\Longrightarrow$(2) : Trivial.
(2)$\Longrightarrow$(3) : Follows from Lemma 1.
(3)$\Longrightarrow$(1) : By Lemma 2, we have for any $t > o$:

$$T_t v + \int_0^t T_s u \, ds \leq v + \int_0^t T_s w \, ds .$$

Moreover, let $g \in E(T)$ such that $W_o u + g = v + W_o w$. Hence for any $t > o$:

$$T_t g + \int_0^t T_s w \, ds + W_o u = T_t v + W_o w + \int_0^t T_s u \, ds. \qquad (*)$$

By assumption we have $\lim_{t \to o} T_t g = o$; moreover, since $W_o u + W_o w < \infty$, we get :

$$\lim_{t \to o} \int_0^t T_s w ds = \lim_{t \to o} \int_0^t T_s \, u \, ds = o .$$

Hence $\lim_{t \to o} T_t v + W_o w = g + W_o u$ holds from (*) and we obtain $\lim_{t \to o} T_t v = v$. This completes the proof.

<u>Definition 2</u>–A function $h \in F^+$ is called P-exact if for any $t > o$ we have :

$$P_t = {}^hP_t + \int_o^t P_{t-s}(h\,{}^hP_s)ds = {}^hP_t + \int_o^t {}^hP_s(hP_{t-s})ds.$$

<u>Theorem 3</u> – Let h be a P-exact function and $(V_r)_{r \geq o}$ the resolvent of $\mathbb{P}$. We suppose that $V_o = \int_o^\infty P_t dt$ is proper. Let $u, v \in F^+$ such that $v + V_o(hv) < \infty$. Then the following statements are equivalent :

(1) $V_o u \underset{E(\mathbb{P})}{\preccurlyeq} v + V_o(hv)$;

(2) $\forall t > o$; $P_t v + \int_o^t P_s\, u\, ds \leq v + \int_o^t P_s(hv)ds$ and $\lim_{t \to o} P_t v = v$;

(3) $\forall r \geq o$; $rV_r v + V_r u \leq v + V_r(hv)$ and $\lim_{r \to \infty} rV_r v = v$;

(4) $V_h u \underset{E({}^h\mathbb{P})}{\preccurlyeq} v$;

(5) $\forall t > o$; ${}^hP_t v + \int_o {}^hP_s u\, ds \leq v$ and $\lim_{t \to o} {}^hP_t\, v = v$;

(6) $\forall r \geq o$; $rV_{r+h} v + V_{r+h} u \leq v$ and $\lim_{r \to \infty} rV_{r+h} v = v$.

<u>Proof</u> – The potential kernel of ${}^h\mathbb{P}$ is V_h and its resolvent is $(V_{h+r})_{r \geq o}$. Hence we have $E(\mathbb{P}) = E_o$ and $E({}^h\mathbb{P}) = E_h$. Thus the equivalence between (1), (3), (4) and (6) holds from Theorem 1. Now, choose in Proposition 3, $T = \mathbb{P}$ and $w = hv$. We obtain the equivalence between (1) and (2). Similarly choose in Proposition 3, $T = {}^h\mathbb{P}$ and $w = o$. We obtain the equivalence between (4) and (5).

<u>Corollary 4</u> – Let h and v be as in Theorem 3. Then the following assertions are equivalent :

(1) $v + V_o(hv) \in E(\mathbb{P})$;

(2) $\forall t > o$; $P_t v \leq v + \int_o^t P_s(hv)ds$ and $\lim_{t \to o} P_t v = v$;

(3) $\forall r \geq o$; $rV_r v \leq v + V_r(hv)$ and $\lim_{r \to \infty} rV_r v = v$;

(4) $v \in E({}^h\mathbb{P})$.

<u>Theorem 4</u> – Let h and v be as in Theorem 3. Let B be a bounded kernel. Then the following statements are equivalent :

(1) $V_o(Bv) \underset{E(\mathbb{P})}{\preccurlyeq} v + V_o(hv)$;

(2) $\forall t > o$; $P_t v + \int_o^t P_s(Bv)ds \leq v + \int_o^t P_s(hv)ds$ and

$\lim_{t \to o} P_t v = v$;

(3) $\forall r \geqq o$; $rV_r v + V_r(Bv) \leqq v + V_r(hv)$ and $\lim_{r \to \infty} rV_r v = v$;

(4) $V_h(Bv) \underset{E({}^h\mathbb{P})}{\preccurlyeq} v$;

(5) $\forall t > o$; ${}^hP_t + \int_o^t {}^hP_s(Bv)ds \leqq v$ and $\lim_{t \to c} {}^hP_t v = v$;

(6) $\forall r \geqq o$; $rV_{r+h}v + V_{r+h}(Bv) \leqq v$ and $\lim_{r \to o} rV_{r+h}v = v$;

(7) $v \in E({}^h\mathbb{P}(B))$.

Proof- Choose $u = Bv$ in Theorem 3. We obtain the equivalence between (1), (2), (3), (4), (5) and (6). The equivalence (5) <=> (7) follows from [2], Theorem 2.5, p.179.

REFERENCES

1. H. Ben Saad
   Generalisation des noyaux V_h et applications.
   Seminaire de Theorie du potentiel Paris, n°7, Lecture Notes in Mathematics 1061
   Springer-Verlag, Berlin, 1984; 14 - 33.

2. N. Boboc and Gh. Bucur
   Perturbations in excessive structures.
   Complex Analysis Fifth Romanian-Finnish Seminar (Part 2), Bucharest 1981, Lecture Notes in Mathematics 1014
   Springer-Verlag, Berlin, 1983; 155 - 187.

3. H. Maagli, M. Selmi
   Perturbation et comparaison de semi-groupes (to appear in Rev.Roum. Math. Pures Appl.).

CAPACITIES ON HARMONIC SPACES WITH ADJOINT STRUCTURE

Fumi-Yuki Maeda

Department of Mathematics, Faculty of Science
Hiroshima University
Hiroshima, Japan

In the classical potential theory, the capacities defined in terms of Green potentials coincide with the capacity defined by Dirichlet integrals; more precisely, for a compact set K in a Greenian domain Ω in $\mathbb{R}^d$,

$$\sup\{\mu(\Omega) \mid G\mu \leqq 1,\ \mathrm{Supp}\ \mu \subset K\} = \inf\{\textstyle\int G\mu\, d\mu \mid G\mu \geqq 1 \text{ on } K\}$$

$$= \inf\{D[f] \mid f\text{: potential on } \Omega \text{ with } f \geqq 1 \text{ on } K\},$$

where $G\mu$ is the Green potential of $\mu \geqq 0$ on Ω and $D[f]$ is the Dirichlet integral of f.

These equalities no longer hold for the potential theory with respect to the heat equation, and thus there arise three different capacities. In fact, in parabolic potential theory or in the boundary value problems for parabolic partial differential equations, various different capacities have been introduced (cf. [3] and [8], in which comparisons of some of them are also discussed).

In the present note, we define three kinds of capacities on a general harmonic space having adjoint structure, and investigate relations between these capacities.

1. DEFINITIONS

We consider a pair $(\mathcal{H}, \mathcal{H}^*)$ of mutually adjoint harmonic structures on a connected locally compact space X with a countable base. By definition, $(X, \mathcal{H})$ and $(X, \mathcal{H}^*)$ are P-harmonic spaces in the sense of [1] and there exists an associated Green function $G(x,y)$ satisfying the following conditions (see [6]):

(G.0) $G(x,y)$ is lower semicontinuous on $X \times X$, continuous off the diagonal;

(G.1) (resp. (G*.1)) For each $y \in X$, $G(\cdot,y)$ (resp. $G(y,\cdot)$) is a potential for $\mathcal{H}$ (resp. $\mathcal{H}^*$) and is harmonic for $\mathcal{H}$ (resp. $\mathcal{H}^*$) on $X \setminus \{y\}$;

This paper is in final form and no version of it will be submitted for publication elsewhere, except the APPENDIX.

(G.2) (resp. (G*.2)) For any continuous potential p for H (resp. H^*), there exists a unique non-negative measure μ on X such that $p = G\mu$ (resp. $p = G^*\mu$), where

$$G\mu(x) = \int G(x,y)d\mu(y) \quad (\text{resp. } G^*\mu(y) = \int G(x,y)d\mu(x)).$$

For simplicity, we assume in this note that the constant function 1 is harmonic for H as well as for H^*.

Associated to G(x,y), there are measure representations $\sigma: R \to M$ and $\sigma^*: R^* \to M$ (see [5] for the definition of a measure representation), where R (resp. R^*) is the sheaf of functions which are locally expressible as differences of continuous H- (resp. H^*-) superharmonic functions, and M is the sheaf of signed measures on X. If $f = G\nu + h$ (resp. $= G^*\nu + h^*$) with a signed measure ν such that $G|\nu|$ (resp. $G^*|\nu|$) is continuous and $h \in H(X)$ (resp. $h^* \in H^*(X)$), then $\sigma(f) = \nu$ (resp. $\sigma^*(f) = \nu$). In particular, $\sigma(1) = \sigma^*(1) = 0$ by our assumption.

For $f \in R(X)$ (resp. $R^*(X)$), its gradient measure δ_f (resp. δ_f^*) is defined by (cf. [5])

$$\delta_f = \frac{1}{2}\{2f\sigma(f) - \sigma(f^2)\} \quad (\text{resp. } \delta_f^* = \frac{1}{2}\{2f\sigma^*(f) - \sigma^*(f^2)\}).$$

We know ([6; Theorem 2.1]) that $\delta_f = \delta_f^*$ for $f \in R(X) \cap R^*(X)$.

Let P_C (resp. P_C^*) be the set of all continuous potentials for H (resp. H^*). For a compact set K in X, let

$$P_C(K) = \{p \in P_C \mid p \geq 1 \text{ on a neighborhood of } K\},$$

$$P_C^*(K) = \{p \in P_C^* \mid p \geq 1 \text{ on a neighborhood of } K\},$$

and we define

$$c_o(K) = \inf\{\mu(X) \mid G\mu \in P_C(K)\}, \quad c_o^*(K) = \inf\{\mu(X) \mid G^*\mu \in P_C^*(K)\};$$

$$C_I(K) = \inf\{\int p d\sigma(p) \mid p \in P_C(K)\}, \quad C_I^*(K) = \inf\{\int p d\sigma^*(p) \mid p \in P_C^*(K)\};$$

$$C_D(K) = \inf\{\delta_p(X) \mid p \in P_C(K)\}, \quad C_D^*(K) = \inf\{\delta_p^*(X) \mid p \in P_C^*(K)\}.$$

It is easy to see that all the above quantities are finite. In fact, in the above definitions, $P_C(K)$ (resp. $P_C^*(K)$) may be replaced by (cf. [6; Theorem 3.1 and Corollary 3.3])

$$P_{BIC}(K) = \{p \in P_C(K) \mid p \text{ is bounded and } \int p d\sigma(p) < +\infty\}$$

$$(\text{resp. } P_{BIC}^*(K) = \{p \in P_C^*(K) \mid p \text{ is bounded and } \int p d\sigma^*(p) < +\infty\}).$$

2. c_o AND c_o^*

In the case where $X = \mathbb{R}^{d+1}$ $(d \geq 1)$ and H is given by solutions of the heat equation, c_o is known as heat capacity ([4]) or thermal capacity ([9], [3]).

In the framework of "parabolic Dirichlet spaces", M. Pierre has shown that $c_o = c_o^*$ (cf. [7; Lemme II.4]). We can show that this is true also for our situation; namely we have

THEOREM 1 — $c_o(K) = c_o^*(K)$ *for any compact set* K; *in fact we have*

$$c_o(K) = c_o^*(K) = \sup\{\mu(K) \mid \mu \geq 0,\ G\mu \leq 1 \text{ on } X,\ \text{Supp } \mu \subset K\}$$
$$(1) \qquad = \sup\{\mu(K) \mid \mu \geq 0,\ G^*\mu \leq 1 \text{ on } X,\ \text{Supp } \mu \subset K\}.$$

PROOF — The proof is more or less standard, and so we give only its outline. Let $\tilde{c}(K)$ and $\tilde{c}^*(K)$ be the third and the fourth quantities in (1). If we show that (i) $c_o(K) \geq \tilde{c}^*(K)$ and (ii) $c_o(K) \leq \tilde{c}(K)$, then together with their adjoint assertions, we obtain the theorem.

(i) is quite easy to show; namely, if $G\nu \in P_C(K)$ and $\mu \geq 0$, Supp $\mu \subset K$, $G^*\mu \leq 1$ on X, then

$$(2) \qquad \nu(X) \geq \int G^*\mu \, d\nu = \int G\nu \, d\mu \geq \mu(X).$$

(ii) For any open set $G \supset K$, there exists $p_G \in P_C(K)$ such that $p_G \leq 1$ and Supp ν_G is compact and contained in G, where $\nu_G = \sigma(p_G)$. We can find a sequence $\{G_n\}$ of open sets such that $\bigcap G_n = K$ and $\{\nu_{G_n}\}$ is vaguely convergent to a non-negative measure μ_o (cf. [6; Lemma 1.3]). Then Supp $\mu_o \subset K$ and $G\mu_o \leq 1$ on X. Thus

$$c_o(K) \leq \lim_{n\to\infty} \nu_{G_n}(X) = \mu_o(X) \leq \tilde{c}(K).$$

REMARK 1 — The relations (2) in the above proof are valid for any $\nu \geq 0$ with $G\nu \geq 1$ on K. Thus, we also have

$$(1)' \qquad c_o(K) = \inf\{\mu(X) \mid \mu \geq 0,\ G\mu \geq 1 \text{ on } K\}$$
$$= \inf\{\mu(X) \mid \mu \geq 0,\ G^*\mu \geq 1 \text{ on } K\}$$

(cf. [9; Lemma 4]).

REMARK 2 — It can occur that $c_o(K) > 0$ while

$$(3) \qquad \sup\{\mu(K) \mid G\mu \in P_C,\ G\mu \leq 1 \text{ on } X,\ \text{Supp } \mu \subset K\} = 0.$$

In fact, for the harmonic space with respect to the heat equation, if K is semi-polar, then (3) holds (cf. [2; 1.XVIII, 12]); thus semi-polar non-(parabolic-)polar compact sets provide examples of such K.

3. RELATIONS BETWEEN c_o AND C_I; C_I AND C_I^*

THEOREM 2 (cf. [8; Theorem 1] and [6; Remark 4.1]) — *For any compact set* K *in* X,

$$C_I(K) \leq c_o(K) \leq 4\, C_I(K) \quad \textit{and} \quad C_I^*(K) \leq c_o(K) \leq 4\, C_I^*(K).$$

PROOF — If $G\mu \in P_C(K)$, then $p = \min(1, G\mu) \in P_C(K)$. Let $\mu' = \sigma(p)$. Then $C_I(K) \leq \int p \, d\mu' \leq \mu'(X) \leq \mu(X)$, where the last inequality follows from [6; Lemma 1.3]. Hence $C_I(K) \leq c_o(K)$.

Given $G\mu \in P_C(K)$, we can find $q \in P_C(K)$ such that Supp $\sigma(q) \subset \{G\mu \geq 1\}$ and $q \leq 1$ on X. Then, by using [6; Proposition 2.2] we have

$$\sigma(q)(X) \leq \int G\mu \, d\sigma(q) \leq \frac{1}{2} \int q \, d\sigma(q) + 2 \int G\mu \, d\mu$$

$$\leq \frac{1}{2}\sigma(q)(X) + 2\int G\mu\, d\mu.$$

Hence, $c_o(K) \leq \sigma(q)(X) \leq 4\int G\mu\, d\mu$, which implies $c_o(K) \leq 4\, C_I(K)$.

COROLLARY 1 - *A compact set* K *is polar if and only if* $C_I(K) = 0$.

COROLLARY 2 - *For any compact set* K *in* X,

$$\frac{1}{4}C_I(K) \leq C_I^*(K) \leq 4\, C_I(K).$$

REMARK 3 - For the harmonic space with respect to the heat equation on $X = \Omega \times (a,b) \subset \mathbb{R}^{d+1}$ (Ω: a domain in $\mathbb{R}^d$),

$$\inf\{\int G\mu\, d\mu \mid \mu \geq 0,\ G\mu \geq 1 \text{ on } K\} = 0$$

for any compact set K in X. For, if we take $t_o \in (a,b)$ such that $K \subset \Omega \times (t_o,b)$ and a non-negative non-zero measure μ with compact support contained in $\Omega \times \{t_o\}$, then $G(\alpha\mu) \geq 1$ on K for some $\alpha > 0$, while $\int G(\alpha\mu)d(\alpha\mu) = 0$.

This shows that the continuity of potentials is essential in the definition of C_I (cf. Remarks 1 and 2 above).

4. RELATIONS AMONG C_I, C_D AND C_D^*

THEOREM 3 - $C_D(K) \leq C_I(K)$ *for any compact set* K *in* X. *The equality does not hold in general; in fact it can occur that* $C_I(K) > 0$ *while* $C_D(K) = 0$.

PROOF - The inequality $C_D(K) \leq C_I(K)$ follows from [6; Theorem 3.1, (ii)]. To show the latter half of the theorem, we consider the harmonic space with respect to the heat equation on $X = \mathbb{R} \times (-\infty,2)$. Let $K = \{0\} \times [0,1]$. The Green function for X is the usual Gauss kernel

$$G((x,t),(y,s)) = \begin{cases} (4\pi(t-s))^{-1/2}\exp\{-(x-y)^2/4(t-s)\}, & 2 > t > s \\ 0, & t \leq s < 2. \end{cases}$$

Define non-negative measures μ_n on X by

$$d\mu_n(x,t) = a_n(1 - |x|/n)\,\chi_{(-n,n)\times(-1,0)}(x,t)dxdt,\quad n = 1,2,\ldots,$$

where χ_A denotes the characteristic function of A and $a_n > 0$ are so chosen that $\inf_{0\leq t\leq 1} G\mu_n(0,t) = 1 + 1/n$. Then, $p_n = G\mu_n \in \mathcal{P}_C(K)$ and $\partial p_n/\partial x$ exists and is continuous on X (cf. [2; 1.XVII.6]). Also we easily see that $a_n \to 1$ $(n \to \infty)$. By direct computation, we have the estimates

$$\left|\frac{\partial p_n}{\partial x}(x,t)\right| \leq \frac{a_n}{n} \quad \text{for any } (x,t) \in X$$

and

$$\left|\frac{\partial p_n}{\partial x}(x,t)\right| \leq a_n \exp\{-(|x|-n)^2/12\},\ \text{if } |x| > 2n,\ (x,t) \in X.$$

Using these estimates, we deduce that

$$\delta_{p_n}(X) = \int_{-1}^{2}\int_{-\infty}^{+\infty}\left(\frac{\partial p_n}{\partial x}\right)^2(x,t)\,dxdt \le \frac{12a_n^2}{n} + \frac{36a_n^2}{n}e^{-n^2/6} \longrightarrow 0 \quad (n \to \infty).$$

Therefore, $C_D(K) = 0$. On the other hand, it is known that $c_o(K) > 0$ (see, e.g., [4; Theorem 2]), so that $C_I(K) > 0$ by Theorem 2.

PROPOSITION 1– *For the harmonic space with respect to the heat equation on* $X = \Omega \times (a, +\infty)$, *where* Ω *is a domain in* $\mathbb{R}^d$ $(d \ge 1)$ *and* $-\infty \le a < +\infty$,

(4) $\qquad C_D(K) = C_I(K)$ *for any compact set* $K \subset X$.

PROOF– In view of [6; Theorem 3.1], it is enough to show that $\sigma(p^2)(X) = 0$ for any $p \in P_C$ such that $\sigma(p)$ is compact in X (cf. (III) of the proof of [6; Proposition 4.1]). Given such p, let $\mu = \sigma(p)$ and Supp $\mu \subset E \times [t_0, t_1]$ with a compact set E in Ω and $a < t_0 < t_1$. The Green function G for X is dominated by the Gauss kernel

$$W((x,t),(y,s)) = \begin{cases} (4\pi(t-s))^{-d/2}\exp\{-|x-y|^2/4(t-s)\}, & t > s \\ 0, & t \le s. \end{cases}$$

It is easy to verify that

$$\int_\Omega [W\mu(x,t)]^2\,dx \le (8\pi)^{-d/2}\mu(X)^2(t-t_1)^{-d/2} \quad \text{for } t > t_1.$$

Since $p \le W\mu$, it follows that $\lim_{t\to+\infty}\int_\Omega p(x,t)^2dx = 0$. In view of [6; Remark 4.1], this implies that $\sigma(p^2)(X) = 0$.

COROLLARY 3– C_D *and* C_D^* *are not equivalent to each other, in general.*

PROOF– Consider the harmonic space and the compact set K given in the proof of Theorem 3. Then, $C_D(K) = 0$, while the adjoint statement of Proposition 1 and Corollary 2 imply that $C_D^*(K) = C_I^*(K) \ge 4^{-1}C_I(K) > 0$.

APPENDIX. REMOVABILITY OF POLAR SETS WITH RESPECT TO DIRICHLET FINITE HARMONIC FUNCTIONS

Here we state a removability result, a detailed proof of which will be given elsewhere.

THEOREM A – *Let* K *be a compact polar set in* X. *If* $u \in H(X\setminus K)$ *and* $\delta_u(X\setminus K) < +\infty$, *then there exists* $h \in H(X)$ *such that* $h|_{X\setminus K} = u$.

REMARK – This result is well-known in the classical case, and is proved by the author for general *self-adjoint* harmonic spaces [Hiroshima Math. J. 14 (1984), 537-546]. It seems that there are no published results of this kind in the non-elliptic case; even for the harmonic space with respect to the heat equation.

In order to prove this theorem, we prepare two lemmas.

LEMMA 1 *Let* U *be an open set in* X. *If* $u \in H(U)$ *and* $\delta_u(U) < +\infty$, *then* $u = u_1 - u_2$ *with* $u_j \in H(U)$ *such that* $u_j \geq 0$ *and* $\delta_{u_j}(U) < +\infty$, $j = 1, 2$.

LEMMA 2 *If* K *is a compact polar set in* X *and* p *is a potential on* X *such that* $p|_{X\setminus K} \in H(X\setminus K)$, *then there exists a non-negative measure* μ *on* X *such that* Supp $\mu \subset K$ *and* $p = G\mu$ *on* X. (Note that $p \notin P_C$ unless $p = 0$.)

Sketch of the proof of THEOREM A. By Lemma 1, we may assume that $u \geq 0$. Then, by [1; Theorem 6.2.1] and Lemma 2, $u = h + G\mu$ on $X \setminus K$ for some $h \in H(X)$ and a non-negative measure μ with Supp $\mu \subset K$. Put $p = G\mu$. We see that $\delta_p(X\setminus K) < +\infty$. Let $\nu = \delta_p|_{X\setminus K} = -\sigma(p^2)|_{X\setminus K}$, which can be regarded as a non-negative measure on X. Then, again by [1; Theorem 6.2.1] and Lemma 2, there is a non-negative measure ν_0 such that Supp $\nu_0 \subset K$ and $G\nu + p^2 = G\nu_0$ on $X \setminus K$. Since $c_0^*(K) = 0$, there is a sequence $\{\lambda_n\}$ such that $G^*\lambda_n \in P_C^*$, $G^*\lambda_n = 1$ on a neighborhood of K for each n and $\lambda_n(X) \to 0$ $(n \to \infty)$. Then

$$\int p^2 d\lambda_n \leq \int G\nu_0 \, d\lambda_n = \int G^*\lambda_n \, d\nu_0 = \nu_0(K),$$

so that

$$\mu(X) = \int G^*\lambda_n \, d\mu = \int p \, d\lambda_n \leq \nu_0(K)^{1/2} \, \lambda_n(X)^{1/2} \longrightarrow 0 \quad (n \to \infty),$$

which implies $\mu = 0$. Thus $u = h$ on $X \setminus K$.

References

[1] C. Constantinescu and A. Cornea, Potential Theory on Harmonic Spaces, Springer-Verlag, Berlin etc., 1972.

[2] J.L. Doob, Classical Potential Theory and Its Probabilistic Counterpart, Springer-Verlag, New York etc., 1984.

[3] R. Gariepy and W.P. Ziemer, Thermal capacity and boundary regularity, J. Diff. Eq. 45 (1982), 374-388.

[4] R. Kaufman and J.-M. Wu, Parabolic potential theory, J. Diff. Eq. 43 (1982), 204-234.

[5] F-Y. Maeda, Dirichlet Integrals on Harmonic Spaces, Lecture Notes Math. 803, Springer-Verlag, Berlin etc., 1980.

[6] F-Y. Maeda, Dirichlet integral and energy of potentials on harmonic spaces with adjoint structure, to appear in Hiroshima Math. J. 17 (1987).

[7] M. Pierre, Représentant précis d'un potentiel parabolique, Sém. Th. Pot., Lecture Notes Math. 814, 186-228, Springer-Verlag, Berlin etc., 1980.

[8] M. Pierre, Parabolic capacity and Sobolev spaces, SIAM J. Math. Anal. 14 (1983), 522-533.

[9] N.A. Watson, Thermal capacity, Proc. London Math. Soc. (3) 37 (1978), 342-362.

BALAYAGE SPACES ON TOPOLOGICAL SUMS

Michael Meyer

Institut für Statistik und Dokumentation
Univeritätsstraße 1
D-4000 Düsseldorf, Federal Republic of Germany

1. INTRODUCTION

Several authors, e.g. N. Bouleau /6/, N. Boboc and Gh. Bucur /4/, H. Ben Saad and K. Janßen /2/, A. Boukricha /5/, studied potentialtheoretic structures on topological sums and their connection with the biharmonic spaces of E.P. Smyrnélis /10/.

In the present paper our main interest is to characterize balayage spaces $(E_1+E_2,\mathcal{W})$ by means of subspaces $(E_i,\mathcal{W}_i)$ and transition kernels $R_i: \underline{\underline{\mathcal{B}(E_{3-i})}}_+ \rightarrow \underline{\underline{\mathcal{B}(E_i)}}_+$ $(i = 1,2)$ and, the other way round, starting from balayage spaces $(E_i,\mathcal{W}_i)$ and appropriate kernels $R_i: \underline{\underline{\mathcal{B}(E_{3-i})}}_+ \rightarrow \underline{\underline{\mathcal{B}(E_i)}}_+$ $(i = 1,2)$ to find necessary and sufficent conditions for the existence of a unique balayage space $(E_1+E_2,\mathcal{W})$ being characterized by $(E_i,\mathcal{W}_i)$ and R_i $(i = 1,2)$ in the desired way and possessing $(E_1,\mathcal{W}_1),(E_2,\mathcal{W}_2)$ as subspaces and R_1,R_2 as transition kernels. Moreover the method of construction allows to obtain certain perturbations of a given balayage space.

2. NOTATIONS

We work within the context of balayage spaces as developed by J. Bliedtner and W. Hansen in /3/ and we take over their notations. Thus, given a sub-Markov resolvent $\mathbb{V}$ and a Hunt process $\mathfrak{X}$, we write $S_{\mathbb{V}},S_{\mathfrak{X}}$ and $E_{\mathbb{V}},E_{\mathfrak{X}}$ for the associated cones of supermedian and excessive functions, respectively, and if $(E,\mathcal{W})$ is a balayage space then $\mathcal{P}$ denotes the cone of continuous potetials in $\mathcal{W}$, $^*\mathcal{H}^+(U)$ the set of positive hyperharmonic functions on U (for an open subset U of E) and H_A the harmonic kernel fulfilling the relation $H_A v = R_v^{\complement A}$ for all $v \in \mathcal{W}$ ($A \in \mathcal{B}(E)$).

Now let $I \subset \mathbb{N}$, $(E_i)_{i\in I}$ be a family of pairwise disjoint locally compact spaces with countable base and $E := \sum_{i\in I} E_i$ their topological sum. For $f: E \longrightarrow F$ (F any set) and a kernel N on $(E, \mathcal{B}(E))$ we introduce the following identifications: $f = (f_i)_{i\in I}$, where $f_j := f|_{E_j}$ ($j \in I$), and $N = (N_{ij})_{i,j\in I}$, where $N_{ij} f_j := N(1_{E_j} f_j)_i$ ($i,j \in I$, $f_j \in \underline{\underline{\mathcal{B}(E_j)}}_+$). Then for $f \in \underline{\underline{\mathcal{B}(E)}}_+$ $Nf = (\sum_{j\in I} N_{ij} f_j)_{i\in I}$ is calculated by matrix multiplication.

3. CHARACTERIZATION OF BALAYAGE SPACES ON TOPOLOGICAL SUMS

Suppose that $I = \{1,2\}$. Let $(E,\mathcal{W})$ be a balayage space and $\mathfrak{X} = (\Omega, \mathcal{M}, \mathcal{M}_t, X_t, \theta_t, P^x)$ a Hunt process on E such that $E_{\mathfrak{X}} = \mathcal{W}$. This implies $1 \in \mathcal{W}$. For $i \in I$ let $\mathfrak{X}^i$ denote the process obtained by killing $\mathfrak{X}$ at time $D_{\complement E_i}$. Then the following holds (c.f. /3/V 1.1, VI 3.20):

3.1. Proposition

For every $i \in I$:

(a) $\mathfrak{X}^i$ is a Hunt process on E_i and $(E_i, \mathcal{W}_i) := (E_i, \mathcal{W}^{E_i}) :=$ $(E_i, {}^*\mathcal{H}^+(E_i)|_{E_i}) = (E_i, E_{\mathfrak{X}^i})$ is a balayage space such that $1 \in \mathcal{W}_i$.

(b) The $\mathcal{W}_i$-fine topology $\tau_{\mathcal{W}_i}$ is the relative $\mathcal{W}$-fine topology $\tau_{\mathcal{W}}$ on E_i.

(c) For every $p \in \mathcal{P}$: $(p - R_p^{E_{3-i}})_i \in \mathcal{P}^i$.

We list some immediate consequences:

3.2. Corollary

(a) $\tau_{\mathcal{W}} = \tau_{\mathcal{W}_1} + \tau_{\mathcal{W}_2}$.

(b) For every $f: E \longrightarrow \overline{\mathbb{R}}$: $f \in \mathcal{W} \iff f \in S_{\mathfrak{X}}$ and $\forall\, i \in I$: $f_i \in \mathcal{W}_i$ $\iff f \in S_{\mathfrak{X}}$ and $\forall\, i \in I$: f_i is l.s.c. with respect to $\tau_{\mathcal{W}_i}$.

Let $\mathbb{U}$ be the resolvent of $\mathfrak{X}$ and $\mathbb{U}^i$ the resolvent of $\mathfrak{X}^i$ for $i \in I$, $U_\lambda := \begin{pmatrix} U_\lambda^1 & 0 \\ 0 & U_\lambda^2 \end{pmatrix}$ ($\lambda \geq 0$) and $T := \inf\{t > 0 \mid \phi \circ X_t \neq \phi \circ X_0\}$, where $\phi := 1_{E_1} + 2 \cdot 1_{E_2}$.

3.3. Proposition

(a) T is an exact and strong terminal stopping time with respect to $(\mathcal{F}_t^o)$ and

$$P_T = \begin{pmatrix} 0 & (H_{E_1})_{12} \\ (H_{E_2})_{21} & 0 \end{pmatrix} .$$

(b) P_T is a strong Feller kernel mapping $\mathcal{P}$ into $\mathcal{P}$.

(c) For all $\alpha \geq 0$: $V_\alpha = \sum_{n \in \mathbb{N}_0} (P_T^\alpha)^n U_\alpha$.

Proof: (a) follows directly from the definition of T and for (c) confer /2/2.. Since $H_{E_i}(\underline{\underline{\mathfrak{B}(E)}}_{b+})|_{E_i} \subset C_b^+(E_i)$ for $i \in I$ P_T is a strong Feller kernel by (a) and from $P_T(\mathcal{W}) \subset \mathcal{W}$, $P_T p \leq p$, we conclude $P_T p \in \mathcal{W} \cap C_{\mathcal{P}}(E) = \mathcal{P}$ for all $p \in \mathcal{P}$.

$$\text{Let } R := \begin{pmatrix} 0 & R_1 \\ R_2 & 0 \end{pmatrix} := P_T .$$

3.4. Proposition

For every $i \in I$: $R_i(\underline{\underline{\mathfrak{B}(E_{3-i})}}_{b+}) \subset \mathcal{P}_b^i$.

Proof: Let $i \in I$, $f_{3-i} \in \underline{\underline{\mathfrak{B}(E_{3-i})}}_{b+}$ and $x \in E_i$. Since $R_i(\underline{\underline{\mathfrak{B}(E_{3-i})}}_{b+}) \subset \mathcal{W}_i \cap C_b(E_i)$ it suffices to show that

$\lim_{n\to\infty} {}^iR^{E_i \setminus K_{n,i}}_{R_i f_{3-i}}(x) = 0$ for some exhaustion $(K_{n,i})_{n\in\mathbb{N}}$ of E_i. Let $D := D_{\complement E_i}$, $D_n := D_{E_i \setminus K_{n,i}}$ $(n \in \mathbb{N})$. Using Hunt's theorem and the strong Markov property for $\mathfrak{X}$ we obtain ${}^iR^{E_i \setminus K_{n,i}}_{R_i f_{3-i}}(x) =$

$P^x(1_{[D_n < D < \zeta]} f_{3-i} \circ X_D)$ for all $n \in \mathbb{N}$, where ζ denotes the life-time of $\mathfrak{X}$. Since $\sup_{n\in\mathbb{N}} D_n \geq \zeta$ P^x-a.s. an application of Lebesgue's convergence theorem finishes the proof.

$(E,\mathcal{W})$ can be characterized by the subspaces $(E_i,\mathcal{W}_i)$ and the transition kernels R_i $(i = 1,2)$ as follows:

3.5. Proposition

$\mathcal{W}_b = \{u \in \underline{\underline{\mathfrak{B}(E)}}_{b+} \mid \forall\, i \in I: u_i, u_i - R_i u_{3-i} \in \mathcal{W}_i\}$. In particular, $1 - R_i 1 \in \mathcal{W}_i$ for all $i \in I$.

Proof: Let $u \in \mathcal{W}_b$ and $(p^n)_{n\in\mathbb{N}}$ a sequence in $\mathcal{P}$ increasing to u. Then for $i \in I$ $u_i \in \mathcal{W}_i$ (3.2.(b)) and $u_i - R_i u_{3-i} = (u - R_u^{E_{3-i}})_i =$

$\lim_{n\to\infty} (p^n - R_{p^n}^{E_{3-i}})_i \in S_{\mathbb{U}^i}$ (by 3.1.(c)). Since $u_i - R_i u_{3-i}$ is l.s.c. we conclude $u_i - R_i u_{3-i} \in \mathcal{W}_i$. Conversely let $u \in \underline{\underline{\mathfrak{B}(E)}}_{b+}$ such that $u_i - R_i u_{3-i} \in \mathcal{W}_i$ for all $i \in I$. Then $u - H_{E_i} u = (u_i - R_i u_{3-i}) 1_{E_i} \in$

${}^*\mathcal{H}^+(E_i)$ for all $i \in I$ and hence $u \in {}^*\mathcal{H}^+(E_1) \cap {}^*\mathcal{H}^+(E_2) = \mathcal{W}$.

4. CONSTRUCTION OF BALAYAGE SPACES ON TOPOLOGICAL SUMS

Let first $I = \mathbb{N}$.

4.1. Proposition

Let $((E_i, \mathcal{W}_i), R_i)_{i\in\mathbb{N}}$ be a sequence of balayage spaces $(E_i, \mathcal{W}_i)$ such that $1 \in \mathcal{W}_i$ and kernels $R_i : \underline{\underline{\mathcal{B}}}(E_{i+1})_+ \to \underline{\underline{\mathcal{B}}}(E_i)_+$ such that $1 - R_i 1 \in \mathcal{W}_i$ and $R_i(\underline{\underline{\mathcal{B}}}(E_{i+1})_{b+}) \subset \mathcal{P}^i_b$ $(i \in \mathbb{N})$. Then:

(a) There exists a unique balayage space $(E, \mathcal{W})$ such that $1 \in \mathcal{W}$ and $\mathcal{W}_b = \{u \in \underline{\underline{\mathcal{B}}}(E)_{b+} \mid \forall\, i \in \mathbb{N} : u_i, u_i - R_i u_{i+1} \in \mathcal{W}_i\}$.

(b) For all $i \in \mathbb{N}$: $R_i = (H_{E_i})_{i\,i+1} = (H_{E \setminus E_{i+1}})_{i\,i+1}$ and $\mathcal{W}_i = \mathcal{W}^{E_i}$.

(c) For all $i \in \mathbb{N}$: $\bigcup_{j \geq i} E_j$ is absorbing.

(d) For every sequence $(\mathbb{U}^i)_{i\in\mathbb{N}}$ of sub-Markov resolvents $\mathbb{U}^i$ on E_i possessing a proper potential kernel U^i_0 and such that $E_{\mathbb{U}^i} = \mathcal{W}_i$ there exists a sub-Markov resolvent $\mathbb{V}$ on E possessing a proper potential kernel V_0 and such that $E_{\mathbb{V}} = \mathcal{W}$, $V_\lambda f = (U^i_\lambda f_i + \sum_{m \geq i+1} [(I_i - \lambda U^i_\lambda) R_i \circ \ldots \circ (I_{m-1} - \lambda U^{m-1}_\lambda) R_{m-1}] U^m_\lambda f_m)_{i\in\mathbb{N}}$ $(f \in \underline{\underline{\mathcal{B}}}(E)_{b+}, \lambda \geq 0)$.

(e) If $q = (q_i)_{i\in\mathbb{N}} \in \mathcal{W}$ such that $\sum_{i\in\mathbb{N}} \|q_i\| < \infty$ and $q_i \in \mathcal{P}^i$ for all $i \in \mathbb{N}$ then $q \in \mathcal{P}$.

4.2. Proof of 4.1.

4.2.1. Preliminaries – Let $\lambda \geq 0$, $i \in \mathbb{N}$ and $\mathbb{U}^i$ be a sub-Markov resolvent on E_i possessing a proper potential kernel $U^i_0 = U^i$ and such that $E_{\mathbb{U}^i} = \mathcal{W}_i$ ($\mathbb{U}^i$ exists by /3/II 8.6). Let R^λ_i denote the sub-Markov kernel fulfilling $R^\lambda_i = (I_i - \lambda U^i_\lambda) R_i$ on $\underline{\underline{\mathcal{B}}}(E_{i+1})_b$ and we define kernels $R^\lambda, U_\lambda, V_\lambda, P^\lambda$ on $(E, \underline{\underline{\mathcal{B}}}(E))$ by $R^\lambda f := (R^\lambda_j f_{j+1})_{j\in\mathbb{N}}$, $U_\lambda f := (U^j_\lambda f_j)_{j\in\mathbb{N}}$, $R := R^0$, $U := U_0$, $P^\lambda := \sum_{n\in\mathbb{N}_0} (R^\lambda)^n$, $V_\lambda := P^\lambda U_\lambda$, $P := P^0$, $V := V_0$ $(f \in \underline{\underline{\mathcal{B}}}(E)_+)$. Then:

(1) $V_\lambda = U_\lambda + R^\lambda V_\lambda$, and for $f \in \underline{\underline{\mathcal{B}}}(E)_+$

(2) $P^\lambda f = (f_j + \sum_{m \geq j+1} R^\lambda_j \circ \ldots \circ R^\lambda_{m-1} f_m)_{j\in\mathbb{N}}$,

$$V_\lambda f = (U^j_\lambda f_j + \sum_{m \geq j+1} R^\lambda_j \circ \ldots \circ R^\lambda_{m-1} U^m_\lambda f_m)_{j\in\mathbb{N}}.$$

Let $\underline{\underline{B}} := \{f \in \underline{\underline{\mathcal{B}}}(E)_b \mid \exists\, n \in \mathbb{N}\ \forall\, m > n : f_m = 0\}$. Then $R^\lambda, I - R^\lambda, R^\lambda : \underline{\underline{B}} \to \underline{\underline{B}}$ and for $\lambda > 0$ $U_\lambda, V_\lambda : \underline{\underline{B}} \to \underline{\underline{B}}$ are linear operators such that $P^\lambda = (I - R^\lambda)^{-1} : \underline{\underline{B}} \to \underline{\underline{B}}$.

4.2.2. Proposition - $\mathbb{V} = (V_\lambda)_{\lambda>0}$ is a sub-Markov resovent on E possessing the proper potential kernel $V = PU$ and such that: (i) $(E_{\mathbb{V}})_b = \{u \in \underline{\underline{\mathcal{B}(E)}}_{b+} \mid \forall\, i \in \mathbb{N}: u_i, u_i - R_i u_{i+1} \in \mathcal{W}_i\}$, $1 \in E_{\mathbb{V}}$, (ii) $\lim_{\lambda\to\infty} \lambda V_\lambda f = f$ for all $f \in \mathcal{K}^+(E)$, (iii) $E_{\mathbb{V}} = S(E_{\mathbb{V}} \cap C_b(E))$.

Proof - By a result of Ben Saad (c.f. /1/) the resolvent equations hold on $\underline{\underline{B}}_+$ and because of $\underline{\underline{\mathcal{B}(E)}}_+ = S(\underline{\underline{B}}_+)$ $\mathbb{V}$ is a resolvent on E. Since $R_i(\underline{\underline{\mathcal{B}(E_{i+1})}}_{b+}) \subset \mathcal{P}^i_b$ we have $\lim_{\lambda\to 0} \lambda U^i_\lambda R_i f_{i+1} = 0$ for all $i \in \mathbb{N}$, $f \in \underline{\underline{\mathcal{B}(E_{i+1})}}_{b+}$ and this implies that $V = PU$ is the potential kernel of $\mathbb{V}$, where V being proper follows directly from (2). Now let $u \in \underline{\underline{\mathcal{B}(E)}}_{b+}$ such that $u_i, u_i - R_i u_{i+1} \in \mathcal{W}_i$ for all $i \in \mathbb{N}$ and define $u^n := (u_1, \ldots, u_n, 0, 0, \ldots)$ for $n \in \mathbb{N}$. Then $(I-R)u^n = ((u_j - R_j u_{j+1})_{j\in\{1,\ldots,n-1\}}, u_n, 0, 0, \ldots) \in E_{\mathbb{U}}$. Since $P(E_{\mathbb{U}}) \subset E_{\mathbb{V}}$, $P(I-R) = I$ on $\underline{\underline{B}}$ and $u^n \in \underline{\underline{B}}_+$ we have $u^n \in E_{\mathbb{V}}$ and hence $u = \sup_{m\in\mathbb{N}} u^m \in E_{\mathbb{V}}$. In particular, $1 \in E_{\mathbb{V}}$. Conversely let $v \in (E_{\mathbb{V}})_b$. Obviously, $v_i \in \mathcal{W}_i$ for all $i \in \mathbb{N}$. Since V is proper there exists a sequence $(g^n)_{n\in\mathbb{N}}$ in $\underline{\underline{\mathcal{B}(E)}}_{b+}$ such that $v = \uparrow\lim_{n\uparrow\infty} Vg^n$. From (1) we conclude $\infty > v = w + Rv$, where $w := \lim_{n\to\infty} Ug^n \in S_{\mathbb{U}}$ and $v, Rv \in E_{\mathbb{U}}$. Hence by regularization we obtain $v_i - R_i v_{i+1} \in \mathcal{W}_i$ for all $i \in \mathbb{N}$. Thus the first part of 4.2.2. included (i) is proved.

Note that $1 \in E_{\mathbb{V}}$ and (1) imply $\lim_{\lambda\to\infty} \lambda R^\lambda V_\lambda f = 0$ for all $f \in \underline{\underline{\mathcal{B}(E)}}_{b+}$. Since $(E_i, \mathcal{W}_i)$ $(i \in \mathbb{N})$ are balayage spaces, $\lim_{\lambda\to\infty} \lambda U_\lambda f = f$ holds for all $f \in \mathcal{K}(E)$. Using again (1) we obtain (ii).

Finally, by a result of V. Dembinski and K. Janßen (c.f. /7/(3.5)) for every $i \in \mathbb{N}$ there exists $g_i \in \underline{\underline{\mathcal{B}(E_i)}}_{b+}$, $0 < g_i \leq 1$, such that $U^i g_i \leq 2^{-i}$ and $U^i(g_i f_i) \in \mathcal{P}^i_b$ for all $f \in \underline{\underline{\mathcal{B}(E_i)}}_{b+}$. Let $g := (g_i)_{i\in\mathbb{N}}$. Then $V(g\cdot)$ is a strong Feller kernel and this implies (iii).

4.2.3. Proposition - Let $(K_n)_{n\in\mathbb{N}}$ be an exhaustion of E and $q = (q_i)_{i\in\mathbb{N}} \in E_{\mathbb{V}}$ such that $\sum_{i\in\mathbb{N}} \|q_i\| < \infty$ and $q_i \in \mathcal{P}^i$ for all $i \in \mathbb{N}$. Then $R_q^{\complement K_n} \underset{n\uparrow\infty}{\searrow} 0$ locally uniformly on E.

Proof - For $j, n \in \mathbb{N}$ let $G_{n,j} := E_j \setminus K_n$. Then for all $j \in \mathbb{N}$:

$$(*) \quad {}^iR_{q_j}^{G_{n,j}} \underset{n\uparrow\infty}{\searrow} 0 \quad \text{locally uniformly on } E_j.$$

Since $\sum_{j\in\mathbb{N}} \|q_j\| < \infty$ we conclude for $m, i \in \mathbb{N}$, $m \geq i+1$, that

$C_b(E_i) \ni \sum_{m \geq i+1} R_i \circ \ldots \circ R_{m-1}({}^m R_{q_m}^{G_{n,m}}) \underset{n\nearrow\infty}{\searrow} 0$, and this convergence is locally uniform on E_i by Dini's lemma. Together with (✱) we obtain by use of (2) that $P(({}^j R_{q_j}^{G_{n,j}})_{j\in\mathbb{N}}) \underset{n\nearrow\infty}{\searrow} 0$ locally uniformly on E. Since for all $v \in E_{\mathcal{W}}$ and $A \subset E$ such that for every $i \in \mathbb{N}$ $A_i := A \cap E_i \in \tau_{\mathcal{W}_i}$ the inequality $R_v^A \leq P(({}^j R_{v_j}^{A_j})_{j\in\mathbb{N}})$ holds, we conclude the desired result.

<u>4.2.4. Corollary</u> -There exists a strictly positive $q \in E_{\mathcal{W}} \cap C_b(E)$ such that $R_q^{\complement K_n} \underset{n\nearrow\infty}{\searrow} 0$ locally uniformly on E for every exhaustion $(K_n)_{n\in\mathbb{N}}$ of E.

<u>Proof</u>- Since $1 \in \mathcal{W}_j$ for all $j \in \mathbb{N}$, we find a sequence $(p_i)_{i\in\mathbb{N}}$ such that $0 < p_i \in \mathcal{P}_b^i$ for all $i \in \mathbb{N}$ and $\sum_{i\in\mathbb{N}} \|p_i\| \leq 1$. Let $q := P((2^{-i} \cdot p_i)_{i\in\mathbb{N}})$. Then by (2) $0 < q \in E_{\mathcal{W}} \cap C_b(E)$, $\sum_{i\in\mathbb{N}} \|q_i\| < \infty$ and $q_i \in \mathcal{P}_b^i$ for all $i \in \mathbb{N}$. An application of <u>4.2.3.</u> now finishes the proof.

<u>4.2.5. Corollary</u> -$(E,\mathcal{W}) := (E, E_{\mathcal{W}})$ is a balayage space.

<u>Proof</u>- Since we have <u>4.2.2.</u> and <u>4.2.4.</u> the assertion follows from /7/(1.4)c) and /3/II 4.7,4.8.

<u>4.2.6. Proposition</u> - Let $i \in \mathbb{N}$, $f_i \in \mathcal{W}_i$ and $A_i \in \mathfrak{B}(E_i)$.

(a) $f := f_i 1_{E_i} + \infty 1_{\complement \bigcup_{j \geq i} E_j} \in \mathcal{W}$.

(b) $\bigcup_{j\geq i} E_j$ is absorbing.

(c) $(H_{A_i})_{ii} = H_{A_i}^i$ and $\mathcal{W}_i = \mathcal{W}^{E_i}$.

(d) $R_i = (H_{E\setminus E_{i+1}})_{i\, i+1} = (H_{E_i})_{i\, i+1}$.

<u>Proof</u> - (a) follows from (2) and (b) is an immediate consequence of (a). For (c) it suffices to prove $(R_u^{B_i \cup (E\setminus E_i)})_i = {}^i R_{v_i}^{B_i}$, where $B_i := E_i \setminus A_i$, $v \in \mathcal{W}$, $u := (v_1, \ldots, v_i, 0, 0, \ldots) \in \mathcal{W}$ (by(b)); now "$\leq$" follows directly from (a) and the converse inequality "$\geq$" is trivial. The second equality in (d) is obtained by utilizing (b) and for the first one we have to show that $R_i v_{i+1} = (R_v^{E_{i+1}})_i$ holds for all $v \in \mathcal{W}$. To this end let $v \in \mathcal{W}$. Note that by (2) $R_i v_{i+1} = P(1_{E_{i+1}} v_{i+1})_i$, $v_{i+1} = P(1_{E_{i+1}} v_{i+1})_{i+1}$, $P(1_{E_{i+1}} v_{i+1}) \in \mathcal{W}$. Thus we have "$\geq$". Finally, if $u \in \mathcal{W}$ such that $u_{i+1} \geq v_{i+1}$ then

$u \geq P(1_{E_{i+1}} v_{i+1})$ (by (2)) and this implies "$\leq$".

Thus 4.1. is proved. Using a general "image theorem" for balayage spaces (c.f. /3/V 8.) we obtain for $I = \{1\}$, $E = E_1$, from 4.1.:

4.3. Corollary

Let $(E,\mathcal{U})$ be a balayage space such that $1 \in \mathcal{U}$ and $R: \mathcal{B}(E)_+ \to \mathcal{B}(E)_+$ a kernel such that $1-R1 \in \mathcal{U}$ and $R(\mathcal{B}(E)_{b+}) \subset \mathcal{P}_b$. Then we have (a)$\Longrightarrow$(b)$\Longrightarrow$(c) for the following conditions:

(a) There exists a sub-Markov resolvent $\mathbb{U}$ on E possessing a proper potential kernel U and such that $E_{\mathbb{U}} = \mathcal{U}$ and $W := \sum_{n \in \mathbb{N}_0} R^n U$ defines a proper kernel on E.

(b) $\mathcal{V}'_b := \{u \in \mathcal{B}(E)_{b+} \mid u, u-Ru \in \mathcal{U}\}$ separates the points of E.

(c) There exists a unique balayage space $(E,\mathcal{V})$ such that $1 \in \mathcal{V}$ and $\mathcal{V}_b = \mathcal{V}'_b$.

Proof - Let $\mathbb{U}$ be any sub-Markov resolvent on E possessing a proper potential kernel $U_0 = U$ and such that $E_{\mathbb{U}} = \mathcal{U}$. For $i \in \mathbb{N}$, $\lambda \geq 0$ we define $\tilde{E}_i := E \times \{i\}$; $\varphi_i: \tilde{E}_i \to E$, $(x,i) \mapsto x$ $(x \in E)$; $\psi_i := \varphi_i^{-1}$; $\tilde{E} := \sum_{j \in \mathbb{N}} \tilde{E}_j$; $\varphi := (\varphi_j)_{j \in \mathbb{N}}: \tilde{E} \to E$; $U^i_\lambda f_i := U_\lambda(f_i \circ \psi_i) \circ \varphi_i$, $R_i := R(f_i \circ \psi_i) \circ \varphi_i$ $(f_i \in \mathcal{B}(\tilde{E}_i)_+)$; $\mathbb{U}^i := (U^i_\lambda)_{\lambda > 0}$; $\mathcal{W}_i := \mathcal{W} \circ \varphi_i = E_{\mathbb{U}^i}$. Then by 4.1.(a),(d) $(\tilde{E},\mathcal{W}) := (\tilde{E}, E_{\mathbb{V}})$ is a balayage space such that $1 \in \mathcal{W}$ and $\mathcal{W}_b = \{u \in \mathcal{B}(\tilde{E})_{b+} \mid \forall\, i \in \mathbb{N}: u_i, u_i - R_i u_{i+1} \in \mathcal{W}_i\}$, where $\mathbb{V}$ is the resolvent associated with $(\mathbb{U}^i, R_i)_{i \in \mathbb{N}}$ by the relation in 4.1.(d). The structure of $\mathbb{V}$ and $\mathcal{W}$ immediately implies that $W_\lambda f := V_\lambda(f \circ \varphi) \circ \psi_1$ $(\lambda > 0,\ f \in \mathcal{B}(E)_+)$ defines a sub-Markov resolvent on E possessing $W = \sum_{n \in \mathbb{N}_0} R^n U$ as potential kernel and such that $(E_{\mathbb{W}})_b = \{v \in \mathcal{B}(E)_{b+} \mid v \circ \varphi \in \mathcal{W}\} = \mathcal{V}'_b$, $W_\lambda f = U_\lambda f + (R^\lambda V_\lambda(f \circ \varphi)) \circ \psi_1$ $(\lambda \geq 0,\ f \in \mathcal{B}(E)_+$, R^λ as in 4.2.1.); therefore, $\lim_{\lambda \to \infty} \lambda W_\lambda f = f$ for all $f \in \mathcal{K}(E)$ (since $1 \in \mathcal{W}$). Moreover, it is easily seen that for every $f: E \to \overline{\mathbb{R}}$ there exists $g: E \to \overline{\mathbb{R}}$ such that $R_{f \circ \varphi} = g \circ \varphi$. Thus (c) follows from (b) by the "image theorem" (c.f. /3/V 8.), and if (a) holds, i.e. W is proper, we conclude (b) from /3/II 4.8.

Thus we see that 4.1. is not so retrictive as it seems; together with the "image theorem" of J. Bliedtner and W. Hansen it covers the situation of certain perturbations of a given balayage space. The general problem of subordination for balayage spaces is discussed in /9/.

Suppose now that $I = \{1,2\}$.

4.4. Theorem

For every $i \in I$ let $(E_i, \mathcal{W}_i)$ be a balayage space such that $1 \in \mathcal{W}_i$ and $R_i : \underline{\underline{\mathcal{B}(E_{3-i})}}_+ \to \underline{\underline{\mathcal{B}(E_i)}}_+$ a kernel such that $1-R_i 1 \in \mathcal{W}_i$ and $R_i(\underline{\underline{\mathcal{B}(E_{3-i})}}_{b+}) \subset \mathcal{P}_b^i$.

1. The following statements are equivalent:

(a) For every $i \in I$ there exists a sub-Markov resolvent $\mathbb{U}^i$ on E_i possessing a proper potential kernel U^i and such that $\mathcal{E}_{\mathbb{U}^i} = \mathcal{W}_i$ and $V = \sum_{n \in \mathbb{N}_0} \begin{pmatrix} 0 & R_1 \\ R_2 & 0 \end{pmatrix}^n \begin{pmatrix} U^1 & 0 \\ 0 & U^2 \end{pmatrix}$ defines a proper kernel on E.

(a') For every $i \in I$ there exists a sub-Markov resolvent $\mathbb{U}^i$ on E_i possessing a bounded potential kernel U^i and such that $\mathcal{E}_{\mathbb{U}^i} = \mathcal{W}_i$ and $V = \sum_{n \in \mathbb{N}_0} \begin{pmatrix} 0 & R_1 \\ R_2 & 0 \end{pmatrix}^n \begin{pmatrix} U^1 & 0 \\ 0 & U^2 \end{pmatrix}$ defines a bounded kernel on E.

(b) $\mathcal{W}_b' := \{u \in \underline{\underline{\mathcal{B}(E)}}_{b+} \mid \forall\, i \in I: u_i, u_i - R_i u_{3-i} \in \mathcal{W}_i\}$ separates the points of E.

(c) There exists a unique balayage space $(E, \mathcal{W})$ such that $1 \in \mathcal{W}$ and $\mathcal{W}_b = \mathcal{W}_b'$.

2. If one of the conditions (a), (a'), (b), (c) holds, then $R_i = (H_{E_i})_{i\,3-i}$ and $\mathcal{W}_i = \mathcal{W}^{E_i}$.

Proof - 1.) (a')$\Longrightarrow$(a): Trivial.

2.) (a)$\Longrightarrow$(b)$\Longrightarrow$(c): This follows from 4.1. and the "image theorem" (/3/V 8.) by the same procedure as in the proof of 4.3., where we now define $\tilde{E}_{2j-1} := E_1 \times \{2j-1\}$, $\tilde{E}_{2j} := E_2 \times \{2j\}$ for $j \in \mathbb{N}$.

3.) (c)$\Longrightarrow$ $R_i = (H_{E_i})_{i\,3-i}$ and $\mathcal{W}_i = \mathcal{W}^{E_i}$ for all $i \in I$:

Therefor let $i \in I$, $A_i \in \mathcal{B}(E_i)$. It suffices to prove

$$(*) \quad (R_u^{E_{3-i} \cup A_i})_i = {}^iR^{A_i}_{(u_i - R_i u_{3-i})} + R_i u_{3-i} \quad \text{for all } u \in \mathcal{W}_b,$$

since (*) implies $(H_{E_i} p)_i = R_i p_{3-i}$ for all $p \in \mathcal{P}$ (by taking $A_i = \emptyset$) and $(H_{B_i})_{ii}(p_i - q_i) = H^i_{B_i}(p_i - q_i)$ for $B_i = E_i \setminus A_i$ and all $p, q \in \mathcal{P}_b$ such that $p_{3-i} = q_{3-i}$. To this end let $u \in \mathcal{W}_b$ and $v_i \in \mathcal{W}_i$ such that $v_i \leq u_i - R_i u_{3-i}$ and $v_i = u_i - R_i u_{3-i}$ on A_i.

Then $w := 1_{E_i}(v_i + R_i u_{3-i}) + 1_{E_{3-i}} u_{3-i} \leqslant u$, $w = u$ on $E_{3-i} \cup A_i$ and $w_1 \in (\mathcal{W}_1)_b$, $w_2 \in (\mathcal{W}_2)_b$. Moreover, from $w_i - R_i w_{3-i} = v_i \in \mathcal{W}_i$ and

$$w_{3-i} - R_{3-i} w_i = \underbrace{(u_{3-i} - R_{3-i} u_i)}_{\in \mathcal{W}_i} + \underbrace{R_{3-i}(\underbrace{(u_i - R_i u_{3-i}) - v_i}_{\geqslant 0})}_{\in \mathcal{W}_i} \in \mathcal{W}_i$$

we conclude that $w \in \mathcal{W}_b$. Thus we have $(R_u^{E_{3-i} \cup A_i})_i \leqslant w_i = v_i + R_i u_{3-i}$ and this implies "$\leqslant$" in (*). Conversely let $v \in \mathcal{W}$ such that $v \leqslant u$, $v_{3-i} = u_{3-i}$ and $v_i = u_i$ on A_i. Then $v \in \mathcal{W}_b$ and hence $\mathcal{W}_i \ni v_i - R_i v_{3-i} = v_i - R_i u_{3-i} \leqslant u_i - R_i u_{3-i}$ and $v_i - R_i u_{3-i} = u_i - R_i u_{3-i}$ on A_i. This implies $v_i - R_i u_{3-i} \geqslant {}^i R^{A_i}_{(u_i - R_i u_{3-i})}$ and we obtain "$\geqslant$" in (*).

4.) (c)$\Longrightarrow$(a'): From the results of J. Bliedtner and W. Hansen (c.f. /3/II 8.6, IV 3.4, 7.5, 7.6) we obtain the existence of a Hunt process $\mathfrak{X}$ on E possessing a bounded potential kernel V and such that $E_{\mathfrak{X}} = \mathcal{W}$. Since $\mathcal{W}_i = \mathcal{W}^{E_i}$ by 3.) $(i = 1,2)$ we conclude from 3.1.(a) and 3.3.(a),(c) that for every $i \in I$ there exists a sub-Markov resovent $\mathbb{U}^i$ on E_i possessing a bounded potential kernel U^i and such that $E_{\mathbb{U}^i} = \mathcal{W}_i$ and

$$V = \sum_{n \in \mathbb{N}_0} \begin{pmatrix} 0 & (H_{E_1})_{12} \\ (H_{E_2})_{21} & 0 \end{pmatrix}^n \begin{pmatrix} U^1 & 0 \\ 0 & U^2 \end{pmatrix}.$$

Since $R_i = (H_{E_i})_{i\,3-i}$ for $i \in I$ by 3.) we see that (a') holds. ┘

4.5. Remark

To give a further probabilistic interpretation of the kernels R_i $(i = 1,2)$ in the situation of 4.4., we note that the following can be shown (c.f. /8/2.,3.):

If for $i \in I$ $\mathfrak{X}^i = (\Omega, \mathfrak{m}^i, \mathfrak{m}^i_t, X^i_t, \theta^i_t, P^x_i)$ is a Hunt process on E_i such that the lifetime ζ_i is finite a.s. and $\mathcal{W}_i = E_{\mathfrak{X}^i}$ there exist sub-Markov kernels $B_i : \underline{\underline{\mathcal{B}}}(E_{3-i})_+ \longrightarrow \underline{\underline{\mathcal{B}}}(E_i)_+$ such that $R_i = Q_i B_i$, where the kernel Q_i, defined by $Q_i f_i(x) := P^x_i(f_i \circ X^i_{\zeta_i -})$ $(x \in E_i$, $f_i \in \underline{\underline{\mathcal{B}}}(E_i)_+)$, is the potential kernel associated with the potential p_i from the Riesz-decomposition of $1 \in \mathcal{W}_i$, i.e.

$p_i(x) = P^x_i([X^i_{\zeta_i -} \in E_i])$ $(x \in E_i$, $i = 1,2)$.

BIBLIOGRAPHY

/1/ H. Ben Saad:
Exemples de noyaux admettant des résolvantes.
Math. Ann. 265 (1983), p. 149-154.

/2/ H. Ben Saad, K. Janßen:
Résolvantes sur les sommes d'espaces topologiques.
Unpublished.

/3/ J. Bliedtner, W. Hansen:
Potential theory.
Springer, Berlin 1986.

/4/ N. Boboc, Gh. Bucur:
Perturbations in excessive structures.
In: "Complex Analysis: Fifth Romanian Finish Seminar", LN in Math. 1014, Springer, Berlin 1983, p. 155-187.

/5/ A. Boukricha:
Espaces biharmoniques.
In: "Théorie du potentiel: Proceedings of the colloque J. Deny", LN in Math. 1096, Springer. Berlin 1984, p. 116-148.

/6/ N. Bouleau:
Espaces biharmoniques et couplage de processus de Markov.
J. Math. pures et appl. 58 (1979), p. 187-240.

/7/ V. Dembinski, K. Janßen:
Standard balayage spaces and standard Markov processes.
In: "Potential Theory Copenhagen 1979", LN in Math. 787, Springer, Berlin 1980, p. 84-105.

/8/ M. Meyer:
Balayage-Räume und Hunt-Prozesse auf abzählbaren topologischen Summen.
Staatsexamensarbeit, Düsseldorf 1986.

/9/ H.-H. Müller:
Subordination for balayage spaces.
To appear.

/10/ E.P. Smyrnélis:
Axiomatique des fonctions biharmoniques.
I: Ann. Inst. Fourier 25 (1975), p. 35-97.
II: Ann. Inst. Fourier 26 (1976), p. 1-47.

SUBORDINATION FOR BALAYAGE SPACES

Hans-Helge Müller

Institut für Statistik und Dokumentation der
Universität Düsseldorf, Universitätsstraße 1
4000 Düsseldorf 1, Federal Republic of Germany

INTRODUCTION

Let X be a lccb space, $\mathbb{U}$ and $\mathbb{V}$ two sub-Markov resolvents on X possessing proper potential kernels U and V such that $\mathbb{V}$ is subordinated to $\mathbb{U}$ (i.e. $V_\lambda \leq U_\lambda \quad \forall \lambda \in]0, \infty[$). Denote by $S_{\mathbb{U}}$, $S_{\mathbb{V}}$ respectively $E_{\mathbb{U}}$, $E_{\mathbb{V}}$ the cone of supermedian respectively excessive functions associated with $\mathbb{U}$ and $\mathbb{V}$. The purpose of this article is: (i) to study the situation where $(X, E_{\mathbb{U}})$ and $(X, E_{\mathbb{V}})$ are both balayage spaces (in the sense of [3]) and P is the sub-Markov kernel on X satisfying $U = V + PU$, and (ii) to give conditions on U, V and P under which $(X, E_{\mathbb{V}})$ respectively $(X, E_{\mathbb{U}})$ form a balayage space, provided $(X, E_{\mathbb{U}})$ respectively $(X, E_{\mathbb{V}})$ is a balayage space. The notations are the notations of [3].

0. PRELIMINARIES

Bliedtner, Hansen, Dembinski and Janßen found necessary and sufficient conditions on a sub-Markov resolvent $\mathbb{W}$ on X possessing a proper potential kernel W such that X with the cone of $\mathbb{W}$-excessive functions is a balayage space. From II 4.7., II 4.8. and II 5.2. in [3] and 1.4 c) and 3.5 in [5] we obtain:

0.1. PROPOSITION: $(X, E_{\mathbb{W}})$ is a balayage space $\iff$

i) $\forall f \in K^+(X): \lim_{\lambda\to\infty} \lambda W_\lambda f = f$ respectively $\forall f \in C_b(X): \lim_{\lambda\to\infty} \lambda W_\lambda f = f$,

ii) $E_{\mathbb{W}} = S(E_{\mathbb{W}} \cap C(X))$ and

iii) $\exists\, p, q \in E_{\mathbb{W}} \cap C(X)$ with $p, q > 0$: $\frac{p}{q} \in C_o(X)$ respectively

$\exists\, p \in E_{\mathbb{W}} \cap C(X)$ with $p > 0$: ${}_{\mathbb{W}}R_p^{\complement K_n} \xrightarrow[n\to\infty]{} 0$ locally uniformly for some exhaustion $(K_n)_{n\in\mathbb{N}}$ of X.

0.2. PROPOSITION: $E_{\mathbb{W}} = S(E_{\mathbb{W}} \cap C(X)) \iff$
$\exists h \in \mathcal{B}^+(X)$ with $0 < h \leq 1$: Wh is a strong-Feller kernel.
If $(X, E_{\mathbb{W}})$ is a balayage space, we have in fact $Wh(\mathcal{B}_b^+(X)) \subset \mathcal{P}_{\mathbb{W},b}$.

1. SUBORDINATION

Let $\mathbb{U}$ and $\mathbb{V}$ be sub-Markov resolvents on X such that the potential kernel U of $\mathbb{U}$ is proper, $\mathbb{V}$ is subordinated to $\mathbb{U}$ and $(X, E_{\mathbb{U}})$ is a balayage space. Combining the results of 19. in [2] with standard arguments in the context of balayage spaces, we obtain a family of kernels associated with the subordination of the resolvents:

1.1. PROPOSITION: There exists a unique family $(P_\lambda)_{\lambda \in [0,\infty[}$ of sub-Markov kernels on X such that $U_\lambda = V_\lambda + P_\lambda U_\lambda \;\; \forall \lambda \in [0,\infty[$. Furthermore, we have: $\forall \lambda \in [0,\infty[: P_\lambda u \leq u \;\; \forall u \in E_{\mathbb{U}_\lambda}$, $P_\lambda + \lambda V_\lambda P = P$, $P_\lambda(\mathcal{B}^+(X)) \subset S_{\mathbb{V}_\lambda}$ and $P_\lambda \underset{\lambda \searrow 0}{\uparrow} P$.

A simple, but a little technical argumentation (using Urysohn's lemma) leads to the following lemma, where $(X, E_{\mathbb{U}})$ is not assumed to be a balayage space:

1.2. LEMMA: Let $v \in E_{\mathbb{V}} \cap C_b(X)$ and $x \in X$ with $v(x) > 0$. Then $\forall f \in K^+(X): \lim_{\lambda\to\infty} \lambda U_\lambda f(x) = f(x)$ implies $\forall f \in C_b(X): \lim_{\lambda\to\infty} \lambda V_\lambda f(x) = f(x)$.

1.3. PROPOSITION: a) $\exists v \in E_{\mathbb{V}} \cap C_b(X): v > 0 \Rightarrow E_{\mathbb{U}} \subset E_{\mathbb{V}}$.
b) $\exists u \in E_{\mathbb{U}} \cap E_{\mathbb{V}}: 0 < u < \infty \Rightarrow \forall \lambda \in [0,\infty[: P_\lambda(\mathcal{B}^+(X)) \subset E_{\mathbb{V}_\lambda}$ and $P_\lambda(E_{\mathbb{U}_\lambda}) \subset E_{\mathbb{U}_\lambda}$.

<u>Proof</u>: a) follows from 0.1. and 1.2. .
b) Since $u \in E_{\mathbb{U}}$, there exists from Hunt's approximation-theorem a sequence $(f_n)_{n\in\mathbb{N}}$ in $\mathcal{B}^+(X)$ such that $u = \uparrow\lim_{n\nearrow\infty} U_\lambda f_n$. $U_\lambda = V_\lambda + P_\lambda U_\lambda$ therefore yields $\infty > u = \lim_{n\to\infty} V_\lambda f_n + P_\lambda u$, where $P_\lambda u$ and $\lim_{n\to\infty} V_\lambda f_n$ are elements of $S_{\mathbb{V}_\lambda}$. Since $u \in E_{\mathbb{V}}$, we conclude $\infty > P_\lambda u \in E_{\mathbb{V}_\lambda}$. Thus we have from $u > 0$ and $P_\lambda(\mathcal{B}^+(X)) \subset S_{\mathbb{V}_\lambda}$: $P_\lambda(\mathcal{B}^+(X)) \subset E_{\mathbb{V}_\lambda}$.
For every $v \in E_{\mathbb{U}_\lambda}$ there exists a sequence $(g_n)_{n\in\mathbb{N}}$ in $\mathcal{B}^+(X)$ such that $v = \uparrow\lim_{n\nearrow\infty} U_\lambda g_n$ and $U_\lambda g_n < \infty$. $\forall \mu \in]0,\infty[: \mu U_{\lambda+\mu} P_\lambda U_\lambda g_n = \mu U_{\lambda+\mu}(U_\lambda g_n - V_\lambda g_n)$
$\leq \mu U_{\lambda+\mu} U_\lambda g_n - \mu V_{\lambda+\mu} V_\lambda g_n = U_\lambda g_n - U_{\lambda+\mu} g_n - V_\lambda g_n + V_{\lambda+\mu} g_n \leq U_\lambda g_n - V_\lambda g_n = P_\lambda U_\lambda g_n$.
This demonstrates $P_\lambda v \in S_{\mathbb{U}_\lambda} \cap E_{\mathbb{V}_\lambda} \subset E_{\mathbb{U}_\lambda}$.

An immediate consequence of 0.2., the condition $E_{\mathbb{U}} = S(E_{\mathbb{U}} \cap C(X))$ and the equation $U = V + PU$ is:

1.4. LEMMA: $E_{\mathbb{V}} = S(E_{\mathbb{V}} \cap C(X)) \iff$
$\exists h \in \mathcal{B}^+(X)$ with $0 < h \leq 1$: PUh is a strong-Feller kernel.

1.5. THEOREM: The following properties are equivalent:

(1) $(X, E_{\mathbb{V}})$ is a balayage space.

(2) $E_{\mathbb{U}} \subset E_{\mathbb{V}}$ and $E_{\mathbb{V}} = S(E_{\mathbb{V}} \cap C(X))$.

(3) $V1 > 0$ and $P(\mathcal{P}_{\mathbb{U},b}) \subset C_b(X)$.

(4) $V1 > 0$ and P is a strong-Feller kernel.

(5) $\exists v \in E_{\mathbb{V}} \cap C_b(X)$: $v > 0$ and all functions in $E_{\mathbb{V}}$ are l.s.c..

Proof: (1) $\Rightarrow$ (5) and (4) $\Rightarrow$ (3) are evident.

(5) $\Rightarrow$ (4): Obviously, $V1 > 0$. From 1.3.a) we obtain $1 \in E_{\mathbb{U}} \subset E_{\mathbb{V}}$, and by 1.3.b) we know $P(\mathcal{B}^+(X)) \subset E_{\mathbb{V}}$. Since $1 = \uparrow\lim_{n\nearrow\infty} Uf_n$ for a sequence $(f_n)_{n\in\mathbb{N}}$ in $\mathcal{B}^+(X)$, $U = V + PU$ implies $1 = \lim_{n\to\infty} Vf_n + P1$, where $\lim_{n\to\infty} Vf_n$ is an element of $S_{\mathbb{V}}$. Hence $\lim_{n\to\infty} Vf_n \in E_{\mathbb{V}}$. Using that all functions in $E_{\mathbb{V}}$ are l.s.c., we conclude $P1 \in C_b(X)$. Moreover, since $P(\mathcal{B}^+(X)) \subset E_{\mathbb{V}}$, P is a strong-Feller kernel.

(3) $\Rightarrow$ (2) follows from 0.1., 0.2., 1.4. and 1.3.a) if we choose v in the form $v = Vh$.

(2) $\Rightarrow$ (1) is a consequence of 0.1. and 1.2..

1.6. THEOREM: If $(X, E_{\mathbb{V}})$ is a balayage space, then the following is true:

a) $E_{\mathbb{U}} = E_{\mathbb{V}} \cap S_{\mathbb{U}}$ and $\mathcal{P}_{\mathbb{U}} \subset \mathcal{P}_{\mathbb{V}}$.

b) $P(\mathcal{B}_b^+(X)) \subset \mathcal{P}_{\mathbb{V},b}$.

c) $\forall \lambda \in [0, \infty[: U_\lambda = \sum_{n\in\mathbb{N}_0} P_\lambda^{\,n} V_\lambda$.

d) $u \in E_{\mathbb{U}} \iff u \in E_{\mathbb{V}} \wedge Pu \prec_{E_{\mathbb{V}}} u \iff u \in E_{\mathbb{V}} \wedge \forall \lambda \in]0, \infty[: \lambda V_\lambda u + P_\lambda u \leq u$.

e) $v \in \mathcal{B}^+(X) \wedge \sum_{n\in\mathbb{N}_0} P^n v \in E_{\mathbb{U}} \wedge \sum_{n\in\mathbb{N}} P^n v < \infty \Rightarrow v \in E_{\mathbb{V}} \Rightarrow \sum_{n\in\mathbb{N}_0} P^n v \in E_{\mathbb{U}}$.

f) $\exists p \in \mathcal{P}_{\mathbb{V},b}$ with $p > 0$: $\sum_{n\in\mathbb{N}_0} P^n p \in \mathcal{P}_{\mathbb{U},b}$ and $\forall q \in \mathcal{P}_{\mathbb{V}}^{\,c}$: $\sum_{n\in\mathbb{N}_0} P^n q \in \mathcal{P}_{\mathbb{U},b}$.

g) The $E_{\mathbb{U}}$-fine- and the $E_{\mathbb{V}}$-fine-topology coincide.

Proof: a) is an immediate consequence of $E_{\mathbb{U}} \subset E_{\mathbb{V}}$.

b) According to 0.2. we choose a function $h \in \mathcal{B}^+(X)$ such that $0 < h \leq 1$ and $0 < Uh \in \mathcal{P}_{\mathbb{U},b} \subset \mathcal{P}_{\mathbb{V},b}$. Further we choose a sequence $(f_n)_{n\in\mathbb{N}}$ in $\mathcal{B}_b^+(X)$ such that $1 = \sum_{n\in\mathbb{N}} f_n$ and such that the support $S(f_n)$ is compact. Then there exists a sequence $(m_n)_{n\in\mathbb{N}}$ in $\mathbb{N}$ such that $f_n \leq m_n Uh$, and consequently $Pf_n \leq m_n PUh \leq m_n Uh \in \mathcal{P}_{\mathbb{V},b}$. Since $P(\mathcal{B}_b^+(X)) \subset E_{\mathbb{V}} \cap C_b(X)$, we conclude $P1 = \sum_{n\in\mathbb{N}} Pf_n \in \mathcal{P}_{\mathbb{V},\sigma,b} = \mathcal{P}_{\mathbb{V},b}$. Moreover, this shows b).

c) i) From $U_\lambda = V_\lambda + P_\lambda U_\lambda$ we obtain inductively $U_\lambda = \sum_{m=0}^{n} P_\lambda^{\,m} V_\lambda + P_\lambda^{\,n+1} U_\lambda$, hence $U_\lambda = \sum_{n\in\mathbb{N}_0} P_\lambda^{\,n} V_\lambda + \downarrow_{n\nearrow\infty} P_\lambda^{\,n} U_\lambda$. Thus we have to show $\downarrow_{n\nearrow\infty} P_\lambda^{\,n} U_\lambda = 0$. Since for every $\lambda \in]0, \infty[$ $P_\lambda \leq P$ and $U_\lambda \leq U$, it is sufficient to demonstrate this for $\lambda = 0$. By 0.2. we choose a function $h \in \mathcal{B}^+(X)$ such that $0 < h \leq 1$ and $Uh \in \mathcal{P}_{\mathbb{U},b}$. If we have seen that $p := \downarrow\lim_{n\nearrow\infty} P^n Uh = 0$, then c) will be proved.

ii) From 1.3.b) we know $P(E_{\mathbb{U}}) \subset E_{\mathbb{U}}$, thus $p \in S_{\mathbb{U}}$. Since $Vh \in E_{\mathbb{V}}$ and $P(\mathcal{B}^+(X)) \subset E_{\mathbb{V}}$, we have $q := \sum_{n\in\mathbb{N}_0} P^n V h \in E_{\mathbb{V}}$. The inclusions $E_{\mathbb{U}} \subset E_{\mathbb{V}}$ and $S_{\mathbb{U}} \subset S_{\mathbb{V}}$ and the equation $Uh = q + p$ therefore imply $p \in E_{\mathbb{V}}$. Hence $p \in E_{\mathbb{V}} \cap S_{\mathbb{U}} = E_{\mathbb{U}}$. All functions in $E_{\mathbb{U}}$ and $E_{\mathbb{V}}$ are l.s.c. , and Uh is continuous, thus $p \in E_{\mathbb{U}} \cap C(X)$. Moreover, $p \le Uh \in \mathcal{P}_{\mathbb{U},b}$, hence $p \in \mathcal{P}_{\mathbb{U},b}$. For every $n \in \mathbb{N}$ $P^n U h$ is dominated by Uh. Therefore, the definition of p and Lebesgue's convergence theorem yields $Pp = p$. According to 1.1. $Pu \le u$ $\forall u \in E_{\mathbb{U}}$. This demonstrates $P(x, \cdot) \in \mathcal{M}_{x,\mathbb{U}}$ and $P(x, p) = p(x)$ for every $x \in X$. From $h > 0$ and $V1 > 0$ we obtain $PUh < Vh + PUh = Uh$, hence $P(x, \cdot) \neq \varepsilon_x$ for every $x \in X$. Thus we have $\delta_{\mathbb{U}}(p) = \emptyset$, and we finally conclude $p = {}_{\mathbb{U}}R_p^{\delta_{\mathbb{U}}(p)} = {}_{\mathbb{U}}R_p^{\emptyset} = 0$.

d) i) Let $u \in E_{\mathbb{U}}$. Then $u \in E_{\mathbb{V}}$, and there exists a sequence $(f_n)_{n\in\mathbb{N}}$ in $\mathcal{B}^+(X)$ such that $u = \uparrow\lim_{n\nearrow\infty} U f_n$. We obtain $u = \liminf_{n\to\infty} V f_n + Pu$, where $\liminf_{n\to\infty} V f_n$ is an element of $S_{\mathbb{V}}$. Since $Pu \in E_{\mathbb{V}}$, the $\mathbb{V}$-excessive regularisation of the last equation yields $u = \widehat{\liminf_{n\to\infty} V f_n}^{\mathbb{V}} + Pu$. Hence $Pu \prec_{E_{\mathbb{V}}} u$.

ii) Let $u \in E_{\mathbb{V}}$ such that $Pu \prec_{E_{\mathbb{V}}} u$. Then we choose a function $v \in E_{\mathbb{V}}$ such that $Pu + v = u$. Let $\lambda \in]0, \infty[$. Adding $Pu + \lambda V_\lambda Pu$ to both sides of the inequality $\lambda V_\lambda v \le v$, we obtain $\lambda V_\lambda u + Pu \le u + \lambda V_\lambda Pu$, i.e. $\lambda V_\lambda u + P_\lambda u \le u$ by 1.1. .

iii) Let $u \in E_{\mathbb{V}}$ such that $\lambda V_\lambda u + P_\lambda u \le u$ $\forall \lambda \in]0, \infty[$. Similar to 1.3. of [4], we obtain inductively $\lambda \sum_{n=0}^{m} P_\lambda^{\,n} V_\lambda u \le u$ $\forall m \in \mathbb{N}_0$. From c) we therefore have $\lambda U_\lambda u \le u$. Hence $u \in E_{\mathbb{V}} \cap S_{\mathbb{U}} = E_{\mathbb{U}}$.

e) i) Let $v \in \mathcal{B}^+(X)$ such that $\sum_{n\in\mathbb{N}_0} P^n v \in E_{\mathbb{U}}$ and $\sum_{n\in\mathbb{N}} P^n v < \infty$. Then we obtain from d): $\infty > \sum_{n\in\mathbb{N}} P^n v = P \sum_{n\in\mathbb{N}_0} P^n v \prec_{E_{\mathbb{V}}} \sum_{n\in\mathbb{N}_0} P^n v = v + \sum_{n\in\mathbb{N}} P^n v$, i.e. $v \in E_{\mathbb{V}}$.

ii) Let $v \in E_{\mathbb{V}}$. We choose a sequence $(f_m)_{m\in\mathbb{N}}$ in $\mathcal{B}^+(X)$ such that $v = \uparrow\lim_{m\nearrow\infty} V f_m$. Then $\sum_{n\in\mathbb{N}_0} P^n v = \uparrow\lim_{m\nearrow\infty} \sum_{n\in\mathbb{N}_0} P^n V f_m = \uparrow\lim_{m\nearrow\infty} U f_m \in E_{\mathbb{U}}$ follows from c).

f) According to 0.2. we choose functions $h_1, h_2 \in \mathcal{B}^+(X)$ such that $0 < h_1, h_2 \le 1$, $Uh_1(\mathcal{B}_b^+(X)) \subset \mathcal{P}_{\mathbb{U},b}$ and $Vh_2(\mathcal{B}_b^+(X)) \subset \mathcal{P}_{\mathbb{V},b}$. Let $h := h_1 h_2$ and $p := Vh$.

i) It follows from c) and $V1 > 0$ that $p \in \mathcal{P}_{\mathbb{V},b}$ such that $p > 0$ and $\sum_{n\in\mathbb{N}_0} P^n p \in \mathcal{P}_{\mathbb{U},b}$.

ii) Let $q \in \mathcal{P}_{\mathbb{V}}^c$. Then $q \le mp$ on the compact carrier $C_{\mathbb{V}}(q)$ for some $m \in \mathbb{N}$. Hence $q = {}_{\mathbb{V}}R_q^{C_{\mathbb{V}}(q)} \le mp$, and we conclude $\sum_{n\in\mathbb{N}_0} P^n q \le m \sum_{n\in\mathbb{N}_0} P^n p \in \mathcal{P}_{\mathbb{U},b}$. From Dini's lemma the series $\sum_{n\in\mathbb{N}_0} P^n mp$ is locally uniformly convergent. Consequently, the series $\sum_{n\in\mathbb{N}_0} P^n q$ is locally uniformly convergent. Using e), we thus obtain $\sum_{n\in\mathbb{N}_0} P^n q \in E_{\mathbb{U}} \cap C(X)$. Since $\sum_{n\in\mathbb{N}_0} P^n q$ is dominated by a function in $\mathcal{P}_{\mathbb{U},b}$, $\sum_{n\in\mathbb{N}_0} P^n q \in \mathcal{P}_{\mathbb{U},b}$ follows.

g) Since $E_{\mathbb{U}} \subset E_{\mathbb{V}}$, the $E_{\mathbb{V}}$-fine-topology is finer than the $E_{\mathbb{U}}$-fine-topology. To prove the converse, let $x \in X$ and $s, t \in E_{\mathbb{V}}$ such that $x \in \{s < t\}$. We choose a function $v \in E_{\mathbb{V}}$ such that $v > 0$ and $\sum_{n\in\mathbb{N}_0} P^n v < \infty$. For every $n \in \mathbb{N}$ let

$t_n := \inf(t, n\,v) \in E_{\mathbb{V}}$. Since $t = \uparrow\lim_{n\nearrow\infty} t_n$, there exists a number $m \in \mathbb{N}$ such that $x \in \{s < t_m\} \subset \{s < t\}$. Let $s_m := \inf(s, t_m) \in E_{\mathbb{V}}$. Then $s_m \le t_m \le m\,v < \infty$ and $x \in \{s_m < t_m\} = \{s < t_m\} \subset \{s < t\}$. We define $s' := \sum_{n\in\mathbb{N}_0} P^n s_m$, $t' := \sum_{n\in\mathbb{N}_0} P^n t_m$, $s'' := P s'$ and $t'' := P t'$. By e) and 1.3.b) these functions are $\mathbb{U}$-excessive, and from the definition we obtain $s_m = s' - s''$ and $t_m = t' - t''$. Hence $s' + t''$, $s'' + t' \in E_{\mathbb{U}}$ and $x \in \{s' + t'' < s'' + t'\} \subset \{s < t\}$. According to $1 \in E_{\mathbb{U}} \subset E_{\mathbb{V}}$ this is sufficient to show that the $E_{\mathbb{U}}$-fine-topology is finer than the $E_{\mathbb{V}}$-fine-topology.

2. SUPERORDINATION

Let $\mathbb{V}$ be a sub-Markov resolvent on X such that the potential kernel V of $\mathbb{V}$ is proper and such that $(X, E_{\mathbb{V}})$ is a balayage space. Further, let P be a kernel on X satisfying $P(\mathcal{B}_b^+(X)) \subset E_{\mathbb{V},b}$. Denote by $(P_\lambda)_{\lambda\in[0,\infty[}$ the unique family of kernels on X such that $P_\lambda + \lambda V_\lambda P = P \quad \forall \lambda \in [0, \infty[$.

2.1. LEMMA: a) $\downarrow_{\lambda\nearrow\infty} P_\lambda = 0$.

b) If P satisfies $P(\mathcal{B}_b^+(X)) \subset \mathcal{P}_{\mathbb{V},b}$, then $P_\lambda \underset{\lambda\searrow 0}{\uparrow} P$.

Proof: a) Since $P(\mathcal{B}_b^+(X)) \subset E_{\mathbb{V},b}$, we have $\uparrow\lim_{\lambda\nearrow\infty} \lambda V_\lambda P f = P f \quad \forall f \in \mathcal{B}_b^+(X)$. Thus the result follows from $P_\lambda + \lambda V_\lambda P = P \quad \forall \lambda \in\,]0, \infty[$.

b) According to 0.2. there exists a strictly positive potential $p \in \mathcal{P}_{\mathbb{V}}$ of the form $p = V h$. From the resolvent equation $\downarrow\lim_{\lambda\searrow 0} \lambda V_\lambda p = 0$ follows. The arguments in the proof of 1.6.f) show $\downarrow\lim_{\lambda\searrow 0} \lambda V_\lambda q = 0 \quad \forall q \in \mathcal{P}_{\mathbb{V}}^c$. Since each $q \in \mathcal{P}_{\mathbb{V}}$ is a countable sum of a sequence in $\mathcal{P}_{\mathbb{V}}^c$, we conclude $\downarrow\lim_{\lambda\searrow 0} \lambda V_\lambda q = 0 \quad \forall q \in \mathcal{P}_{\mathbb{V}}$. Thus $P(\mathcal{B}_b^+(X)) \subset \mathcal{P}_{\mathbb{V},b}$ yields $\downarrow\lim_{\lambda\searrow 0} \lambda V_\lambda P f = 0 \quad \forall f \in \mathcal{B}_b^+(X)$. Hence the result is a consequence of $P_\lambda + \lambda V_\lambda P = P \quad \forall \lambda \in\,]0, \infty[$.

2.2. PROPOSITION: Let $\mathbb{U}$ be a sub-Markov resolvent on X such that the potential kernel U of $\mathbb{U}$ is proper. If P satisfies $P(\mathcal{B}_b^+(X)) \subset E_{\mathbb{V}} \cap C_b(X)$ and $U = V + P U$, then the following is true:

a) $\forall \lambda \in [0, \infty[: U_\lambda = V_\lambda + P_\lambda U_\lambda$. In particular, $\mathbb{V}$ is subordinated to $\mathbb{U}$.

b) $\forall f \in C_b(X)$: $\lim_{\lambda\to\infty} \lambda U_\lambda f = f$.

c) $E_{\mathbb{U}} = S(E_{\mathbb{U}} \cap C(X))$.

d) $E_{\mathbb{U}} = E_{\mathbb{V}} \cap S_{\mathbb{U}}$.

Proof: a) follows from 19. in [2].

b) Since $(X, E_{\mathbb{V}})$ is a balayage space, we obtain from 0.1. $\forall f \in C_b(X)$: $\lim_{\lambda\to\infty} \lambda V_\lambda f = f$. Hence 2.1.a) and the estimation $|\lambda U_\lambda f - \lambda V_\lambda f| = |P_\lambda \lambda U_\lambda f| \le P_\lambda \lambda U_\lambda \|f\| \le P_\lambda \|f\| = \|f\|\, P_\lambda 1$ demonstrate b).

c) Since the potential kernel U of the sub-Markov resolvent $\mathbb{U}$ is proper, there exists a function $g \in \mathcal{B}^+(X)$ such that $0 < g \leq 1$ and $U g \leq 1$. According to 0.2. we choose a function $h \in \mathcal{B}^+(X)$ such that $0 < h \leq 1$ and such that $V h$ is a strong-Feller kernel. Furthermore, P is a strong-Feller kernel. From $U = V + P U$ we therefore obtain that $U g h$ is a strong-Feller kernel, where $g h \in \mathcal{B}^+(X)$ and $0 < g h \leq 1$. Hence c) follows by 0.2. .

d) It is enough to show $E_{\mathbb{U}} \subset E_{\mathbb{V}}$. Since we know from c) that each function in $E_{\mathbb{U}}$ is l.s.c. , this is an immediate consequence of $S_{\mathbb{U}} \subset S_{\mathbb{V}}$ since $(X, E_{\mathbb{V}})$ is a balayage space.

In the following lemma we do not assume $(X, E_{\mathbb{V}})$ to be a balayage space.

2.3. LEMMA: Let Q be a kernel on X and $v \in S_{\mathbb{V}} \cap C(X)$ such that $Q v < \infty$ and $\{v > 0\} \supset \{V 1 > 0\}$. Then there exists a function $h \in \mathcal{B}^+(X)$ such that $0 < h \leq 1$, $V h \leq 1$ and $Q V h < \infty$.

Proof: Since the potential kernel V of the sub-Markov resolvent $\mathbb{V}$ is proper, there exists a function $g \in \mathcal{B}^+(X)$ such that $0 < g \leq 1$ and $V g \leq 1$. Since v is continuous, there exists an exhaustion $(K_n)_{n \in \mathbb{N}}$ of $\{v > 0\}$. For each $n \in \mathbb{N}$ there exists a number $m_n \in \mathbb{N}$ such that $V g 1_{K_n} \leq V g \leq m_n v$ on the compact set K_n. Hence $V g 1_{K_n} \leq m_n v$ since $m_n v \in S_{\mathbb{V}}$. In addition $V 1_{\{v = 0\}} = 0$ is a consequence of $\{v > 0\} \supset \{V 1 > 0\}$ and $V 1_{\{V 1 = 0\}} \leq V 1 = 0$ on $\{1_{\{V 1 = 0\}} > 0\} = \{V 1 = 0\}$. Now let $h := g\,(\sum_{n \in \mathbb{N}} \frac{1}{2^n m_n} 1_{K_n} + 1_{\{v = 0\}})$. Then $h \in \mathcal{B}^+(X)$ such that $0 < h \leq g \leq 1$, $V h \leq V g \leq 1$, $V h \leq v$ and $Q V h \leq Q v < \infty$.

From now on we suppose that P satisfies $P(\mathcal{B}_b^+(X)) \subset \mathcal{P}_{\mathbb{V},b}$ and that there exists a function $v \in E_{\mathbb{V}} \cap C(X)$ such that $v > 0$ and $\sum_{n \in \mathbb{N}_0} P^n v < \infty$. Let $(U_\lambda)_{\lambda \in [0, \infty[}$ be the family of kernels defined by $U_\lambda := \sum_{n \in \mathbb{N}_0} P_\lambda^{\,n} V_\lambda$ for each $\lambda \in [0, \infty[$. In particular, $U_\lambda = V_\lambda + P_\lambda U_\lambda$ $\forall \lambda \in [0, \infty[$.

2.4. PROPOSITION: $\mathbb{U}$ is a resolvent, U is proper and U is the potential kernel of $\mathbb{U}$.

Proof: Applying 2.3. to $Q = \sum_{n \in \mathbb{N}_0} P^n$, we see that U is proper. Since V is the potential kernel of $\mathbb{V}$, we obtain from 2.1.b) and the definition of $\mathbb{U}$ and U that U is the potential kernel of $\mathbb{U}$. We now prove the resolvent equation for $\mathbb{U}$:

i) The resolvent equation of $\mathbb{V}$ and the definition of the family $(P_\lambda)_{\lambda \in [0, \infty[}$ demonstrate $\forall \lambda, \mu \in [0, \infty[$ with $\mu \leq \lambda$: $P_\mu = P_\lambda + (\lambda - \mu) V_\lambda P_\mu$.

ii) Inductively we obtain from the resolvent equation of $\mathbb{V}$ and i) $\forall k \in \mathbb{N}_0$ $\forall f \in \mathcal{B}^+(X)$ with $U f < \infty$: $P_\mu^{\,k} V_\mu f - P_\lambda^{\,k} V_\lambda f$

$= (\lambda - \mu) \sum_{\substack{m,n\in[\![0,k]\!]\\ m+n=k}} P_\lambda^m V_\lambda P_\mu^n V_\mu f = (\lambda - \mu) \sum_{\substack{m,n\in[\![0,k]\!]\\ m+n=k}} P_\mu^n V_\mu P_\lambda^m V_\lambda f.$

iii) According to the definition of the family $(U_\lambda)_{\lambda\in[0,\infty[}$ we conclude with the equation in ii): $U_\mu f - U_\lambda f = (\lambda-\mu) U_\lambda U_\mu f = (\lambda-\mu) U_\mu U_\lambda f$. Hence the resolvent equation for $\mathbb{U}$ follows.

2.5. THEOREM: If in addition $P1 \prec_{E_{\mathbb{V}}} 1$ holds, then $\mathbb{U}$ is a sub-Markov resolvent and $(X, E_{\mathbb{U}})$ is a balayage space.

Proof: i) According to the proof of 1.6.d) we obtain from $1 \in E_{\mathbb{V}}$, $P1 \prec_{E_{\mathbb{V}}} 1$ and $U_\lambda = \sum_{n\in\mathbb{N}_0} P_\lambda^n V_\lambda$ that $1 \in E_{\mathbb{U}}$. In particular, $\mathbb{U}$ is a sub-Markov resolvent. Furthermore, the statements of 2.2. and 1.6.d) and e) are valid for $\mathbb{U}$, $\mathbb{V}$, U, V and P.

ii) Let $X\times\mathbb{N}$ be the countable topological sum of the family of spaces $(X\times\{n\})_{n\in\mathbb{N}}$ and π the projection $\pi: X\times\mathbb{N} \to X$, $(x, n) \mapsto x$. We identify $X\times\{n\}$ with X and write $(f_n)_{n\in\mathbb{N}}$ for a function $f: X\times\mathbb{N} \to [-\infty, \infty]$ and $(Q_{ij})_{(i,j)\in\mathbb{N}\times\mathbb{N}}$ for a kernel $Q: \mathcal{B}^+(X\times\mathbb{N}) \to \mathcal{B}^+(X\times\mathbb{N})$ with the usual intuitive interpretation. For each $\lambda \in [0, \infty[$ and $i, j \in \mathbb{N}$ let $(\tilde{V}_\lambda)_{ij} := \delta_{ij} V_\lambda$, $(\tilde{P}_\lambda)_{ij} := \delta_{i+1,j} P_\lambda$ and $\tilde{U}_\lambda := \sum_{n\in\mathbb{N}_0} \tilde{P}_\lambda^n \tilde{V}_\lambda$. Obviously, $\tilde{\mathbb{V}}$ is a sub-Markov resolvent on $X\times\mathbb{N}$ such that $\tilde{V}$ is proper, $\tilde{V}$ is the potential kernel of $\tilde{\mathbb{V}}$, $S_{\tilde{\mathbb{V}}} = \{(w_n)_{n\in\mathbb{N}} \mid w_n \in S_{\mathbb{V}} \ \forall n \in \mathbb{N}\}$, $E_{\tilde{\mathbb{V}}} = \{(w_n)_{n\in\mathbb{N}} \mid w_n \in E_{\mathbb{V}} \ \forall n \in \mathbb{N}\}$, $(X\times\mathbb{N}, E_{\tilde{\mathbb{V}}})$ is a balayage space, $\mathcal{P}_{\tilde{\mathbb{V}}} = \{(w_n)_{n\in\mathbb{N}} \mid w_n \in \mathcal{P}_{\mathbb{V}} \ \forall n \in \mathbb{N}\}$, $\tilde{P}(\mathcal{B}_b^+(X\times\mathbb{N})) \subset \mathcal{P}_{\tilde{\mathbb{V}},b}$, $\tilde{P}1 \prec_{E_{\tilde{\mathbb{V}}}} 1$, $v\circ\pi \in E_{\tilde{\mathbb{V}}} \cap C(X\times\mathbb{N})$, $v\circ\pi > 0$, $\sum_{n\in\mathbb{N}_0} \tilde{P}^n (v\circ\pi) < \infty$ and such that $\tilde{P}_\lambda + \lambda \tilde{V}_\lambda \tilde{P} = \tilde{P}$ $\forall \lambda \in [0, \infty[$. Thus we have seen that $\tilde{\mathbb{U}}$ is a sub-Markov resolvent on $X\times\mathbb{N}$ such that $\tilde{U}$ is the proper potential kernel of $\tilde{\mathbb{U}}$. Moreover, the statements of 2.2. and 1.6.d) and e) are valid for $\tilde{\mathbb{U}}$, $\tilde{\mathbb{V}}$, $\tilde{U}$, $\tilde{V}$ and $\tilde{P}$.

iii) According to 0.2. we choose a function $h \in \mathcal{B}^+(X)$ such that $0 < h \le 1$ and $Uh \in C_b(X)$. By 0.2. we obtain from $Uh = \sum_{n\in\mathbb{N}_0} P^n Vh$ that $Vh \in \mathcal{P}_{\mathbb{V},b}$. Since $P(\mathcal{B}_b^+(X)) \subset \mathcal{P}_{\mathbb{V},b}$, we therefore conclude $Uh \in E_{\mathbb{U}} \cap \mathcal{P}_{\mathbb{V},\sigma,b} = E_{\mathbb{U}} \cap \mathcal{P}_{\mathbb{V},b}$. Let $p := \sum_{n\in\mathbb{N}_0} \tilde{P}^n (2^{-m} Vh)_{m\in\mathbb{N}} = (2^{-m} \sum_{n\in\mathbb{N}_0} (\frac{1}{2}P)^n Vh)_{m\in\mathbb{N}}$. Obviously, $p > 0$. From 1.6.e) and $(2^{-m} Vh)_{m\in\mathbb{N}} \in \mathcal{P}_{\tilde{\mathbb{V}},b}$ we obtain $p \in E_{\tilde{\mathbb{U}}} \cap \mathcal{P}_{\tilde{\mathbb{V}},\sigma,b} = E_{\tilde{\mathbb{U}}} \cap \mathcal{P}_{\tilde{\mathbb{V}},b}$. We choose an exhaustion $(K_n)_{n\in\mathbb{N}}$ of X. Then $\tilde{K}_n := K_n \times [\![1, n]\!]$ defines an exhaustion $(\tilde{K}_n)_{n\in\mathbb{N}}$ of $X\times\mathbb{N}$. From 4.2.3. in [6] we obtain ${}_{\tilde{\mathbb{U}}}R_p^{\complement\tilde{K}_n} \xrightarrow[n\to\infty]{} 0$ locally uniformly. Indeed, if we define for each $n \in \mathbb{N}$ $f_n = ((f_n)_m)_{m\in\mathbb{N}}$ by $(f_n)_m := 2^{-m} {}_{\mathbb{V}}R_{Uh}^{\complement K_n}$ for $m \in [\![1, n]\!]$ and $(f_n)_m := 2^{-m} Uh$ for $m \in [\![n+1, \infty[\![$, then $f_n \in E_{\tilde{\mathbb{V}}}$ and $\sum_{i\in\mathbb{N}_0} \tilde{P}^i f_n \in E_{\tilde{\mathbb{U}}}$ by 1.6.e), where $(\sum_{i\in\mathbb{N}_0} \tilde{P}^i f_n)_m = 2^{-m} (\sum_{i\in[\![0,n-m]\!]} (\frac{1}{2}P)^i \, {}_{\mathbb{V}}R_{Uh}^{\complement K_n} + \sum_{i\in[\![n-m+1,\infty[\![} (\frac{1}{2}P)^i Uh)$ for $m \in [\![1, n]\!]$ and $(\sum_{i\in\mathbb{N}_0} \tilde{P}^i f_n)_m = 2^{-m} \sum_{i\in\mathbb{N}_0} (\frac{1}{2}P)^i Uh$ for $m \in [\![n+1, \infty[\![$. This construction yields $p \le (2^{-m} Uh)_{m\in\mathbb{N}} \le \sum_{i\in\mathbb{N}_0} \tilde{P}^i f_n$ on $\complement\tilde{K}_n$, hence ${}_{\tilde{\mathbb{U}}}R_p^{\complement\tilde{K}_n} \le \sum_{i\in\mathbb{N}_0} \tilde{P}^i f_n$.

Further, $(\sum_{i\in\mathbb{N}_0} \tilde{P}^i f_n)_m \in C_b(X)$ for $m \in [\![n+1, \infty[\![$ and $(\sum_{i\in\mathbb{N}_0} \tilde{P}^i f_n)_m - 2^{-m}\, {}_{\mathbb{V}}R^{\complement K_n}_{Uh}$ $\in C_b(X)$ for $m \in [\![1, n]\!]$ since the series in the explicit presentation is uniformly convergent. Since $Uh \in \mathcal{P}_{\mathbb{V},b}$, we have ${}_{\mathbb{V}}R^{\complement K_n}_{Uh} \xrightarrow[n\to\infty]{} 0$ locally uniformly and by Lebesgue's convergence theorem $\lim_{n\to\infty} \sum_{i\in\mathbb{N}} (\frac{1}{2}P)^i\, {}_{\mathbb{V}}R^{\complement K_n}_{Uh} = 0$. Therefore, for each $m \in \mathbb{N}$ $(\sum_{i\in[\![1,n-m]\!]} (\frac{1}{2}P)^i\, {}_{\mathbb{V}}R^{\complement K_n}_{Uh} + \sum_{i\in[\![n-m+1,\infty[\![} (\frac{1}{2}P)^i Uh)_{n\in[\![m,\infty[\![}$ decreases to 0 if n increases to infinity, and by Dini's lemma this convergence is a locally uniform convergence. Thus we have $\sum_{i\in\mathbb{N}_0} \tilde{P}^i f_n \xrightarrow[n\to\infty]{} 0$ locally uniformly, hence ${}_{\tilde{\mathbb{U}}}R^{\complement \tilde{K}_n}_p \xrightarrow[n\to\infty]{} 0$ locally uniformly is demonstrated. Finally, $(X\times\mathbb{N}, E_{\tilde{\mathbb{U}}})$ is a balayage space by ii) and 0.1. .

iv) As a consequence of the definition of $\tilde{\mathbb{U}}$ in ii) we obtain for each $f\in\mathcal{B}^+(X\times\mathbb{N})$: $f \in S_{\tilde{\mathbb{U}}} \iff \forall\lambda \in]0,\infty[\ \forall m\in\mathbb{N}: \lambda\sum_{n\in\mathbb{N}_0} P_\lambda^n V_\lambda f_{n+m} \le f_m$. Hence we have from 2.2.d) for every $u\in\mathcal{B}^+(X)$: $u\circ\pi \in E_{\tilde{\mathbb{U}}} \iff u \in E_{\mathbb{U}}$. Furthermore, this characterization of $\tilde{\mathbb{U}}$-supermedian functions and 2.2.d) demonstrate for each $g\in\mathcal{B}^+(X)$: ${}_{\tilde{\mathbb{U}}}R_{g\circ\pi} = {}_{\mathbb{U}}R_g \circ\pi$. It is finally a consequence of i), iii), 2.2.b), 2.4. and the image theorem V 8.5. in [3] that $(X, E_{\mathbb{U}})$ is a balayage space.

It follows from 1.1. and 1.6. that we obtain all sub-Markov resolvents $\mathbb{U}$ such that the potential kernel U of $\mathbb{U}$ is proper, $\mathbb{V}$ is subordinated to $\mathbb{U}$ and such that $(X, E_{\mathbb{U}})$ is a balayage space in this way.

REFERENCES

1. H. Ben Saad: Exemples de noyaux admettant des résolvantes. Mathematische Annalen 265, 149-154 (1983).
2. H. Ben Saad: Généralisation de noyaux V_h et applications. In: Séminaire de Théorie du Potentiel No. 7. Lecture Notes in Mathematics 1061, 14-39, Springer, Berlin (1984).
3. J. Bliedtner, W. Hansen: Potential Theory - An Analytic and Probabilistic Approach to Balayage. Universitext, Springer, Berlin - Heidelberg - New York - Tokyo (1986).
4. N. Boboc, Gh. Bucur: Perturbations in excessive structures. In: Complex Analysis, Fifth Romanian - Finnish Seminar. Lecture Notes in Mathematics 1014, 155-187, Springer, Berlin (1983).
5. V. Dembinski, K. Janßen: Standard Balayage Spaces and Standard Markov Processes. In: Potential Theory Copenhagen 1979. Lecture Notes in Mathematics 787, 84-105, Springer, Berlin (1980).
6. M. Meyer: Balayage Spaces on Topological Sums. To appear.

ON THE NEW PROBLEM IN THE POTENTIAL GRAVITY THEORY

Jiří Nedoma

General Computing Centre, Czech.Acad.Sci
Pod vodárenskou věží 2
182 07 Praha 8 - Libeň, P.O.B.5, Czechoslovakia

ABSTRACT

The results presented in this paper were obtained in the course of an investigation of a simple model problem of the time-dependent Earth's gravity field, but they are equally applicable to other problems governed by thermo-magnetohydrodynamics and geophysics. The author has studied the problem of the existence, uniqueness and regularity of the solution of the variational problem of the coupled gravity field, velocity field in thermo-Bingham's fluid (by which the Earth's interior is simulated) and a quasi-stationary magnetic field.

INTRODUCTION

While, the global geodynamic model is discussed in Nedoma (1986), in the present paper the mathematical model of the global gravity field coupled with geodynamic processes and the thermo-magnetohydrodynamic processes in the Earth's interior based on thermo-Bingham's rheology is given. According to Nedoma (1986), the dynamic N-dimensional problem (N=2,3) of the coupled global time-dependent gravity field model, simulated by a thermo-Bingham imcompressible fluid under the effect of a quasi-stationary magnetic field, can be formulated in the following way :

Let G be a bounded domain in R^N with a sufficiently smooth boundary dG. Let $I = \langle t_0, t_1 \rangle$, $G_t = G\times I$, $dG_t = dG\times I$. The problem is to find a group of four functions (v, B, T, ϕ), i.e. vector functions $v(x,t)$-the velocity of the medium and $B(x,t)$-the magnetic induction, scalar functions $T(x,t)$-the temperature and $\phi(x,t)$-the gravity potential, which satisfy the following initial-boundary value problem :

Let $v = du/dt$ be known, then

$$u(x,t) = \int_{t_0}^{t_1} v(x,s)\,ds + u_0(x), \quad u_0(x) = u(x,t_0) .$$

We then solve problem (P) :

$$\rho(x,t)\, Dv_i/Dt = d\tau_{ij}/dx_j + f_i$$

$$\text{in } G_t ,$$

$$(1) \quad -dB/dt = \mathrm{Rot}\,[v,B] + \mathrm{Rot}(\eta_o \,\mathrm{Rot}\, B),\ \eta_o=(\mu\sigma)^{-1} \quad \text{in } G_t ,$$

$$\rho b_{ij}T_oD_{ij} + \rho c_e DT/Dt = W - d/dx_j(K_{ij}\ dT/dx_j) - \sigma^{-1}_{ij}\ j_i j_j$$

$$\Delta\phi = -4\pi G_c \rho p K^{-1} + 2\omega^2 \quad \text{in } G_t ,$$

with conditions

$$(2) \quad \mathrm{Div}\ v = 0,\ \mathrm{Div}\ B = 0,\ g = \mathrm{grad}\ \phi ,$$

the boundary-value conditions on dG

$$(3) \quad v = v_1,\ [n,\mathrm{Rot}\ B] = 0,\ B\,n = 0,\ T = T_1,\ \phi = \phi_1$$

and initial conditions

$$(4) \quad v(x,t_o)=v_o(x),\ B(x,t_o)=B_o(x),\ T(x,t_o)=T_o(x),\ \phi(x,t_o)= = \phi_o(x),\quad x\in G ,$$

where $D/Dt = d/dt+v_k\ d/dx_k$, ρ is the density, τ_{ij} is the stress tensor and $\tau_{ij} = {}^B\tau_{ij} + {}^T\tau_{ij} + {}^M\tau_{ij}$, ${}^B\tau_{ij} = -p\delta_{ij} + \hat{g}D_{ij} \cdot D_{II}^{-1/2} + 2\hat{\mu}D_{ij}$ is the stress tensor in Bingham's fluid, p is the spherical part of the stress tensor, $\hat{g}, \hat{\mu}$ are the thres - holds of plasticity and viscosity of Bingham's fluid, $D_{II}=1/2\ D_{ij}D_{ij}+\varepsilon$, $\varepsilon > 0$, $D_{ij} = 1/2(dv_i/dx_j+dv_j/dx_i)$, ${}^T\tau_{ij} = b_{ij}(T-T_o)$ is the thermal stress, b_{ij} the coefficient of thermal expansion, T_o the initial temperature at $t=t_o$, ${}^M\tau_{ij} = H_iB_j+H_jB_i - (HB)\delta_{ij}$ Maxwell's stress, H the intensity of the magnetic field, f the body forces, μ permeability, $\sigma=(\sigma_{ij})$ electric conductivity, c_e specific heat, W thermal sources, K_{ij} thermal conductivity, $j = \sigma(E+[v,B])$ the electric current (defined by the Ohm's law), E intensity of the electric field, G_c gravitational constant, K bulk modulus, ω angular velocity of the Earth and g the gravity acceleration vector.

If the data related to the presented problem are suf - ficiently smooth, on applying the well-known transformations, it leads to a problem with homogeneous boundary and initial conditions. Further we shall solve problem (1), (2) with homogeneous boundary conditions on dG

$$(3) \quad v = 0,\ [n,\mathrm{Rot}\ B] = 0,\ B\,n = 0,\ T = 0,\ \phi = 0$$

and initial conditions

$$(4) \quad v(x,t_o) = 0,\ B(x,t_o) = 0,\ T(x,t_o) = 0,\ \phi(x,t_o) = 0 .$$

VARIATIONAL SOLUTION AND SUMMARY OF THE RESULTS

Let us put $W = W - \sigma^{-1}_{ij}\ j_i j_j$,

$(u,v)= \int_G \rho\, u_i v_i\ dx$, $|u|^2=(u,u)$, $(f,g)= \int_G \rho\, fg\ dx$, $|f|^2= =(f,f)$, $\|u\|^2 = \int_G e_{ij}(u)e_{ij}(u)\ dx + |u|^2$, $e_{ij} = 1/2(du_i/dx_j+ +du_j/dx_i)$, $\|f\|^2 = \int_G df/dx_i\ df/dx_i\ dx + |f|^2$ and let

$a(t;v,w) = 2 \int_G D_{ij}(v)D_{ij}(w)\ dx$, $b_M(t;B,C)=(\sigma\mu^{-2}\mathrm{Rot}\ B,\mathrm{Rot}\ C)$,

$A(t;T,z) = \int_G K_{ij}\, dT/dx_i dz/dx_j\, dx$, $a_g(t;\phi,\varphi) =$

$= \int_G d\phi/dx_i\, d\varphi/dx_j\, dx$, $j(v) = 2\int_G (D_{II}(v))^{1/2}\, dx$,

$b_o(t;v,g,z) = \int_G \rho c_e v_k\, dg/dx_k\, z\, dx$, $b(t;u,v,w) =$

$= \int_G \rho u_i dv_j/dx_i\, w_j\, dx$, $b_s(t;T,v) = \int_G d/dx_j (b_{ij}T) v_i dx$,

$b_p(t;v,g) = \int_G \rho T_o\, b_{ij}\, dv_i/dx_j\, g\, dx$ be the bilinear and trilinear forms. Let

$V = \{vIv \in [H^1_o(G)]^N, \mathrm{Div}\ v=0\}$; ${}^1V = \{CIC \in [H^1(G)]^N, \mathrm{Div}\ C=0, nC=0\}$,

${}^2V = \{zIz \in H^1(G), z=0 \text{ on } dG\}$, ${}^3V = \{\varphi I \varphi \in H^1(G), \varphi=0 \text{ on } dG\}$.

Multiplying (1) in turn by $w-v(t), C-B(t), z-T(t), \varphi-\phi(t)$, integrating over G, applying the divergence theorems and the boundary conditions and, since the stress-strain relation in Bingham's medium can be rewritten as follows

$${}^D\tau_{II}^{1/2} < \hat{g} \iff D_{ij} = 0 ,$$
$${}^D\tau_{II}^{1/2} = \hat{g} \iff D_{ij} = (2\hat{\mu})^{-1}(1-\hat{g}\,{}^D\tau_{II}^{-1/2})\,{}^B\tau^D_{ij}$$

where ${}^B\tau^D_{ij}$ is the deviator of the stress tensor

${}^B\tau_{ij}$, ${}^B\tau_{II} = 1/2\,{}^B\tau^D_{ij}\,{}^B\tau^D_{ij}$, we obtain the following variational formulation

(5) $(dv(t)/dt, w-v(t)) + \mu a(t;v(t), w-v(t)) + b(t;v(t), v(t), w-v(t)) +$

$+ b_s(t;T(t), w-v(t)) + \hat{g} j(w) - \hat{g}\, j(v(t)) - (d/dx_i(\mu^{-1}B_i(t)B(t)),$

$w-v(t)) \geq (f(t), w-v(t)) \quad \forall\, w \in V$,

$(\mu^{-1}dB(t)/dt, C-B(t)) + b_M(t;B(t), C-B(t)) - (\mu^{-1}\mathrm{Rot}\,[v(t),$

$B(t)], C-B(t)) \geq 0 \quad \forall\, C \in {}^1V$,

$(dT/dt, z-T(t)) + b_o(t;v(t), T(t), z-T(t)) + b_p(t;v'(t), z-T(t)) +$

$A(t;T(t), z-T(t)) \geq (W(t), z - T(t)) \quad \forall\, z \in {}^2V$,

$a_g(t;\phi(t), \varphi-\phi(t)) \geq (\psi(t), \varphi-\phi(t)) \quad \forall\, \varphi \in {}^3V$

and with the initial conditions (4).

Let $\hat{V} = \{vIv \in [D(G)]^N, \mathrm{Div}\ v=0\}$, (where D(G) is the space of all functions from C^∞ with a compact support in G), V_s being a closure of $\hat{V}$ in $[H^s(G)]^N$ $(s=N/2)$, then V_s is a Hilbert space with the norm $IIvII^2_s = ((v,v))_s$ if $((v,w))_s = (v_i, w_i)_{H^s(G)}$. Further $V_1=V$, $IIvII_1=IIvII$, $V_o=H$, $IIvII_o = IVI$, and for $s>1$ $V_s \subset V \subset H \subset V' \subset V'_s$, where V', V'_s are dual spaces of V and V_s and $H=H'$. Further let

${}^1\hat{V} = \{CIC \in [D(G)]^N, \mathrm{Div}\ C=0, nC=0 \text{ on } dG\}$, ${}^1V_s$ being a closure of ${}^1\hat{V}$ in $H^s(G)^N$ and then ${}^1V_s$ is a Hilbert space with the norm $IICII^2_s = ((C,C))_s$ if $((C,B))_s = (C_i, B_i)_s$. Let ${}^1V_1 = {}^1V$, ${}^1V_o = {}^1H$ and for $s>1$ ${}^1V_s \subset {}^1V \subset {}^1H \subset {}^1V' \subset {}^1V'_s$, where ${}^1V'$, ${}^1V'_s$ are dual spaces of ${}^1V$, ${}^1V_s$ and ${}^1H = {}^1H'$. Further let

${}^2\hat{V} = \{ z \mid z \in D(G),\ z=0 \text{ on } dG \}$, ${}^3\hat{V} = \{ \varphi \mid \varphi \in D(G),\ \varphi=0 \text{ on } dG \}$, ${}^2V$ being a closure of ${}^2\hat{V}$ in $H^s(G)$, ${}^3V_s$ being a closure of ${}^3\hat{V}$ in $H^s(G)$. Both spaces are Hilbert spaces with scalar products $(z,y)_s$ and $(\varphi,\psi)_s$. Let ${}^2V_1 = {}^2V$, ${}^2V_o = {}^2H$, ${}^3V_1 = {}^3V$, ${}^3V_o = {}^3H$ and for $s > 1$ ${}^2V_s \subset {}^2V \subset {}^2H \subset {}^2V' \subset {}^2V_s'$ and ${}^3V_s \subset {}^3V \subset {}^3H \subset {}^3V' \subset {}^3V_s'$, where ${}^2V'$, ${}^2V_s'$, ${}^3V'$, ${}^3V_s'$ are dual spaces of ${}^2V$, ${}^2V_s$, ${}^3V$, ${}^3V_s$ and ${}^2H = {}^2H'$, ${}^3H = {}^3H'$. Further let

$\underline{W} = \{ v \mid v \in L^2(I;V_s),\ v' \in L^2(I;H),\ v(t_o)=0 \}$, ${}^1W = \{ C \mid C \in L^2(I,{}^1V_s),\ C' \in L^2(I;{}^1H),\ C(t_o)=0 \}$, ${}^2W = \{ z \mid z \in L^2(I;{}^2V_s),\ z' \in L^2(I;{}^2H),\ z(t_o)=0 \}$, ${}^3W = \{ \varphi \mid \varphi \in L^2(I;{}^3V_s),\ \varphi' \in L^2(I;{}^3H),\ \varphi(t_o)=0 \}$.

As a result we have the following theorem

<u>Theorem 1</u> : Let $N=3$, $s=N/2$

(7) $f \in L^2(I;V')$, $W \in L^2(I;{}^2V')$, $\psi \in L^2(I;{}^3V')$

There then exist functions v, B, T, ϕ such that

(8) $v \in L^2(I;V) \cap L^\infty(I;H)$, $dv/dt \in L^2(I;V_s')$,
$B \in L^2(I;{}^1V) \cap L^\infty(I;{}^1H)$, $dB/dt \in L^2(I;{}^1V_s')$,
$T \in L^2(I;{}^2V) \cap L^\infty(I;{}^2V)$, $dT/dt \in L^2(I;{}^2V_s')$,
$\phi \in L^2(I;{}^3V)$, $d\phi/dt \in L^2(I;{}^3V_s')$,

(9) $v(t_o)=0$, $B(t_o)=0$, $T(t_o)=0$, $\phi(t_o)=0$

and v, B, T, ϕ satisfy

$$\int_{t_o}^{t_1} \{ (dv(t)/dt, w-v(t)) + \hat{\mu} a(t;v(t),w-v(t)) + (\mu^{-1} dB(t)/dt,$$
$$C-B(T)) + b_M(t,B(t),C-B(t)) + (dT(t)/dt, z-T(t)) + A(t;T(t),$$
$$z-T(t)) + a_g(t;\phi(t), \varphi-\phi(t)) + b(t;v(t),v(t),w) + \hat{g}\, j(w) - \tag{10}$$
$$-\hat{g}\, j(v(t)) + b_s(t;T(t),w-v(t)) + b_o(t;v(t),T(t),z-T(t)) +$$
$$+b_p(t;v'(t),z-T(t)) - (d/dx_i(\mu^{-1}B_i(t)B(T)),w) -$$
$$-(\mu^{-1}\operatorname{Rot}[v(t),B(t)],C) - (f(t),w-v(t)) - (W(t),z-T(t)) -$$
$$-(\psi(t), \varphi-\phi(t)) \}\, dt = 0 \quad \forall v \in \underline{W}, C \in {}^1\underline{W}, z \in {}^2\underline{W}, \varphi \in {}^3\underline{W}$$

and such that

$v \in L^2(I;V_s)$, $dv/dt \in L^2(I;H)$, $v(t_o)=0$,
$C \in L^2(I;{}^1V_s)$, $dC/dt \in L^2(I;{}^1H)$, $C(t_o)=0$,
$z \in L^2(I;{}^2V_s)$, $dz/dt \in L^2(I;{}^2H)$, $z(t_o)=0$,
$\varphi \in L^2(I;{}^3V_s)$, $d\varphi/dt \in L^2(I;{}^3H)$, $\varphi(t_o)=0$.

Outline of the proof : Double regularization will be used.Let

(11) $j_\varepsilon(w)=2/(1+\varepsilon)\int_G (D_{II}(w))^{(1+\varepsilon)/2}\,dx,\quad \varepsilon>0$

for which

(12) $(j'_\varepsilon(v),w)=\int_G D_{II}(v)^{(\varepsilon-1)/2} D_{ij}(v)\, D_{ij}(w)\,dx$.

Further we add terms with viscosity regularization i.e. the terms $\eta((v_{\varepsilon\eta},w))_s$, $\eta((B_{\varepsilon\eta},C))_s$, $\eta((T_{\varepsilon\eta},z))_s$, $\eta((\phi_{\varepsilon\eta},\varphi))_s$.

Then for N=2 and s=1 ${}^iV_s={}^iV$, i=0,1,2,3 and the terms with viscosity regularization are of the same order as the bilinear forms and thus can be omitted.

We now solve the double regularized problem $(P_r)_v$:

(13)
$$\begin{aligned}
&(v'_{\varepsilon\eta}(t),w)+\hat{\mu}\,a(t;v_{\varepsilon\eta}(t),w)+b(t;v_{\varepsilon\eta}(t),v_{\varepsilon\eta}(t),w)+\\
&+b_s(t;T_{\varepsilon\eta}(t),w)+\hat{g}(j'_\varepsilon(v_{\varepsilon\eta}),w)+\eta((v_{\varepsilon\eta},w))_s-\\
&-(d/dx_i(\mu^{-1}B_{i\varepsilon\eta}(t)B_{\varepsilon\eta}(t),w)=(f(t),w)\quad \forall\, w\in V_s,\\
&(\mu^{-1}\,dB_{\varepsilon\eta}(t)/dt,C)+b_M(t;B_{\varepsilon\eta}(t),C)+\eta((B_{\varepsilon\eta}(t),C))_s-\\
&-(\mu^{-1}\mathrm{Rot}[v_{\varepsilon\eta}(t),B_{\varepsilon\eta}(t)],C)=0\quad \forall\, C\in {}^1V_s\\
&(dT_{\varepsilon\eta}(t)/dt,z)+b_o(t;v_{\varepsilon\eta}(t),T_{\varepsilon\eta}(t),z)+b_p(t;v'_{\varepsilon\eta}(t),z)+\\
&+A(t;T_{\varepsilon\eta}(t),z)+\eta((T_{\varepsilon\eta}(t),z))_s=(W(t),z)\quad \forall\, z\in {}^2V_s,\\
&a_g(t;\phi_{\varepsilon\eta}(t),\varphi)+\eta((\phi_{\varepsilon\eta}(t),\varphi))_s=(\psi(t),\varphi)\quad \forall\,\varphi\in {}^3V_s
\end{aligned}$$

with initial conditions

(14) $v_{\varepsilon\eta}(x,t_o)=0,\ B_{\varepsilon\eta}(x,t_o)=0,\ T_{\varepsilon\eta}(x,t_o)=0,\ \phi_{\varepsilon\eta}(x,t_o)=0.$

The purpose of the proof :

a) the technique of the Galerkin approximation will be used to prove the existence of the solution of (13)-(14),
b) a priori estimates I and II independent of ε and η will be given,
c) limitation processes for the Galerkin approximation and for $\varepsilon\to 0$ and $\eta\to 0$. add a) Let $\{w_i\}$, $\{C_i\}$, $\{z_i\}$, $\{\varphi_i\}$ be separable bases of V_s, ${}^1V_s$, ${}^2V_s$, ${}^3V_s$. Let us put $v_m=v_{\varepsilon\eta m}$, $C_m=C_{\varepsilon\eta m}$, $z_m=z_{\varepsilon\eta m}$, $\varphi_m=\varphi_{\varepsilon\eta m}$. Let W^m, ${}^1W^m$, ${}^2W_m$, ${}^3V_m$ be spaces built by $w_1,\dots.w_m$; $C_1,\dots.C_m$, $z_1,\dots.z_m$; $\varphi_1,\dots.\varphi_m$. The approximate solution $\{w_m,C_m,z_m,\varphi_m\}$ of order m then satisfies

(15)
$$\begin{aligned}
&(v'_m,w_j)+\hat{\mu}\,a(t;v_m,w_j)+b(t;v_m,v_m,w_j)+b_s(t;T_m,w_j)+\\
&+\hat{g}(j'_\varepsilon(v_m),w_j)+\eta((v_m,w_j))_s-(d/dx_i(\mu^{-1}B_{im}B_m),w_j)=\\
&=(f,w_j)\quad \forall w_j\in W^m,\quad \forall\, t\in I,\quad 1\le j\le m,\\
&(\mu^{-1}B'_m,C_j)+b_M(t;B_m,C_j)+\eta((B_m,C_j))_s-(\mu^{-1}\mathrm{Rot}[v_m,B_m],C_j)=\\
&=0\quad \forall\, C_j\in {}^1W^m,\quad \forall\, t\in I,\quad 1\le j\le m,\\
&(T'_m,z_j)+b_o(t;v_m,T_m,z_j)+b_p(t;v'_m,z_j)+A(t;T_m,z_j)+
\end{aligned}$$

$+\eta((T_m,z_j))_s = (W,z_j) \quad \forall z_j \in {}^2W^m, \ \forall t \in I, \ 1 \le j \le m,$

$a_g(t;\phi_m,\varphi_j)+\eta((\phi_m,\varphi_j))_s=(\psi,\varphi_j) \quad \forall \varphi_j \in {}^3W^m, \ \forall t \in I, \ 1 \le j \le m$

and initial conditions

(16) $v_m(x,t_o)=0, \ B_m(x,t_o)=0, \ T_m(x,t_o)=0, \ \phi_m(x,t_o)=0.$

Since $\{w_j\}$, $\{C_j\}$, $\{z_j\}$, $\{\varphi_j\}$ are linearly independent, system (15)-(16) represents a regular system of ordinary differential and algebraic equations and thus (15)-(16) uniquely define $\{v_m,B_m,T_m,\phi_m\}$ in the interval $I_m=\langle t_o,t_m\rangle$

add b) (15)-(16) yield

$$\begin{aligned}
(17)\quad & 1/2\ D/Dt\ \{|v_m(t)|^2+\mu^{-1}|B(t)|^2+|T_m(t)|^2\}+\mu\ \hat{a}(t;v_m(t),\\
& v_m(t))+b_M(t;B_m(t),B_m(t))+A(t;T_m(t),T_m(t))+a_g(t;\phi_m(t),\\
& \phi_m(t))+\hat{g}(j_\varepsilon'(v_m(t),v_m(t))+\eta\|v_m(t)\|_s^2+\eta\|B_m(t)\|_s^2+\\
& +\eta\|T_m(t)\|_s^2+\eta\|\phi_m(t)\|_s^2+b_s(t;T_m(t),v_m(t))+\\
& +b_p(t;v_m'(t),T_m(t)) = (f(t),v_m(t))+(W(t),T_m(t))+\\
& +(\psi(t),\phi_m(t)) \ .
\end{aligned}$$

Then, integrating (17) over t from t_o to t, and since

$|(d/dx_i(\mu^{-1}B_iB),w)| \le C\ \|B\|_{[L^p(G)]^N} \sum_i \|P_iw\|_{L^N(G)},$

$1/p=1/2-1/(2N),$

$|(\mu^{-1}Rot[v,B],C)| \le C\ \|v\|_{[L^p(G)]^N}\|B\|_{[L^p(G)]^N}\|C\|_s,$

$|(d/dx_i(\mu^{-1}B_iB),w)| \le c\|B\|\ |B|\ \|w\|_s,$

$|(\mu^{-1}\ Rot[v,B],C)| \le c\ \|v\|^{1/2}|v|^{1/2}\|B\|^{1/2}|B|^{1/2}\|C\|_s,$

$(d/dx_i(\mu^{-1}B_{im}B_m),v_m)+(\mu^{-1}Rot[v_m,B_m],B_m) = 0,$

$b(t;v_m(t),v_m(t),v_m(t))=0, \ b_o(t;v_m(t),T_m(t),T_m(t))=0$

$|b_s(t;T-T_o,v)+b_p(t;v,T)| \le c(1+|T_m(t)|)|v_m(t)|,$

$(j_\varepsilon'(v),v) \ge 0, \ b_M(t;B,B) \ge c\ \|B\|^2 \quad \forall B \in {}^1V, \ a(T;v,v) \ge$

$\ge \alpha\|v\|^2 \quad \forall v \in V, \ A(t;z,z) \ge \alpha_o\|z\|^2 \ \forall z \in {}^2V,$

$a_g(t;\varphi,\varphi) \ge \alpha_1\|\varphi\|^2 \quad \forall \varphi \in {}^3V,$

the application of Gronwall's lemma and the Lax-Milgram theorem yields

$|v_m(t)| \le c, \ |B_m(t)| \le c, \ |T_m(t)| \le c,$

$\int_{t_o}^{t_1}\|v_m(s)\|^2ds \le c, \int_{t_o}^{t_1}\|B_m(s)\|^2ds \le c, \int_{t_o}^{t_1}\|T_m(s)\|^2ds \le c,$

$$\int_{t_o}^{t_1} \|\phi(s)\|^2 ds \le c \ , \quad \eta\int_{t_o}^{t_1} \|v_m\|_s^2 ds \le c \ , \quad \eta\int_{t_o}^{t_1} \|B_m(s)\|^2 ds \le c,$$

$$\eta\int_{t_o}^{t_1} \|T_m(s)\|^2 ds \le c \ , \quad \int_{t_o}^{t_1} \|\phi(s)\|^2 ds \le c \ , \quad c=\text{const. independent of } m \ ,$$

which imply that

$$\begin{aligned}
&v_m \in X \subset L^2(I;V) \cap L^{\infty}(I;H), \ B_m \in {}^1X \subset L^2(I;{}^1V) \cap L^{\infty}(I;{}^1H),\\
&T_m \in {}^2X \subset L^2(I;{}^2V) \cap L^{\infty}(I;{}^2H), \ \phi_m \in {}^3X \subset L^2(I;{}^3V),\\
(18)\quad &\eta^{1/2} v_m \in X_o \subset L^2(I;V^s), \ \eta^{1/2} B_m \in {}^1X_o \subset L^2(I;{}^1V_s),\\
&\eta^{1/2} T_m \in {}^2X_o \subset L^2(I;{}^2V_s), \ \eta^{1/2} \phi_m \in {}^3X_o \subset L^2(I;{}^3V_s),
\end{aligned}$$

where $X, \ldots, {}^3X_o$ are bounded subsets.

To finish the second part of the proof we shall verify that

$$\begin{aligned}
(19)\quad &v_m' \in Y_o \subset L^2(I;V_s'), \ B_m' \in {}^1Y_o \subset L^2(I,{}^1V_s'), \ T_m' \in {}^2Y_o \subset L^2(I;{}^2V_s'),\\
&\phi_m' \in {}^3Y_o \subset L^2(I;{}^3V_s') \ ,
\end{aligned}$$

where $Y_o, {}^1Y_o, {}^2Y_o, {}^3Y_o$ are bounded subsets.

It can be proved that (15) is equivalent to the following system

$$\begin{aligned}
(20)\quad &(v_m' + \hat{\mu} A_B v_m + \eta\Lambda v_m + \hat{g}\, j_\varepsilon'(v_m) + h_m - f, w_j) = 0, \ 1 \le j \le m \ ,\\
&(\mu^{-1} B_m' + A_m B_m - \mathcal{B}_m + \eta\, \Sigma_s B_m, C_j) = 0 \ , \quad 1 \le j \le m \ ,\\
&(T_m' + A_T T_m + \eta\, \mathcal{T}_s T_m + g_m - W, z_j) = 0 \ , \ 1 \le j \le m \ ,\\
&(A_G \phi_m + \eta\, \mathcal{X}_s \phi_m - \psi, \varphi_j) = 0 \ , \quad 1 \le j \le m \ ,
\end{aligned}$$

where $a(t;v,w) = (A_B v,w)$, $A_B \in \mathcal{L}(V,V')$, $b_M(t;B,C) = (A_M B,C)$, $A_M \in \mathcal{L}({}^1V, {}^1V')$, $A(t;T,z) = (A_T T,z)$, $A_T \in \mathcal{L}({}^2V, {}^2V')$, $a_g(t;\phi,\varphi) = (A_G\phi,\varphi)$, $A_G \in \mathcal{L}({}^3V,{}^3V')$, $((v,w))_s = (\Lambda_s v,w)$, $\Lambda_s \in \mathcal{L}(V_s,V_s')$, $((B,C))_s = (\Sigma_s B,C)$, $\Sigma_s \in \mathcal{L}({}^1V_s, {}^1V_s')$, $((T,z))_s = (\mathcal{T}_s T,z)$, $\mathcal{T}_s \in \mathcal{L}({}^2V_s, {}^2V_s')$, $((\phi,\varphi))_s = (\mathcal{X}_s\phi,\varphi)$, $\mathcal{X}_s \in \mathcal{L}({}^3V_s, {}^3V_s')$, because the bilinear forms $a(t;v,w)$, $b_M(t;B,C)$, $A(t;T,z)$, $a_g(t;\phi,\varphi)$ are continuous on V, ${}^1V$, ${}^2V$, ${}^3V$. Further

$b(t;v_m,v_m,w) - (d/dx_i(\mu^{-1} B_{im} B_m),w) + b_s(t;T_m,w) = (h_m(t),w)$,

$h_m \in K_m \subset L^2(I;V_s')$, $b_o(t;v_m,T_m,z_m) + b_p(t;v_m',z) = (g_m(t),z)$,

$g_m \in {}^2K_m \subset L^2(I;{}^2V_s')$, $(\mu^{-1} \operatorname{Rot}[v_m,B_m],C) =$

$= (\mathcal{B}_m,C)$, $\mathcal{B}_m \in {}^1K_m \subset L^2(I;{}^1V_s')$.

The technique of orthogonal projection yields

$$v_m' = P_m(f - \hat{\mu} A_B v_m - \eta \Lambda_s v_m - \hat{g} j_\varepsilon'(v_m) - h_m) = P_m p_m ,$$

$$\mu^{-1} B_m' = Q_m(-A_M B_m + \mathcal{B}_m - \eta \Sigma_s B_m) = Q_m q_m ,$$

$$T_m' = R_m(W - A_T T_m - \eta \mathcal{T}_s T_m + g_m) = R_m r_m ,$$

$$0 = S_m (- A_g \phi_m - \eta \chi_s \phi_m + \psi) .$$

Then $p_m \in \mathcal{P}_p \subset L^2(I;V_s')$, $q_m \in \mathcal{P}_Q \subset L^2(I;{}^1V_s')$, $r_m \in \mathcal{P}_R \subset L^2(I;{}^2V_s')$, where $\mathcal{P}_p$, $\mathcal{P}_Q$, $\mathcal{P}_R$ are bounded subsets. Further

$$\| P_m p \|_{V_s'} \le c \| P \|_{V_s'} = c \| \Lambda_s^{-1} p \|_{V_s}$$

$$\| Q_m q \|_{{}^1V_s'} \le c \| q \|_{{}^1V_s'} = c \| \Sigma_s^{-1} q \|_{{}^1V_s} ,$$

$$\| R_m r \|_{{}^2V_s'} \le c \| r \|_{{}^2V_s'} = c \| \mathcal{T}_s^{-1} r \|_{{}^2V_s} ,$$

which together with the Lax-Milgram theorem prove (19).

The limitation over m (Galerkin) and $\varepsilon \to 0$, $\eta \to 0$ finish the proof of the existence of the solution.

To prove the uniqueness of the solution, we assume that N=2 and that (v,B,T,ϕ) and (v_1,B_1,T_1,ϕ_1) are two solutions of the problem satisfying (5).
Let us put $(U,\hat{B},\hat{T},\hat{\phi}) = (v-v_1,B-B_1,T-T_1,\phi-\phi_1)$. Then, after some modifications, we have

$$D/Dt \{ |U(t)|^2 + \mu^{-1}|\hat{B}(t)|^2 + |\hat{T}(T)|^2 + |\hat{\phi}(t)|^2 \} + \hat{\mu}\, a(t;U(t),U(t)) +$$

$$+ b_M(t;\hat{B}(t),\hat{B}(t)) + A(t;\hat{T}(t),\hat{T}(t)) + b_g(t;\hat{\phi}(t),\hat{\phi}(t)) \le$$

$$\le c\, g_0(t) (|U(t)|^2 + |\hat{B}(t)|^2 + |\hat{T}(t)|^2 + |\hat{\phi}(t)|^2) ,$$

where $g_0(t)$ is an integrable function in t. Integrating over t from t_0 to t_1, we then obtain

$$|U(t)|^2 + |\hat{B}(t)|^2 + |\hat{T}(t)|^2 + |\hat{\phi}(t)|^2 \le c \int_{t_0}^{t_1} g_0(s) (|U(s)|^2 + |\hat{B}(s)|^2 + |\hat{T}(s)|^2 + |\hat{\phi}(s)|^2)\, ds.$$

Hence $U(t) = 0$, $\hat{B}(t) = 0$, $\hat{T}(t) = 0$, $\hat{\phi}(t) = 0$, which completes the proof.

REFERENCES

Duvaut, G., Lions, J.L., 1972, "Les inéquations en mécanique et en physique," Dunod, Paris.

Duvaut, G., Lions, J.L., 1972, Inéquations en Thermoélasticité et Magnetohydrodynamique, Archive for Rational Mechanics and Analysis, 46:241-279.

Nedoma, J., 1986, On the geodynamic model of the Earth and the plate tectonic hypothesis I-III. Gerlands Beitr. Geophysik, 95,1:36-62, 95,2:89-105.

STOCHASTIC DIFFERENTIATION FORMULA AND CONVERGENCE THEOREM FOR FINELY HYPOHARMONIC FUNCTIONS

Nguyen Xuan-Loc

Department of Math. Statistics
Lieu Giai. Ba Dinh
Hanoi, R.S. Vietnam

As a consequence of a convergence theorem for finely hyperharmonic functions given by us in [4e] , we give here an elementary proof of the following fact :

Let $K \subset C$ *be a compact set in the complex plane and let* (f_n) *be a Cauchy sequence in* $(C_c(K), \sup|\ |)$ *of functions holomorphic in the neighborhoods of K. Then for any point z in the fine interior K' of K there is a compact fine neighborhood V of z in K' such that : The process* $(\partial_z f_n(Z^\tau))$ *is locally a Cauchy sequence in* $L^2([0,\infty[\times\Omega, \mu_{z'})$ *for any z' in K', where* $\mu_{z'}$ *denotes the Doléans measure on* $[0,+\infty[\times\Omega$ *associated with the complex standard Brownian motion started at z' and stopped at the fine boundary of K'.*

The above result is then applied to show the existence of the *fine* complex differentiation as follows (see Coro.3.) :

Let $U \subset C$ *be finely open then there is a mapping* ∂_U *, called the fine complex differentiation, from the space of finely harmonic functions in U into the algebra of quasi-everywhere finely holomorphic functions in U such that,*

$$\partial_U(H_f(U)) \subset H_f(U),$$

i.e., the algebra $H_f(U)$ *of finely holomorphic functions in U is an invariant sub-vector space for the linear operator* ∂_U *.*

Furthermore we could also answer a question on compactness property of the fine Green operator raised earlier in [4b].
From now on we are concerned with the potential theory in a Green domain of the complex plane C. Note that the strong domination principle (D) is verified here and the Markov process associated with this special case is nothing else than the restriction of the standard complex Brownian motion in the given domain.
Let us recall that a complex valued standard process with state space C is called a *standard complex Brownian motion* and is denoted by,

$$(1) \qquad Z := (\Omega, F, (F_t), Z_t = X_t + iY_t, \Theta_t, P^z),$$

provided that there is assigned to each complex number $z \in C$ a probability law P^z on (Ω, F) such that $X = (X_t)$ and $Y = (Y_t)$ are identified with two independent 1-dimensional Wiener processes on (Ω, F_t, P^z) and $P^z(Z_0 = z) = 1$.
Let $K \subset C$ be a compact set with non-empty fine interior $K' \neq \emptyset$. The first exit-time of τ_K for the set K' is defined as,

$$(2) \qquad \tau_K(\omega) := \begin{cases} \inf\{s \mid Z_s(\omega) \notin K'\} \\ +\infty \qquad \text{if such s doesn't exist.} \end{cases}$$

The complex Brownian function stopped at a random time T will be denoted by $Z^T = (Z^T(.,s))$, where $Z^T(.,s) = Z(.,s \wedge T)$.
Without loss of generalities we suppose in the rest of the paper that any function holomorphic in a neighborhood of a compact set is understood to be at the same time smooth and having compact support contained in a fixed open disk $D(0,R) \supset K$. The complex derivative $\frac{1}{2}(\partial_x - i\partial_y)$ will be denoted by ∂.
The following result will be used frequently in the sequel.
<u>Proposition A.</u>a) *Let $K \subset C$ be a compact set with non-empty fine interior K' and let (f_n) be a sequence of functions holomorphic in the neighborhoods of K and converging uniformly in K. For each couple of integers p and n, define,*

$$(3)\begin{cases} C_p^n(z) := \sum_{K=0}^{\infty} |u_k^n(z)| + |u_\infty^n(z)| \quad \text{where} \\ u_k^n(z) = \int_{|\xi-z|=2^{-(K+1)}} \frac{\|f_n\|^{-1} f_n(\xi)\,d\xi}{(\xi-z)^{p+1}} - \int_{|\xi-z|=2^{-K}} \frac{\|f_n\|^{-1} f_n(\xi)\,d\xi}{(\xi-z)^{p+1}}, \quad u_\infty^n(z) = \int_{|\xi-z|=1} \frac{\|f_n\|^{-1} f_n(\xi)\,d\xi}{(\xi-z)^{p+1}} \end{cases}$$

then the function C_p^n $(n,p=1,2..)$ *are finely subharmonic in* K'.

b) *Both* $\lim_n \partial f_n(z)$ *and* $\lim_n C_p^n(z) := C_p(z)$ *exist for any* $z \in K'$ *and furthermore* $C_p(z)$ *is a finely subharmonic function in* K' *such that,*

$$(4) \qquad \lim_n \partial^p f_n(z) \leq C_p(z) \qquad , \quad \forall z \in K' .$$

Proof. The proof is given in details in [4e]. Let us only mention that since the partial sum of the series in the right hand side of (3) forms a specifically increasing sequence of finely subharmonic functions in K' and furthermore since $C_p(z)$ is finite in K', it follows from our convergence theorem for finely hypoharmonic function (see [4e],Pro.1) that C_p is also finely subharmonic in K'. Note also that $\lim_n \partial^p f_n$ exists by the bounded point derivation theorem (see [3]and [4e]).

A *property* of processes (for instance being a martingale, belonging to some class of processes so on...) is said to *hold locally* for a process X if there exists an increasing sequence of (F_t)-stopping times (T_n) with $\lim_n T_n = +\infty$ a.s. such that each process X^{T_n} n=1,2, ... has that property. (T_n) is said to reduce X.

Theorem 1. Let $K \subset C$ *be a compact set with non-empty fine interior* K'. *Let* (f_n) *be a sequence of functions, holomorphic in the neighborhoods of K anf converging uniformly in K. Then for any z in K' and any integer p=1,2,... there is a compact fine neighborhood* V_z *of z in K' such the following sequence of processes,*

$$(5) \qquad \{\partial^p f_n(Z^{\tau V_z})\}$$ *is locally a Cauchy sequence in* $L^2([0,+\infty]\times\Omega, \mu_z)$ *for any z in the fine interior of* V_z , *where* μ_z

denotes the Doléans measure associated with the stopped process starting at z.

Proof. We recall that the Doléans measure is defined on the σ-field of predictable sets in $[0,+\infty]\times\Omega$ as follows,

(6) $$\mu_z(A) \equiv E_z\{\int_0^\infty \chi_A(\cdot,s)\,d(s\wedge\tau_V)\}\ ,\ A\subset[0,+\infty]\times\Omega.$$

p=1. Since the f_n (n=1,2,...) are holomorphic in the neighborhoods of K they can be represented as complex Itô-integrals of the forms:

(7) $$f_n(Z^{\tau_V}) - f_n(Z_o^{\tau_V}) = \int \partial f_n(Z^{\tau_V})\,d(Z^{\tau_V}-Z_o^{\tau_V}) \quad \text{a.s.}P^y$$

where V is any compact subset of K' and y is any point of the fine interior of V. Now by taking the L^2-norm of both sides of (7) it turns out that,

(8) $$E_y\{|f_n(Z^{V})-f_n(Z_o^{V})|^2\} = \int_{[0,+\infty[\times\Omega}|\partial f_n|^2(Z^{\tau_V})\,d\mu_y \quad .$$

Since (f_n) is a Cauchy sequence with respect to the supremum norm in K, it has the same property on any compact set of K', hence it is clear from (8) that $(\partial_z f_n)$ is a Cauchy sequence in $L^2([0\ +\infty]\times\Omega;\mu_y)$.

p>2. For p=1,2,... given, let us consider the sequence of finely subharmonic function (C_p^n) and the function C_p given by the Proposition A. Let $\varepsilon>0$ be fixed constant and put,

$$\begin{aligned} O_n &:= \{\sup_n C_p^n < C_p+\varepsilon \mid n\geq n_0\} \\ &= \cap\{C_p^n < C_p+\varepsilon \mid n\geq n_0\} \end{aligned}$$

Since finely hypoharmonic functions are finely continuous the O_n form an increasing sequence of G_δ-sets, furthermore since $(C_p^{\cdot})$ converges pointwise to C_p the union of the O_n turns out to be K'; we get the followings :

- the exit-time τ_n of O_n increases a.s. to $\tau_{K'}$ (the exit-time of K')and for any $n\geq n_0$, any y in K',

(9) $$|\partial^p f_n(Z(\cdot,s\wedge\tau_n\wedge\tau_V))|\leq C_p(Z(\cdot,s\wedge\tau\wedge\tau_V))+\varepsilon \quad \text{a.s.}P^y$$

On the other hand since $\partial^p f_n$ (n=1,2,...) are also holomorphic in the neighborhoods of K, they could be represented as (see (7)):

$$(10)\quad \begin{cases} \partial^{p-1} f_n(z^{\tau}v) - \partial^{p-1} f_n(z^{\tau}V) = \int \partial^p f_n(z^{\tau}V)\, d(z^{\tau}V - z_o^{\tau}V) \\ \text{where } V \text{ is any compact set of } K'. \end{cases}$$

Now if z is given in K' then for any $\varepsilon>0$ there are a compact fine neighborhood V of z in K' and a constant $M_p(z)$ (depending on the sequence (f_n)) such that,

$$|C_p(y) + \varepsilon|^2 \leq M_p(z), \quad \forall y \in V.$$

Since $(\partial^p f_n)$ converges pointwise in K' by the bounded point derivation theorem (see [4e]), by taking account of (9) we get the following desired equality by Lebesgue bounded convergence theorem :

$$\lim_{n,m} \int_{[0,\infty[\times\Omega} |\partial^{p-1} f_n - \partial^{p-1} f_m|^2 (z^{\tau}V) . \mu_y = 0,$$

for every point y belonging to the fine interior of V.

q.e.d.

Since the left hand side of (11) has the following potential theoretic interpretation (i.e. is identical with):

$$\lim_{n,m} G^{f}(|\partial^p (f_n - f_m)|^2)(y) := \int u_f(y,\xi)\, |\partial^p (f_n - f_m)|^2(\xi)\, d\xi = 0,$$

where $u_f(y,\xi)$ *denotes the fine Green kernel in the fine interior of* V, we obtain the following corollary:

<u>*Corollary 2*</u>. *Let* $K \subset C$ *be compact set with non empty fine interior* K', *then every point* y *in* K' *has a compact fine neighborhood* V *in* K' *such that if we consider the fine complex differential* ∂^p *as an operator defined in the algebra* $R_f(V)$ *and having values in* $L^2(V; u_f(y,\xi)d\xi)$ *as* $\partial^p f$ *with* $f \in R_f(V)$) *then* ∂^p *is a compact operator.*

Note that the Corollary 2. above answers exactely to a question raised by us in [5b].

Corollary 3. *Let $U \subset C$ be a fine domain and let $H_f(U)$ be the vector space of finely harmonic functions in U, then for $p=1,2,\ldots$ there is an operator ∂^p, called the p-th fine complex differential such that :*

a) $\partial^p f$ is quasi-everywhere finely holomorphic in U for any f belonging to $H_f(U)$,

b) $\partial^p f$ is finely holomorphic in U provided that f is and furthermore,

(12) $$\partial^p f(z^{\tau_V}) - \partial^p f(z_0^{\tau_V}) = \int \partial^{p+1} f(z^{\tau_V})\, d(z^{\tau_V} - z_0^{\tau_V})$$

where V is understood to be a fine neighborhood of any given point z in U and where the equality holds a.s.P^y, $\forall y \in V$.

Proof. a) the proof is already given in [4c].

b) the assertation (5) in the theorem 1. is nothing else than a condition which insures that $(\partial^{p-1} f_n)$ is a Cauchy sequence in the Hilbert space of holomorphic martingales (see for instance [4b] or [4d]), hence we get as n tending to ∞ the following:

$$\partial^{p-1} f(z^{\tau_V}) - \partial^{p-1} f(z^{\tau_V}) = \int U(.)\, d(z^{\tau_V} - z^{\tau_V}) \quad \text{a.s.} P^y.$$

where the process $U(.)$ is the limit of $\partial^p f_n(z^{\tau_V})$. The above equality is nothing else the characterization for fine holomorphy of a function f in the neighborhood V (see [4b]).

q.e.d.

Remark. Corollary 2. was a key result in the theory of finely holomorphic functions which has attracted the attention of many authors (see [1],[4b] and [5]).

REFERENCES

[1] A. Debiard and B.Gaveau, Potentiel fin et algèbre de fonctions, analytiques,II, J. Funct. Ana., t.17, 1974, p.296-310.

[2] B. Fuglede, Fine topology and finely holomorphic functions, Proc. 18-th Scand. Congr. Math., Birhauser, 1980,p.22-38.

[3] H. Hallstrom, On bounded point derivation and analytic capacity J. Funct. Ana.,t.4,1969,p.153-165.

[4] a. Nguyen Xuan-Loc, Fine (BMP) and dual processes, Z. Wahr. Ver. Geb.,27,1973,p. 233-56.

b. ---------------, Sur la théorie des fonctions finement holomorphes, I, Bull. Soc. Math., 2è ser.,102,1978, p. 337-64.

c. --------------, Sur la théorie des functions finement holomorphes , II, Proc. Complex Ana.,Lect. Notes747, p.289-300.

d --------------, Smooth conformal martingales and fine holomorphy, Proc. Summer School in Potential Theo.(ed. by O. Martio), Uni. Jyväskylä, 1987, Report 34, p. 35-82.

e The relation between bounded point derivation and the fine derivation (to appear in Proc. 13-th Nevanlinna Coll. 1987).

[5] T. Lyons, A theorem in fine potential theory and applications to finely holomorphic functions, J. Funct. Ana.,37, 1980, p. 19-26.

MORPHISMS OF STONIAN CONES

Eugen Popa

Seminarul Matematic
Universitatea "Al.I.Cuza"
Iaşi, Romania

INTRODUCTION

The main result of this paper puts forward a class of morphisms of H-cones, possessing image: namely those commuting with arbitrary infimum. Applying this result to the canonical embedding in the bidual, one gets a decomposition of any H-cone. In the particular case when the cone has the specific order, this decomposition gives a well-known theorem of Dixmier [1], on the structure of stonian compact spaces.

All the notations and notions about H-cones will be as in [2]. For the notion of a stonian cone, see [2],[3]. The definition of a regular stonian cone, as well as the notation $Ch_T\ X$, are explained in [3]. The notion of morphism of H-cone is used as in [2],[4].

A map $\varphi: S \longrightarrow T$ between two ordered sets is called bimonotone if: $\varphi(s) \leq \varphi(t)$ if and only if $s \leq t$. The notion of stonian compact space and hyperstonian space are defined by Dixmier [1]. For the notion of normal measure, see [2]. If S and T are stonian cones on X,Y, a map $\varphi: Y \longrightarrow X$ is called H-map if: φ is continuous and for every $s \in S$ we have $s \circ \varphi \in \tilde{T}$.

THE MAIN RESULTS

Theorem 1 - Let S and T be H-cones and $\varphi \in \mathrm{Hom}\,(S,T)$ be such that:

(*) for every family $(s_i)_{i \in I}$ from $D(\varphi)$, we have:

$$\varphi(\bigwedge_{i \in I} s_i) = \bigwedge_{i \in I} \varphi(s_i).$$

Then, there exist the balayages B_1 on $D(\varphi)$ and B_2 on S, such that:

a) The restriction of φ to $B_1(D(\varphi))$ is a bimonotone morphism.

b) $S_2 = B_2(S)$ is naturally solid in S and $\varphi|_{S_2} = 0$.

Corollary 2 - Let S and T be stonian cones on the compact spaces X and Y. Let $\varphi: Y \longrightarrow X$ be a H-map such that $\varphi(Y)$ is open in X. Then:

a) φ is an open map and induces a finite morphism, posessing the property ($*$) from Theorem 1.

b) Let us suppose, moreover, that S is a regular stonian cone and that $Ch_T X$ is connected. Then φ induces an injective morphism of cones.

Theorem 3 - Let S be an H-cone with weak unit. There exist balayages B^+ and B^- on S, such that:

a) The restriction of the canonical morphism $S \longrightarrow S^{**}$ to $S^+ = B^+(S)$ is bimonotone.

b) $(S^+)^*$ separates S^+; $(S^-)^* = \{0\}$.

Corollary 4 - (Dixmier [1].) Let X be a stonian compact space. There exist then the open and closed sets X_1 and X_2 such that: $X = X_1 \cup X_2$; $X_1 \cap X_2 = \emptyset$; the only normal measure on X_1 is 0; X_2 is a hyperstonian space.

Comments - a) As it was proved in [5], the condition ($*$) is equivalent, when S and T are standard H-cones of functions, with the fact that semi-polar sets are preserved by inverse image. Under some supplementary assumptions, see [6], ($*$) is also equivalent with the fact that φ is finely open.

b) The hypothesis from Corollary 2 b) can be reformulated as: " S is a standard H-cone possessing a nearly saturated, finely connected representation space".

PROOFS OF THE MAIN RESULTS

Proof of Theorem 1 - Let us define $B_1: D(\varphi) \longrightarrow D(\varphi)$ by

$B_1 s = \bigwedge \{u \in D(\varphi);\ \varphi(u) = \varphi(s)\}$.

From ($*$) it follows the key property:

$$\varphi(s) = \varphi(B_1 s) \text{ whenever } s \in D(\varphi).$$

Lemma 5 - B_1 is a balayage on $D(\varphi)$.

Proof - Since $\varphi(B_1 s + B_1 t) = \varphi(s) + \varphi(t)$, we get $B_1 s + B_1 t \geq B_1(s + t)$. From the Riesz decomposition property, there exist $u, v \in D(\varphi)$ such that $B_1(s + t) = u + v$, $u \preceq B_1 s$, $v \preceq B_1 t$. Hence

$$\varphi(s) + \varphi(t) = \varphi(s + t) = \varphi(B_1(s + t)) =$$
$$= \varphi(u) + \varphi(v) \preceq \varphi(B_1 s) + \varphi(B_1 t) = \varphi(s) + \varphi(t).$$

This shows that $\varphi(u) = \varphi(s)$ and $\varphi(v) = \varphi(t)$, hence $B_1 s = u$, $B_1 t = v$, B_1 is additive.

Let now $s \preceq t$. Then:

$$\varphi(B_1 s \wedge B_1 t) = \varphi(B_1 s) \wedge \varphi(B_1 t) = \varphi(s) \wedge \varphi(t) = \varphi(s)$$

proves that $B_1 s \preceq B_1 s \wedge B_1 t$, B_1 is increasing. Let $(s_i)_{i \in I}$ be an increasing family from $D(\varphi)$ such that $s = \bigvee_{i \in I} s_i \in D(\varphi)$. $(B_1 s_i)_{i \in I}$ is an increasing and dominated family. From:

$$\varphi(\bigvee_{i\in I} B_1 s_i) = \bigvee_{i\in I}\varphi(B_1 s_i) = \bigvee_{i\in I}\varphi(s_i) = \varphi(s)$$

we get $B_1 s \leq \bigvee_{i\in I} B_1 s_i$, B_1 is continuous. The fact that B_1 is a contraction is a direct consequence of the definition.

Finally, since $\varphi(B_1(B_1 s)) = \varphi(B_1 s) = \varphi(s)$, we get $B_1 s \leq B_1(B_1 s)$, and B_1 is a balayage on $D(\varphi)$.

We prove next that $\varphi|_{S_1}$ is bimonotone. Let $\varphi(B_1 s) \leq \varphi(B_1 t)$. It results: $\varphi(s) \leq \varphi(t)$ and, as above $B_1 s \leq B_1 t$.

Let us define now $B_2 : S \longrightarrow S$ by:

$B_2 s = \bigvee\{u \in S;\ \varphi(u) = 0,\ u \leq s\}$.

The family from the second member being increasing and dominated in S, the definition is correct and B_2 results additive and continuous. Clearly, B_2 is increasing and contraction. Since $\varphi(B_2 s) = 0$, we get B_2 a balayage on S and $\varphi|_{S_2} = 0$. Clearly S_2 is naturally solid in S.

Proof of Corollary 2 – a) Clearly, φ is an open map, hence φ^{-1} preserves the dense sets. This implies that φ induces a finite morphism. Next, φ^{-1} preserves also the sets of the first category, hence induces a morphism possessing the property (*). For b), we use the following:

Lemma 6 – Let S, T be stonian cones on X, Y and $\varphi : Y \longrightarrow X$ be a H-map with the property (*). Then $b(B_1) = \varphi(Y)$.

Proof – The equality $\varphi(s) = \varphi(B_1 s)$, for every $s \in D(\varphi)$, can be written now: $s(\varphi(y)) = (B_1 s)(\varphi(y))$, for every $y \in Y$, hence: $\varphi(Y) \subseteq b(B_1)$. Conversely, if $x \notin \varphi(Y)$, there exist $s, t \in D(\varphi)$ such that $s = t$ on $\varphi(Y)$ and $s(x) < t(x)$. Then $\varphi(s) = s \circ \varphi = t \circ \varphi = \varphi(t)$, hence $B_1 s = B_1 t$. On the other hand, $t(x) > s(x) \geq B_1 s(x) = B_1 t(x)$, hence $x \notin b(B_1)$.

We prove now b). From the hypothesis and the lemma above, it follows that $b(B_1) \cap Ch_T X$ is open and closed in $Ch_T X$. Hence $b(B_1) \cap Ch_T X$ is $\emptyset$ or $Ch_T X$. Since

$$B_1 = B^{b(B_1) \cap Ch_T X},$$

it follows that $B_1 = 0$ or $B_1 = I$. In the first case $\varphi = 0$, which is impossible; in the second case, φ results surjective, hence the induced morphism is injective.

Proof of Theorem 3 – Using [2], Theorem 1 can be applied to the canonical morphism $S \longrightarrow S^{**}$. It remains to prove that $(S^+)^*$ separates S^+ and $(S^-)^* = \{0\}$. If $s, t \in S^+$ are such that $s \neq t$, then $\tilde{s} \neq \tilde{t}$, hence there exists $\mu \in S^*$ such that $\mu(s) \neq \mu(t)$. Since S^+ is the image of a balayage, it follows that $\mu|_{S^+} \in (S^+)^*$. Let $u \in S$ be a weak unit and denote $u_0 = B^- u$. $u_0 \in S^-$ is a weak unit in S^-. Indeed, for each $s \in S^-$ we have:

$s \wedge n.u_0 = s \wedge \bigvee\{t \in S^-;\ t \leq n.u\} =$

$= \bigvee \{s \wedge t;\ t \in S^-, t \leq n.u\} = s \wedge n.u.$

Let now $\mu \in (S^-)^*$. Since $\tilde{u}_0 \in (S^-)^{**}$, it follows that there exists an increasing family $(\mu_i)_{i \in I}$ from $(S^-)^*$ such that $\mu = \bigvee_{i \in I} \mu_i$ and $\mu_i(u_0) < \infty$. We can suppose $\mu(u_0) < \infty$. Denoting:

$$\bar{\mu}(s) = \sup\{\mu(s \wedge t);\ t \in S^-\},\ s \in S,$$

we get $\bar{\mu} \in S^*$. Indeed, from

$$(s+s') \wedge t \leq s \wedge t + s' \wedge t \leq (s+s') \wedge 2.t$$

it results that $\bar{\mu}$ is additive; $\bar{\mu}$ is clearly increasing. From $(\bigvee_{i \in I} s_i) \wedge t = \bigvee_{i \in I}(s_i \wedge t)$ it follows the continuity of $\bar{\mu}$. Finally, from $\mu(u_0) < \infty$, it follows that $\bar{\mu} \in S^*$. But: $\mu = \bar{\mu}|_{S^-} = 0$, hence $(S^-)^* = \{0\}$.

Proof of Corollary 4 Let $S = \mathcal{C}_+(X)$ be endowed with the specific order. S is an H-cone. Applying Theorem 3, we get the balayages B^+ and B^-. Let us denote $X_1 = b(B^-)$, $X_2 = b(B^+)$.

Lemma 7 Any basic set (in X, with respect to the stonian cone S) is open.

Proof Let $A \subseteq X$ be basic, hence closed. For each $x \notin A$, there is $s \in S$ such that $s = 1$ on A and $s(x) = 0$. Hence: $\{x \in X;\ B^A 1(x) = 0\} = X \setminus A$, and A results open.

From [2], each normal measure on X_1 induces an H-integral on S^-. Hence X_1 does not carry normal measures. Let now $f \in \mathcal{C}(X_2)$, $f \neq 0$. Clearly we can suppose $f \geq 0$ on X_2. There exists $\mu \in (S^+)^*$ such that $\mu(f) > 0$. Moreover, μ can be chosen finite, hence μ defines a normal measure on X_2. From $B^+ \leq (B^-)'$ we get $b(B^+) \subseteq \overline{X \setminus b(B^-)}$, hence $X_1 \cap X_2 = \emptyset$. On the other hand, we get also:

$$B^+s + B^-s \leq (B^-)'s + B^-s = s,\ s \in S.$$

Hence there is $u \in S$ such that $B^+s + B^-s + u = s$. For any $\mu \in S^*$ we have now $\mu(B^+s) = \mu(s)$ and $\mu(B^-s) = 0$. It follows that $\mu(u) = 0$, whenever $\mu \in S^*$, which shows that $u \in S^-$. Since $B^-s + u \leq s$ and $B^-s + u \in S^-$, we get $B^-s + u \leq B^-s$, and $u = 0$. We get finally $B^+ = (B^-)'$, hence $X_1 \cup X_2 = X$.

H-MAPS BETWEEN STONIAN CONES

In this part we give, without proofs, some results about those morphisms which are induced by continuous maps between the base stonian spaces.

Proposition 8 Let S, T be stonian cones on the compact spaces X, Y. There exists then a canonical bijection between: i) those morphisms $\varphi \in \mathrm{Hom}(S,T)$ satisfying: $\varphi(1) = 1$ and

$\varphi(s \wedge t) = \varphi(s) \wedge \varphi(t)$ whenever $s, t \in S$. ii) The continuous maps $\tilde{\varphi}: Y \longrightarrow X$ such that $s \circ \tilde{\varphi} \in \tilde{T}$ for every $s \in \tilde{S}$ and $\tilde{\varphi}^{-1}(A)$ is of the first category in Y whenever $A \in \mathcal{N}_S$.

φ and $\tilde{\varphi}$ are related by $\varphi(s) = s \circ \tilde{\varphi}$, $s \in \tilde{S}$.

Theorem 9 - Let S be a regular stonian cone on X, T be a stonian cone on Y. Let $\tilde{\varphi}: Y \longrightarrow X$ be a continuous map such that $s \circ \tilde{\varphi} \in \tilde{T}$ for every $s \in S_1$. Then, there exists a unique morphism $\varphi \in \mathrm{Hom}\,(S,T)$, such that $\varphi(s) = s \circ \tilde{\varphi}$ for every $s \in S_1$.

Remarks - 1) If there exists an isomorphism between S and T which preserves 1, then X and Y are homeomorphic. 2) If S and T can be represented as H-cones of functions on the sets X' and Y', then it results from [4] that any H-map from Y' to X' extends to a unique continuous map from Y to X. Conversely, any continuous map $\tilde{\varphi}: Y \longrightarrow X$ as in ii) has the property: $\tilde{\varphi}(Y') \subseteq X'$ and $\tilde{\varphi}|_{Y'}$ is an H-map.

REFERENCES

[1] J. Dixmier, Sur certains espaces considérés par M.H. Stone, Summa Bras. Math. 151, 1951.

[2] N. Boboc, Gh. Bucur, A. Cornea, Order and Convexity in Potential Theory: H-cones, Lecture Notes in Mathematics, vol. 853, Springer-Verlag, 1981.

[3] N. Boboc, Gh. Bucur, Stonian Structures in Potential Theory, Preprint INCREST, Bucureşti, 52/1984.

[4] E. Popa, Morphisms of H-cones, An.Şt. Univ. Iaşi, t. XXIX, f. 2 (1983).

[5] E. Popa, Morphisms of H-cones and Semi-polar Sets, An. Şt. Univ. Iaşi, t. XXX, f. 3 (1984).

[6] E. Popa, Finely Open Morphisms of H-cones, Math. Ann. 270 (1985).

ON THE TRANSITION FUNCTION OF THE INFINITE DIMENSIONAL ORNSTEIN-UHLENBECK PROCESS GIVEN BY THE FREE QUANTUM FIELD

Michael Röckner

University of Edinburgh, Department of Mathematics
James Clerk Maxwell Building, King's Buildings
Mayfield Road, Edinburgh EH9 3JZ, Scotland

1. INTRODUCTION AND BASIC NOTIONS

In this paper we want to investigate a particular semigroup $(\pi_t)_{t>0}$ of probability kernels defined on some infinite dimensional Banach space B_α contained in $\mathcal{S}'(\mathbb{R}^{d-1})$ (i.e. the tempered distributions on $\mathbb{R}^{d-1}$), $d \geq 2$, $\alpha \in]\frac{d-2}{2},\infty[$. This semigroup $(\pi_t)_{t>0}$ is the transition function of the infinite dimensional Ornstein-Uhlenbeck process given by the free field of Euclidean quantum field theory on $\mathbb{R}^d$ (cf. [1], [2]). We will establish that $(\pi_t)_{t>0}$ has the Feller property, determine its generator (on a suitable domain) and its Dirichlet form. Furthermore, we will characterise the associated entrance space in the sense of Dynkin (cf. [3], [4], [5]).

We emphasize that $(\pi_t)_{t>0}$ could also be defined on $\mathcal{S}'(\mathbb{R}^{d-1})$ and it is, in fact, much easier to study $(\pi_t)_{t>0}$ and the associated Ornstein-Uhlenbeck process on this state space (cf. 1.3(iii) below). $\mathcal{S}'(\mathbb{R}^{d-1})$, however, is so large that the set of continuous nonlinear functionals on it is rather poor. Since non-linear functionals of this process are of great interest in quantum field theory, $\mathcal{S}'(\mathbb{R}^{d-1})$ is not suitable as the state space. Therefore, it is important to "localize" this process as well as possible. The Banach space B_α introduced in [6] (cf. (1.3) below) which is much smaller than $\mathcal{S}'(\mathbb{R}^{d-1})$ appears to be a much better and more suitable state space to work with.

Because of its importance in Euclidean field theory we think that $(\pi_t)_{t>0}$ presents a fundamental example of a potential theory on an infinite dimensional (hence non-locally compact) topological vector space. Due to the fact that the free field is defined in terms of the operator $\Delta-1$ on $\mathbb{R}^d$ we will see that this "infinite dimensional potential theory on B_α" is linked very closely to the "finite dimensional potential theory on $\mathbb{R}^d$" given by the operator $\Delta-1$ (e.g. the entrance space associated with $(\pi_t)_{t>0}$ can be characterised in terms of $(\Delta-1)$-harmonic functions). For more information about general infinite dimensional Ornstein-Uhlenbeck processes we refer to [7], [8] and [9].

In order to give the precise definitions of B_α and $(\pi_t)_{t>0}$ we need to introduce the following notations.

Let dx denote the Lebesgue measure and let $\mathfrak{F}\xi$ denote the Fourier transform of $\xi \in \mathcal{S}'(\mathbb{R}^{d-1})$, i.e. if g is a test function in $\mathcal{S}(\mathbb{R}^{d-1})$ (equipped with the usual topology)

$$\mathfrak{F}g(y) = \int_{\mathbb{R}^{d-1}} e^{iy\cdot x}\, g(x)\, dx\,,\ y \in \mathbb{R}^{d-1}\,.$$

Let P_t, $t>0$, be the Poisson kernel for the half space $D_0 := \{(x,s) \in \mathbb{R}^d : x \in \mathbb{R}^{d-1},\ s>0\}$, i.e. its Fourier transform is given by

$$(1.1) \qquad (\mathfrak{F}P_t)(x) = e^{-t\sqrt{1+|x|^2}}\,,\ x \in \mathbb{R}^{d-1}\,,\ t>0.$$

Hence it is easy to see that for $x \in \mathbb{R}^{d-1}$

$$(1.2) \qquad P_t(x) = \frac{t}{\sqrt{4\pi}} \int_0^\infty \frac{e^{-t^2/4s}}{s^{3/2}}\, p_s^{d-1}(x)\, ds$$

$$= \int_0^\infty p_s^{d-1}(x)\, \eta_t^1\,(ds)$$

where $p_s^{d-1} : \mathbb{R}^{d-1} \to \mathbb{R}_+$ is the density of the transition function of (d-1)-dimensional Brownian motion with killing 1, i.e. $p_s^{d-1}(x) = \frac{e^{-s}}{\sqrt{4\pi s}^{\,d-1}} e^{-|x|^2/4s}$ $x \in \mathbb{R}^{d-1}$, $s>0$. (We should mention that we consider Brownian motion with a time scale such that Δ_{d-1} instead of $\frac{1}{2}\Delta_{d-1}$ is its generator.) $(\eta_t^1)_{t>0}$ in

(1.2) denotes the one-sided stable semigroup on $\mathbb{R}_+\setminus\{0\}$ (cf. [10], Sect.V, 3). Of course, if $\mu^{D_o}_{(x,t)}$ denotes the harmonic measure of D_o at $(x,s)\in D_o$ then

$$\int f(z)\mu^{D_o}_{(x,t)}(dz) = \int f(z)\, P_t(x-z)\, dz = (P_t * f)(x)$$

for $x\in\mathbb{R}^{d-1}$, $t>0$, and f a positive, Borel-measurable function on $\mathbb{R}^{d-1}$ $(=\mathbb{R}^{d-1}\times\{0\})$. (Here $*$ means convolution with respect to $\mathbb{R}^{d-1}$.) Let for $\alpha\in\mathbb{R}$, $H^{\alpha}(\mathbb{R}^{d-1})$ denote the Sobolev space of order α on $\mathbb{R}^{d-1}$, i.e.

$$H^{\alpha}(\mathbb{R}^{d-1}) := \{\xi\in \mathcal{S}'(\mathbb{R}^{d-1}) : \int_{\mathbb{R}^{d-1}} |\mathcal{F}\xi|^2(x)\ (1+|x|^2)^{\alpha} < \infty\}$$

with corresponding inner product $\langle\ ,\ \rangle_{H^{\alpha}}$ and norm $\|\ \|_{H^{\alpha}}$.

Fix a bounded infinitely differentiable function ρ on $\mathbb{R}^{d-1}$ which is strictly positive and dx-square integrable such that ρ and ρ^{-1} is an $\mathcal{S}'(\mathbb{R}^{d-1})$ multiplier. Assume that for each $t\in\mathbb{R}$

$$P_t * \rho^2 \leqslant \rho^2.$$

Furthermore, fix $p_o\in[2,\infty[$ and $p_1,\ldots,p_N\in[1,\infty[$, $N\in\mathbb{N}$. Then we define for $\alpha > 0$

(1.3) $$B_{\alpha} := \{\xi\in\mathcal{S}'(\mathbb{R}^{d-1}) : \rho\cdot\xi\in H^{-\alpha}(\mathbb{R}^{d-1}) \text{ and for } 0\leqslant i\leqslant N$$
$$\int_0^{\infty}\int_{\mathbb{R}^{d-1}} (t^{\alpha}|P_t*\xi|)^{p_i}\, \rho^2\, t^{-1}\, dx\, dt < +\infty\},$$

with norm

$$\|\xi\|_{B_{\alpha}} := \left[\|\rho\xi\|^2_{H^{-\alpha}} + \sum_{i=0}^{N}\left(\int_0^{\infty}\int_{\mathbb{R}^{d-1}} (t^{\alpha}|P_t*\xi|)^{p_i}\, \rho^2\, t^{-1}\, dx\, dt\right)^{2/p_i}\right]^{\frac{1}{2}}.$$

B_{α} should be thought of as a space of distributions for which when given as boundary data on $\mathbb{R}^{d-1}\times\{0\}$ one can solve a Dirichlet problem for the half-space D_o such that the boundary behaviour of the solutions is controlled in a certain way (cf. also [6, Sect. 6]). In addition, the traces of the solutions on $\{(x,t)\in\mathbb{R}^d : x\in\mathbb{R}^{d-1}\}$ for $t>0$ are in the weighted L^2-space $L^2(\mathbb{R}^{d-1}, \rho^2 dx)$ (cf. the proof of 3.3 in [6]). Some complications arise since for our purpose (cf. 1.2, 1.3(i) and 1.4 below) we need to consider distributions in the scaled Sobolev space $\rho^{-1}H^{-\alpha}(\mathbb{R}^{d-1})$. Nevertheless, we have that B_α is invariant under P_t. More precisely, we have the following proposition which was proven in [6, Sect. 3]:

1.1. Proposition – For any $\alpha > 0$, B_α is a separable Banach space such that

(1.4) $\quad P_t * \xi \in B_\alpha$ for any $\xi \in B_\alpha$.

Note that $P_t * \xi$ is well defined for $\xi \in B_\alpha$ since $B_\alpha \subset \mathcal{S}'(\mathbb{R}^{d-1})$ and $P_t \in \mathcal{S}(\mathbb{R}^{d-1})$ by (1.1). Let B'_α be the (topological) dual of B_α equipped with the weak*-topology. By [6, 3.6] and the Hahn-Banach theorem it follows that each $g \in \mathcal{S}(\mathbb{R}^{d-1})$ can be considered as an element in B'_α given by $\langle\ ,g\rangle$ where $\langle\ ,\ \rangle$ is the dualisation between $\mathcal{S}'(\mathbb{R}^{d-1})$ and $\mathcal{S}(\mathbb{R}^{d-1})$. Moreover, this imbedding is dense in B'_α. Using this we can restate one of the main results in [6]. From now on we denote the Borel σ-field of a topological space X by $\mathcal{B}(X)$.

1.2. Theorem – If $\alpha > \frac{d-2}{2}$, there exist unique measures $\mathbb{P}$, $\mathbb{P}_t$, $t \in \mathbb{R}$, on $(B_\alpha, \mathcal{B}(B_\alpha))$ such that for every $g \in \mathcal{S}(\mathbb{R}^{d-1})$

$$(1.5) \qquad \int_{B_\alpha} \exp(i\langle \xi, g\rangle)\, \mathbb{P}(d\xi) = \exp(-\tfrac{1}{2}\|g\|^2_{H^{-\frac{1}{2}}})$$

and

$$(1.6) \qquad \int_{B_\alpha} \exp(i\langle \xi, g\rangle)\, \mathbb{P}_t(d\xi) = \exp(-\tfrac{1}{2}\, {}_{H^{-\frac{1}{2}}}\langle g - P_{2t} * g, g\rangle_{H^{-\frac{1}{2}}}).$$

Proof – [6, 4.2]

1.3 Remark–(i) It is well known that in 1.2 ρ can not be taken to be the constant function 1. Also $\alpha > \frac{d-2}{2}$ is necessary (cf. [11] and [6, 5.5]).

(ii) (1.5), (1.6) just say that the functions $\xi \mapsto \langle \xi, g\rangle$, $g \in \mathcal{S}(\mathbb{R}^{d-1})$, are Gaussian distributed on $\mathbb{R}$ under $\mathbb{P}$, $\mathbb{P}_t$, $t \in \mathbb{R}$, with mean zero and covariance $\|g\|^2_{H^{-\frac{1}{2}}}$ resp. ${}_{H^{-\frac{1}{2}}}\langle g - P_{2t} * g, g\rangle_{H^{-\frac{1}{2}}}$.

(iii) The existence of $\mathbb{P}$, $\mathbb{P}_t$ satisfying (1.5) resp. (1.6) as measures on $\mathcal{S}'(\mathbb{R}^{d-1})$ (equipped with the σ-field of its weak*-Borel sets is a simple consequence of the Bochner/Minlos theorem (cf. [12, Chap. IV. 3]). So the main point in 1.2 is that they can actually be constructed on a much smaller space, namely the Banach space B_α, if α is big enough. Fundamental work on problems of this kind was done by L. Gross (cf. [13], [14]). We also refer to [15] and [16].

Now we are prepared to define the semigroup $(\pi_t)_{t>0}$:
Let $t > 0$. For $\xi \in B_\alpha$, $B \in \mathcal{B}(B_\alpha)$ we define

(1.7) $$\pi_t(\xi,B) := \int_{B_\alpha} 1_B(\eta + P_t{*}\xi)\, \mathbb{P}_t(d\eta)$$

(where 1_B means indicator function of B). Note that π_t is well-defined because of (1.4). $(\pi_t)_{t>0}$ defines a Markovian semigroup on $(B_\alpha, \mathcal{B}(B_\alpha))$ (cf. [17, §64]), i.e. a family of probability kernels on $(B_\alpha, \mathcal{B}(B_\alpha))$ satisfying

(1.8) $$\pi_{t+s} = \pi_t \pi_s \quad \text{for all } s,t > 0,$$

i.e. $\pi_{t+s}(\xi,F) = \pi_t(\xi, \pi_s(\cdot,F))$ for every $\xi \in B$ and every $\mathcal{B}(B_\alpha)$-measurable bounded function F on B_α. Here we set as usual $\pi_t(\xi,F) := \int F(\eta)\pi_t(\xi,d\eta)$. For a proof of (1.8) see [6, Sect. 4].

As mentioned above $(\pi_t)_{t>0}$ is important because of its relevance in mathematical physics. This is the content of the following theorem. But first we need to introduce some notations.

Let $\Omega := C(\mathbb{R}, B_\alpha)$ denote the space of all continuous functions from $\mathbb{R}$ to B_α. As usual, for $t \in \mathbb{R}$, let $X_t : \Omega \to B_\alpha$ be defined by $X_t(\omega) := \omega(t)$, $\omega \in \Omega$. Furthermore, let $\mathcal{F}_t$, $t \in \mathbb{R} \cup \{+\infty\}$, denote the σ-field on Ω which is generated by $\{X_s : s \in]-\infty, t]\}$ and set $\mathcal{F} := \mathcal{F}_\infty$. Let P be some probability measure on $(\Omega, \mathcal{F})$ and $(\tilde{\Omega}, \tilde{\mathcal{F}})$ a measurable space. Given an $\mathcal{F}/\tilde{\mathcal{F}}$-measurable map $Y : \Omega \to \tilde{\Omega}$ let Y(P) denote the image measure on $(\tilde{\Omega}, \tilde{\mathcal{F}})$. For a sub-$\sigma$-field Σ of $\mathcal{F}$ let $E_P(\cdot \mid \Sigma)$ denote the conditional expectation of P given Σ.

<u>1.4. Theorem</u> – If $\alpha > \frac{d-2}{2}$ then there exists a unique measure P_o on $(\Omega, \mathcal{F})$ such that $X_t(P_o) = \mathbb{P}$ for every $t \in \mathbb{R}$ and for all $s,t \in \mathbb{R}, s < t$,

$$E_{P_o}(F \circ X_t \mid \mathcal{F}_s) = \pi_{t-s}(X_s, F) \quad P_o\text{-a.e.},$$

for every bounded $\mathcal{B}(B_\alpha)$-measurable function F on B_α. Moreover, $\mathbf{P}_t = (X_t - P_t * X_o)(P_o)$ for every $t \in \mathbf{R}$.

Proof.– [6, 4.2, 4.6 and 5.4]

1.5 Remark–(i) P_o (i.e. $X(P_o)$ in the notation of [6]) is the free field of Euclidean quantum field theory (cf. [1], [18] and [19]). We note that P_o is usually constructed on $\mathcal{D}'(\mathbb{R}^d)$, i.e. the space of Schwarz distributions on $\mathbf{R}^d$, but in [6] it was proven that it is actually supported on $C(\mathbf{R}, B_\alpha)$ (which, of course, can be considered as a subspace of $\mathcal{D}'(\mathbf{R}^d)$, cf. [6, Sect.5]). For the case of other spaces replacing B_α we refer to [20] and [16].
(ii) It is easy to check that $(X_t)_{t\in\mathbb{R}}$ is a (jointly) Gaussian process under P_o (cf. [6, 4.1]).

In the following sections $(\pi_t)_{t>0}$ will be in the centre of our considerations. The probability measure $\mathbf{P}$ on $(B_\alpha, \mathcal{B}(B_\alpha))$ and its relation with $(\pi_t)_{t>0}$ will play an important role in section 3 in connection with the Dirichlet form of $(\pi_t)_{t>0}$. The free field P_o will be essential in section 4 on the "entrance space" of $(\pi_t)_{t>0}$. Let us start with the Feller property of $(\pi_t)_{t>0}$ and the related question about the strong Markov property. We assume from now on that $\alpha > \frac{d-2}{2}$.

2. FELLER PROPERTY AND THE ASSOCIATED STRONG MARKOV PROCESS

Define for $t \geq 0$, $T_t : B_\alpha \to B_\alpha$ by

$$T_t\xi := \begin{cases} P_t * \xi & \text{if } t > 0 \\ \xi & \text{if } t = 0 \end{cases}, \quad \xi \in B_\alpha. \tag{2.1}$$

2.1 Theorem – $(\pi_t)_{t>0}$ has the Feller property, i.e. for any $t > 0$ and any bounded continuous function F on B_α the function $\xi \mapsto \pi_t(\xi, F)$ is continuous on B_α.

By Lebesgue's theorem of dominated convergence, 2.1 is an immediate consequence of:

2.2. Proposition – For any $t \geq 0$, T_t is a continuous linear operator on B_α.

Proof – Let $t \in \mathbb{R}$. Since T_t is defined on all of B_α, convergence in B_α implies convergence in $\mathcal{S}'(\mathbb{R}^{d-1})$ and T_t is continuous on $\mathcal{S}'(\mathbb{R}^{d-1})$, the assertion follows by the closed graph theorem.

Using the usual "Kolmogorov scheme" one can construct a canonical Markov process associated with the semigroup $(\pi_t)_{t>0}$ (cf. e.g. [17, §62, 64, 65]). One would like to prove that this Markov process has indeed the strong Markov property. If we knew that the Markov process had right continuous sample paths on $[0,\infty[$ then the strong Markov property would follow from 2.1 by [21, Satz 5.10]. A necessary condition for the right continuity of the sample paths on $[0,\infty[$ is that

(2.2) $$\lim_{t \downarrow 0} \pi_t(\xi, F) = F(\xi)$$

for every $\xi \in B_\alpha$ and any bounded continuous function on B_α which in turn is implied by

(2.3) $$\lim_{t \downarrow 0} T_t \xi = \xi \text{ with respect to } \| \ \|_{B_\alpha} \text{ for every } \xi \in B_\alpha .$$

For a general space B_α (as defined in (1.3)) without further assumptions on ρ, (2.3) seems to be hard to prove. Therefore, we assume from now on that in addition

(2.4) $$\rho \in \mathcal{W}$$

i.e. ρ is a smooth tempered weight function in the sense of [22, Definition 4.1] to which we refer for the definition of $\mathcal{W}$. We only mention here that if $\beta > \frac{d-1}{2}$ and $\gamma \in]0,+\infty[$ is large enough then e.g.

$$\rho(x) := (\gamma + |x|^2)^{-\beta/2}, \quad x \in \mathbb{R}^{d-1}$$

satisfies all our assumptions. Now we have the following.

2.3. Proposition – There exists $C \in]0,\infty[$ such that for every $t \geq 0$ and $\xi \in B_\alpha$, $\|T_t \xi\|_{B_\alpha} \leq C \|\xi\|_{B_\alpha}$. Furthermore, the map $t \mapsto T_t \xi$ from $[0,\infty[$ to B_α is continuous.

Proof - The first part of the assertion is a consequence of [22, Corollary to Lemma 4.1 and Proposition 4.3] and the proof of 3.3 in [6]. The second part then follows since $H^{\frac{1}{2}} \subset B_\alpha$ continuously and densely (cf. [6, 3.4 and 4.4]) and $\lim_{t\downarrow 0} \|T_t\xi - \xi\|_{H^{\frac{1}{2}}} = 0$ for every $\xi \in H^{\frac{1}{2}}$. □

Now we want to present an alternative, direct construction of a Markov process associated with $(\pi_t)_{t>0}$ without using the usual "Kolmogorov scheme" mentioned above. This is possible in our particular situation.

Define $Y : \Omega \to \Omega$, $T : B_\alpha \to \Omega$ componentwise by

$$Y_t := \begin{cases} X_t - T_t X_\circ & , \text{ if } t \geq 0 \\ 0 & , \text{ if } t < 0 \end{cases}$$

and for $\xi \in B_\alpha$

$$(T\xi)_t := \begin{cases} T_t\xi & , \text{ if } t \geq 0 \\ \xi & , \text{ if } t < 0 \,. \end{cases}$$

By 2.3, Y and T actually take values in Ω. Moreover, Y is $\mathcal{F}/\mathcal{F}$-measurable and T is $\mathcal{B}(B_\alpha)/\mathcal{F}$-measurable. Define for $\xi \in B_\alpha$ the measure P^ξ on $(\Omega, \mathcal{F})$ by

(2.5) $$P^\xi(A) = Y(P_\circ)(A - T\xi) \quad , \quad A \in \mathcal{F}$$

(i.e. P^ξ is defined in terms of the free quantum field $P_\circ$ and the solution for the Dirichlet problem for $D_\circ$ and boundary data ξ on $\mathbb{R}^{d-1} \times \{0\}$). We define $\pi_\circ(\xi, d\eta)$, $\xi \in B_\alpha$, to be the Dirac measure on $\mathcal{B}(B_\alpha)$ in ξ.

2.4 Theorem - For all $\xi \in B_\alpha$, $0 \leq t_1 < t_2 \ldots < t_n < +\infty$, $n \in \mathbb{N}$, and $B_1, \ldots, B_n \in \mathcal{B}(B_\alpha)$

(2.6) $$P^\xi(X_{t_1} \in B_1, \ldots, X_{t_n} \in B_n)$$
$$= \int_{B_1} \pi_{t_1}(\xi, d\xi_1) \int_{B_2} \pi_{t_2 - t_1}(\xi_1, d\xi_2) \ldots \int_{B_n} \pi_{t_n - t_{n-1}}(\xi_{n-1}, d\xi_n) \,.$$

In particular, $(\Omega, \mathcal{F}, (X_t)_{t\geq 0}, (P^\xi)_{\xi\in B_\alpha})$ is a strong Markov process with state space B_α having continuous sample paths and transition function $(\pi_t)_{t\geq 0}$.

Proof — Using [6, 4.1] it is easy to calculate the Fourier transform of the left hand side of (2.6), i.e. for $\beta_1,\ldots,\beta_n \in \mathbb{R}$ and $g_1,\ldots,g_n \in \mathcal{S}(\mathbb{R}^{d-1})$

$$(2.7) \qquad \int \exp\Big(i \sum_{k=1}^{n} \beta_k \langle X_{t_k}, g_k\rangle\Big)\, dP^{\xi}$$

$$= \exp\Big[i\langle \xi, \sum_{k=1}^{n} \beta_k T_{t_k} g_k\rangle - \frac{1}{2}\sum_{k,l=1}^{n} \beta_l \beta_k \big({}_{H^{-\frac{1}{2}}}\langle g_l, T_{|t_l - t_k|} g_k\rangle_{H^{-\frac{1}{2}}} - {}_{H^{-\frac{1}{2}}}\langle g_l, T_{t_l + t_k} g_k\rangle_{H^{-\frac{1}{2}}}\big)\Big].$$

Then (2.6) is a simple consequence by induction and 1.2. The rest of the proof is standard (cf. [17, §65]). The strong Markov property follows by [21, Satz 5.10].

3. THE GENERATOR AND THE DIRICHLET FORM ASSOCIATED WITH $(\pi_t)_{t>0}$

Now we intend to determine the generator of $(\pi_t)_{t>0}$ on the following domain of functions on B_α.

$$(3.1) \qquad \mathcal{F}C_b^2 := \{F : B_\alpha \to \mathbb{R}: \text{ there exist } g_1,\ldots g_n \in \mathcal{S}(\mathbb{R}^{d-1}) \text{ and } F^* \in C_b^2(\mathbb{R}^{d-1}) \text{ such that } F(\xi) = F^*(\langle \xi, g_1\rangle, \ldots, \langle \xi, g_n\rangle),\ \xi \in B_\alpha\}$$

(where $C_b^2(\mathbb{R}^{d-1})$ = all twice differentiable functions on $\mathbb{R}^{d-1}$ with bounded continuous derivatives).

Given a bilinear map $V : B_\alpha \times B_\alpha \to \mathbb{R}$ and a Hilbert space $(H, \|\ \|_H)$ such that $H \subset B_\alpha$ and $V_{|H\times H}$ is continuous, we define

$$(3.2) \qquad \mathrm{Trace}_H V := \sum_{k=1}^{\infty} V(h_k, h_k)$$

for some orthonormal basis $(h_k)_{k\in\mathbb{N}}$ of H. (3.2) is, of course, independent of the special choice of $(h_k)_{k\in\mathbb{N}}$.

In the following DF, D^2F etc. shall denote Frechet derivatives of real valued functions F on some Banach space. Note that by the chain rule for $F = F^*(\langle\cdot, g_1\rangle, \ldots, \langle\cdot, g_n\rangle) \in \mathcal{F}C_b^2$ and $\xi, \eta, \eta' \in B_\alpha$ we have

$$(3.3) \qquad DF(\xi)(\eta) = DF^*(\langle \xi, g_1\rangle, \ldots, \langle \xi, g_n\rangle)\ (\langle \eta, g_1\rangle, \ldots, \langle \eta, g_n\rangle)$$

and

(3.4) $D^2F(\xi)(\eta,\eta')$

$$= D^2F^*(\langle\xi,g_1\rangle,\dots,\langle\xi,g_n\rangle)((\langle\eta,g_1\rangle,\dots,\langle\eta,g_n\rangle),(\langle\eta',g_1\rangle,\dots,\langle\eta',g_n\rangle)).$$

Hence both $DF(\cdot)(\cdot)$ and $D^2F(\cdot)(\cdot,\cdot)$ extend to continuous functions on $(\mathcal{S}'(\mathbb{R}^d))^2$, $(\mathcal{S}'(\mathbb{R}^d))^3$ respectively. Now the proof of the following theorem is fairly standard (cf. e.g. [23, Theorem 1.4] where the simpler case of $e^{-t}Id_{B_\alpha}$ replacing T_t is treated).

<u>3.1 Theorem</u> — Let $F \in \mathcal{F}C_b^2$. Then

(3.5) $$L_oF(\xi) := \lim_{t\downarrow 0}\frac{1}{t}(\pi_t(\xi,F) - F(\xi)) = \text{Trace}_{L^2(\mathbb{R}^{d-1},dx)} D^2F(\xi) - DF(\xi)(\sqrt{-\Delta_{d-1}+1}\,\xi)$$

for each $\xi \in B_\alpha$. In particular, if $F(\xi) = F^*(\langle\xi,g_1\rangle,\dots,\langle\xi,g_n\rangle)$, $\xi\in B_\alpha$, for some $F^*\in C_b^2(\mathbb{R}^{d-1})$ and an orthonormal system $g_1,\dots,g_n$ of $L^2(\mathbb{R}^{d-1},dx)$ then

(3.6) $$L_oF(\xi) = \sum_{j=1}^n \left[\left(\frac{d^2}{dt^2}F(\xi+tg_j)\right)\Big|_{t=0} - \left(\frac{d}{dt}F(\xi+tg_j)\right)\Big|_{t=0}\langle\xi,\sqrt{-\Delta_{d-1}+1}\,g_j\rangle\right].$$

<u>Proof</u> — Let $\xi\in B_\alpha$ and $t>0$ then by Taylor's formula

$$\frac{1}{t}(\pi_t(\xi,F) - F(\xi)) = \frac{1}{t}\int (F(\xi+(T_t\xi-\xi+\eta)) - F(\xi))\,\mathbb{P}_t(d\eta)$$

$$= \frac{1}{t}\Big[DF(\xi)(T_t\xi-\xi) + \frac{1}{2}D^2F(\xi)(T_t\xi-\xi,\,T_t\xi-\xi)$$

$$+ \frac{1}{2}\int D^2F(\xi)(\eta,\eta)\,\mathbb{P}_t(d\eta) + \int R_2(\xi,T_t\xi-\xi+\eta)\,\mathbb{P}_t(d\eta)\Big]$$

where in the last step we used the fact that the Bochner integral $\int \eta\,\mathbb{P}_t(d\eta)$ is equal to zero. Let $F^*,g_1,\dots,g_n$ be as in the theorem, then by (3.4) we have for the remainder R_2 that for $\eta_1,\eta_2\in B_\alpha$

$$R_2(\eta_1,\eta_2) = \int_0^1 (1-s)\,(D^2F^*(G(\eta_1+s\eta_2)) - D^2F^*(G(\eta_1)))(G(\eta_2),G(\eta_2))ds$$

where $G(\eta) := (\langle\eta,g_1\rangle,\dots,\langle\eta,g_n\rangle)$, $\eta\in B_\alpha$. Therefore, the modulus of the above integral of the remainder is by 1.3(ii) and 1.4 dominated by

$$\left[\int\int_0^1 \|D^2F^*(G(\xi+s(T_t\xi-\xi+X_t-T_tX_o))) - D^2F^*(G(\xi))\|^2_{\mathbb{R}^{n^2}}ds\,dP_o\right]^{1/2}$$

$$\times\frac{8}{t}\sum_{j=1}^n\left[\langle\xi,T_tg_j-g_j\rangle^2 + {}_{H^{-\frac12}}\langle g_j-T_{2t}g_j,g_j\rangle_{H^{-\frac12}}\right]$$

which converges to zero if $t\downarrow 0$. By the remarks preceding 3.1 the same is true for the second term above while the first converges to $-DF(\xi)(\sqrt{-\Delta_{d-1}+1}\,\xi)$.

Now we consider the third term. Extend $g_1,\dots,g_n$ to an orthonormal basis $(g_j)_{j\in\mathbb{N}}$ of $L^2(\mathbb{R}^{d-1},dx)$. Since $L^2(\mathbb{R}^{d-1},dx)\subset B_\alpha$ (cf. the proof of [6, 3.3]) and because of (3.4), $\mathrm{Trace}_{L^2(\mathbb{R}^{d-1},dx)} D^2F(\xi)$ exists for every $\xi\in B_\alpha$. Define for $i,j\in\mathbb{N}$ and $\xi\in B_\alpha$, $a_{ij}(\xi):= D^2F(\xi)(g_i,g_j)$, then by (3.4)

$$\frac{1}{2t}\int D^2F(\xi)(\eta,\eta)\,\mathbb{P}_t(d\eta) = \frac{1}{2t}\sum_{i,j=1}^{n} a_{ij}(\xi)\int\langle \eta,g_i\rangle\langle\eta.g_j\rangle\,\mathbb{P}_t(d\eta)$$
$$=\frac{1}{2t}\sum_{i,j=1}^{n} a_{ij}(\xi)\ {}_{H^{-\frac12}}\langle g_i - T_{2t}g_i,g_j\rangle_{H^{-\frac12}}\ .$$

If $t\downarrow 0$, this converges to

$$\sum_{i,j=1}^{n} a_{ij}(\xi)\ {}_{H^{-\frac12}}\langle\sqrt{-\Delta_{d-1}+1}\ g_i,g_j\rangle_{H^{-\frac12}} = \sum_{i=1}^{n} a_{ii}(\xi) = \mathrm{Trace}_{L^2(\mathbb{R}^{d-1},dx)} D^2F(\xi).$$

The last part of the theorem is now obvious by (3.3) and (3.4).

Now let's turn to the Dirichlet form associated with $(\pi_t)_{t>0}$ on $L^2(B_\alpha,d\mathbb{P})$. First we have:

<u>3.2 Proposition</u> – $(\pi_t)_{t>0}$ is $\mathbb{P}$-symmetric, i.e. for every $t>0$

$$\int \pi_t(\xi,F)\,G(\xi)\ \mathbb{P}(d\xi) = \int F(\xi)\ \pi_t(\xi,G)\ \mathbb{P}(d\xi)$$

for all bounded $\mathcal{B}(B_\alpha)$-measurable functions F, G.

<u>Proof</u> – For $f,g\in\mathcal{S}(\mathbb{R}^{d-1})$ and $t>0$ we have by (1.5) - (1.7)

$$\int\pi_t(\xi,\exp(i\langle\cdot,f\rangle))\exp(i\langle\xi,g\rangle)\mathbb{P}(d\xi) = \exp\left[-\frac{1}{2}(\|g\|^2_{H^{-\frac12}} + \|f\|^2_{H^{-\frac12}} - 2\ {}_{H^{-\frac12}}\langle g,T_tf\rangle_{H^{-\frac12}})\right]$$

As in [24, §1.4] one now easily sees that for each $t>0$

$$\tilde{S}_tF := \pi_t(\cdot,F),\qquad F\ \mathcal{B}(B_\alpha)\text{-measurable, bounded,}$$

defines an operator on $L^2(B_\alpha,\mathbb{P})$ which extends uniquely to a symmetric contraction S_t on $L^2(B_\alpha,\mathbb{P})$ such that $(S_t)_{t>0}$ is a semigroup of operators on $L^2(B_\alpha,\mathbb{P})$ which are sub-Markovian (i.e. $0\leq S_tu\leq 1$ $\mathbb{P}$-a.e. if $0\leq u\leq 1$ $\mathbb{P}$ - a.e., $u\in L^2(B_\alpha,\mathbb{P})$). Define

$$(3.7)\qquad \begin{cases} D(\mathcal{E}) := \{u\in L^2(B_\alpha,\mathbb{P}):\ \lim\limits_{t\to0}\frac{1}{t}(u-S_tu,u) < +\infty\} \\ \mathcal{E}(u,v) = \lim\limits_{t\to0}\frac{1}{t}(u-S_tu,v)\ ,\quad u,v\in D(\mathcal{E}). \end{cases}$$

Note that $\frac{1}{t}(u-S_tu,u)$ is increasing as $t\downarrow0$ for every $u\in L^2(B_\alpha,\mathbb{P})$. Now we have by [24, Lemma 1.3.4] and [25, Theorem 1.1]:

3.3 Theorem — $(\mathcal{E}, D(\mathcal{E}))$ is a Dirichlet form on $L^2(B_\alpha, \mathbb{P})$, i.e. a closed non-negative, symmetric bilinear form, densely defined on $L^2(B_\alpha, \mathbb{P})$ such that for every $u \in D(\mathcal{E})$, $|u| \in D(\mathcal{E})$ and

$$\mathcal{E}(|u|,|u|) \leqslant \mathcal{E}(u,u).$$

By 3.1 we know that for $u,v \in \mathcal{F}C_b^2$

$$\mathcal{E}(u,v) = \int (-L_0 u)\, v\, d\mathbb{P} = \int_{H^0} \langle Du, Dv \rangle_{H^0}\, d\mathbb{P} \tag{3.8}$$

where the second equality follows by the well-known integration by parts formula proved in [2] and the fact that Du may be considered as an element in $H^0(\mathbb{R}^{d-1}) = L^2(\mathbb{R}^{d-1}, dx)$ by (3.3).

3.4 Remark — It is, of course, not clear at this point whether $\mathcal{E}$ is determined by (3.8), i.e. whether $\mathcal{F}C_b^2$ is a form core for $\mathcal{E}$. This and related questions will be discussed in a forthcoming paper.

Starting with the semigroup $(\pi_t)_{t>0}$ we have thus recovered the Dirichlet form studied extensively by S. Albeverio and R. Høegh-Krohn in [2], [26] and [27] as a quadratic form on $L^2(\mathcal{S}'(\mathbb{R}^{d-1});\mathbb{P})$. For the general framework of Dirichlet forms on Banach spaces see [28].

4. THE ENTRANCE SPACE OF $(\pi_t)_{t>0}$

In this section we want to investigate the structure of the set of all probability measures on $(\Omega, \mathcal{F})$ having $(\pi_t)_{t>0}$ as transition function. Following Dynkin (see [3], [4], [5] and also [29]) we define:

4.1 Definition — $\mathcal{M}$ is the set of all probability measures P on $(\Omega, \mathcal{F})$ such that for all $s,t \in \mathbb{R}, t > s$, F bounded, $\mathcal{B}(B_\alpha)$-measurable

$$E_p(F \circ X_t | \mathcal{F}_s) = \pi_{t-s}(X_s, F) \quad P\text{-a.s.} \tag{4.1}$$

It is easy to see that (4.1) is equivalent to

$$(4.2) \quad P(X_{t_1} \in B_1, \ldots, X_{t_n} \in B_n) = \int_{B_1} X_{t_1}(P)(d\xi_1) \int_{B_2} \pi_{t_2-t_1}(\xi_1, d\xi_2) \cdots \int_{B_n} \pi_{t_n-t_{n-1}}(\xi_{n-1}, d\xi_n)$$

for $t_1 < t_2 < \ldots < t_n$, $B_i \in \mathcal{B}(B_\alpha)$, $n \in \mathbb{N}$.

4.2 Remark — Let $P \in \mathcal{M}$ and set $\mu_s := X_s(P)$, $s \in \mathbb{R}$, then

$$(4.3) \quad \int \pi_t(\xi, B) \ \mu_s(d\xi) = \mu_{t+s}(B) \text{ for every } t \in \mathbb{R},\ B \in \mathcal{B}(B_\alpha).$$

$(\mu_s)_{s \in \mathbb{R}}$ is called the entrance law of P (see [4]). Clearly, by (4.2) P is uniquely determined by its entrance law.

The set $\mathcal{M}$ is obviously convex; let $\mathcal{M}_e$ denote the set of its extreme points. It is well known and easy to prove by the reverse martingale convergence theorem (cf. e.g. [17, 60.9]) and (4.2) that for $P \in \mathcal{M}$

(4.4) $P \in \mathcal{M}_e$ if and only if $P(A) = 1$ or 0 for $A \in \mathcal{F}_{-\infty} := \bigcap_{s \in \mathbb{R}} \mathcal{F}_s$.

In our particular case $\mathcal{M}_e$ can be characterised. To this end we define

(4.5) $\mathcal{H}_0 := \{h : \mathbb{R}^d \to \mathbb{R} : h(\cdot, t) \in B_\alpha$ and $T_{t-s} h(\cdot, s) = h(\cdot, t)$ for all $t, s \in \mathbb{R}, t > s\}$

(where we denote points in $\mathbb{R}^d$ by (x,s), $x \in \mathbb{R}^{d-1}$, $s \in \mathbb{R}$). Clearly, $\mathcal{H}_0$ is a space of $(\Delta - 1)$-harmonic functions. We identify $h \in \mathcal{H}_0$ with the map $t \mapsto h(\cdot, t)$ from $\mathbb{R}$ to B_α. Then it follows by 2.3 that $\mathcal{H}_0 \subset \Omega (= C(\mathbb{R}, B_\alpha))$. For $\omega \in \Omega$ let P_ω denote the probability measure on $(\Omega, \mathcal{F})$ defined by

$$(4.6) \quad P_\omega(A) = P_0(A - \omega), \ A \in \mathcal{F}.$$

4.3 Theorem — $\mathcal{M}_e = \{P_h : h \in \mathcal{H}_0\}$.

The proof is done in several steps. Since $P_0 \in \mathcal{M}$ by 1.4, it follows immediately from the definition of $\mathcal{H}_0$ that $P_h \in \mathcal{M}$. If we can show that P_0 is trivial on $\mathcal{F}_{-\infty}$ then the same is obviously true for P_h. So, by the following proposition and (4.4) we have shown that $\{P_h : h \in \mathcal{H}_0\} \subset \mathcal{M}_e$.

4.4 Proposition — $P_0(A) = 1$ or 0 on $\mathcal{F}_{-\infty}$.

Proof — We have to show that for every $\mathcal{F}$-measurable bounded function F on Ω

$$(4.7) \quad E_{P_0}(F \mid \mathcal{F}_{-\infty}) = \int F \, dP_0.$$

We may assume that $F = \exp[i \sum_{k=1}^{n} \beta_k \langle X_{t_k}, g_k \rangle]$ for some $\beta_k \in \mathbb{R}$, $-\infty < t_1 < \ldots < t_n < +\infty$, $g_k \in \mathcal{S}(\mathbb{R}^{d-1})$, $n \in \mathbb{N}$. By the reverse martingale convergence theorem we have that P_o – a.s.

$$(4.8) \quad E_{P_o}(F \mid \mathcal{F}_{-\infty}) = \lim_{n\to\infty} E_{P_o}(F \mid \mathcal{F}_{-n}) = \lim_{n\to\infty} \int \exp(i \sum_{k=1}^{n} \beta_k \langle X_{t_k+n}, g_k \rangle)\, dP^{X-n}.$$

where the last equality is a consequence of (2.6), (4.2) and the invariance of P_o under "time shifts". Since by [6, 4.1]

$$\int F\, dP_o = \exp[-\tfrac{1}{2} \sum_{k,l=1}^{n} \beta_k \beta_l \,{}_{H^{-\frac{1}{2}}}\langle g_l, T_{|t_l - t_k|} g_k \rangle_{H^{-\frac{1}{2}}}],$$

(4.7) now follows from (4.8) by (2.7).

Now let us prove that $\mathcal{M}_e \subset \{P_h : h \in \mathcal{H}_o\}$: let $P \in \mathcal{M}_e$. Let $t \in \mathbb{R}$ and let $\mathcal{S}_1$, $\mathcal{L}_1$ be countable dense subsets of $\mathcal{S}(\mathbb{R}^{d-1})$ resp. $L^1(\mathbb{R}, ds)$ (= all complex valued Lebesgue integrable functions on $\mathbb{R}$). By the reverse martingale convergence theorem and Fubini's theorem there exists $\omega \in \Omega$ such that for all $g \in \mathcal{S}_1$, $f \in \mathcal{L}_1$

$$\int f(s) \int \exp(is\langle X_t, g\rangle)\, dP\, ds = \lim_{n\to\infty} \int f(s)\, \pi_{t+n}(X_{-n}(\omega), \exp(is\langle \cdot, g\rangle))\, ds.$$

Now similarly as in the proof of 8.3 in [30] it follows by [30, 8.2] that $\langle X_t, g\rangle$ is Gaussian distributed on $\mathbb{R}$ with variance $\|g\|^2_{H^{-\frac{1}{2}}}$ for every $g \in \mathcal{S}_1$. By [31, Lemma 1.5] it follows that the same is true for ${}_{B_\alpha}\langle X_t, g\rangle_{B'_\alpha}$ for every $g \in B'_\alpha$ (where ${}_{B_\alpha}\langle\ ,\ \rangle_{B'_\alpha}$ is the dualisation between B_α and B'_α) and that $h_t : g \mapsto \int {}_{B_\alpha}\langle X_t, g\rangle_{B'_\alpha} dP$ is continuous on B'_α (with respect to the weak*topology). Hence $h_t \in B_\alpha$. Furthermore, for all $t, s \in \mathbb{R}$, $s < t$, $g \in \mathcal{S}(\mathbb{R}^{d-1})$

$$\langle h_t, g\rangle = \int \langle X_t, g\rangle dP = \int E_P(\langle X_t, g\rangle \mid \mathcal{F}_s)\, dP = \int \pi_{t-s}(X_s, \langle \cdot, g\rangle)\, dP$$

$$= \int \langle T_{t-s} X_s, g\rangle dP = \int \langle X_s, T_{t-s} g\rangle dP = \langle T_{t-s} h_s, g\rangle.$$

Hence the map $h : t \mapsto h_t$ from $\mathbb{R}$ to B_α is in $\mathcal{H}_o$. Clearly, $X_t(P) = X_t(P_h)$ for every $t \in \mathbb{R}$; hence $P = P_h$ by (4.2).

Let $\mathcal{B}(\mathcal{M})$ be the σ-algebra on $\mathcal{M}$ generated by all mappings $P \mapsto P(A)$ on $\mathcal{M}$ where A runs through $\mathcal{F}$. Since $(\Omega, \mathcal{F})$ is a standard Borel space (cf.[32, V Def. 2.2]), it follows that a regular conditional probability exists given $\mathcal{F}_{-\infty}$ (cf. [32, V.7.1]). Hence we can apply Dynkin's well-known representation theorem (cf. [4, Theorem 2.1]) to derive our final result on the structure of $\mathcal{M}$:

4.5 Theorem – $\mathcal{M}_e \in \mathcal{B}(\mathcal{M})$. If $\mathcal{B}_o$ is the σ-field induced by the map $h \mapsto P_h$ on $\mathcal{H}_o$, then every $P \in \mathcal{M}$ can be uniquely represented in the form

$$P(A) = \int_{\mathcal{H}_o} P_h(A)m(dh), \quad A \in \mathcal{F}, \tag{4.8}$$

where m is a probability measure on $(\mathcal{H}_o, \mathcal{B}_o)$.

4.6 Remark-(i) For results related to 4.5 we refer to [33], [34,Sect.2], [30, Sect. 8] and [35].

(ii) 4.4 and 4.5 are the higher dimensional analogues of results about the one dimensional Ornstein-Uhlenbeck process obtained in [29].

(iii) Concerning the question how big $\mathcal{H}_o$ really is, we note that this depends strongly on the choice of our density function ρ in the definition of B_α. But as in the one dimensional case the (Δ-1)-harmonic functions $(x,s) \mapsto \beta e^{-t}$, $x \in \mathbb{R}^{d-1}$, $t \in \mathbb{R}$, for $\beta \in \mathbb{R}$, are all contained in $\mathcal{H}_o$.

ACKNOWLEDGEMENTS

It is a pleasure to thank Shigeo Kusuoka for valuable discussions concerning proposition 2.3 in this paper and for bringing reference [22] to our attention.

REFERENCES

[1] E. Nelson, The free Markov field, J. Funct. Anal. 12, 221-227 (1973).
[2] S. Albeverio, R. Høegh-Krohn, Dirichlet forms and diffusion processes on rigged Hilbert spaces, Z. Wahrscheinlichkeitstheorie verw. Gebiete 40, 1-57 (1977).
[3] E. B. Dynkin, Entrance and exit spaces for a Markov process, Actes Congres Intern. Math. 1970, Tome 2, 507-512 (1971).
[4] E. B. Dynkin, The initial and final behaviour of trajectories of a Markov process, Russian Math. Surveys 26, 4, 165-185 (1971).
[5] E. B. Dynkin, Sufficient statistics and expreme points, Ann. Prob. 9, No. 5, 706-730 (1978).
[6] M. Röckner, Traces of harmonic functions and a new path space for the free quantum field, to appear in J. Funct. Anal.
[7] K. Ito, Infinite dimensional Ornstein-Uhlenbeck processes, Taniguchi

Symp. SA, Katata 1982, K. Ito, ed., 197-224, North Holland, Amsterdam, Oxford, New York (1984).

[8] K. Ito, Foundations of stochastic differential equations in infinite dimensional spaces, CBMS-NSF Regional Conference Series in Appl. Math. Vol. 47, SIAM, Philadelphia (1984).

[9] A. Antoniades, R. Carmona, Eigenfunction expansions for infinite dimensional Ornstein-Uhlenbeck processes, Probab. Th. Rel. Fields 74, 31-54 (1987).

[10] J. Bliedtner, W. Hansen, Potential theory, Springer, Berlin, Heidelberg, New York, Tokyo (1986).

[11] J. T. Cannon, Continuous sample paths in quantum field theory, Commun. Math. Phys. 35, 215-233 (1974).

[12] I. M. Gel'fand and N. J. Vilenkin, Generalized functions, Vol. 4., Some applications of harmonic analysis, Academic Press, New York (1964).

[13] L. Gross, Measurable functions on Hilbert space, Trans. Amer. Math. Soc. 105, 372-390 (1962).

[14] L. Gross, Abstract Wiener spaces, Proc. 5th Berkeley Symp. Math. Stat. Prob. 2, 31-42 (1965).

[15] H. Kuo, Gaussian measures in Banach spaces, Lecture Notes in Math. 463, 1-224, Springer, Berlin, Heidelberg, New York (1975).

[16] R. Carmona, Measurable norms and some Banach space valued Gaussian processes, Duke Math. J. 44 no. 1, 109-127 (1977).

[17] H. Bauer, Wahrscheinlichkeitstheorie und Grundzüge der Masstheorie, de Gruyter, Berlin, New York (1978).

[18] B. Simon, The $P(\Phi)_2$ Euclidean (quantum) field theory, Princeton University Press, Princeton, N.J. (1974).

[19] J. Glimm, A. Jaffe, Quantum physics: A functional integral point of view, Springer, New York, Heidelberg, Berlin (1981).

[20] R. L. Dobrushin, R. A. Minos, An investigation of the properties of generalized Gaussian random fields, Sel. Math. Sov. 1, 215-263 (1981). Originally published in Zadachy mekhaniky i matematicheskoy fiziky, Moscow: Nauka, 117-165 (1976).

[21] E. B. Dynkin, Die Grundlagen der Theorie der Markoffschen Prozesse, Springer, Berlin, Göttingen, Heidelberg (1961).

[22] S. Kusuoka, The support property of a Gaussian white noise and its application, J. Fac. Sci. Univ. Tokyo, Sec. I.A. 29, no 2, 387-400 (1982).

[23] S. Watanabe, Lectures on stochastic differential equations and Malliavin calculus, Springer, Berlin, Heidelberg, New York, Tokyo (1984).

[24] M. Fukushima, Dirichlet forms and Markov processes, North Holland, Amsterdam, Oxford, New York (1980).

[25] N. Bouleau, F. Hirsch, Formes de Dirichlet générales et densité des variables aléatoires réelles sur l'espace de Wiener, J. Funct. Anal. 69, 229-259 (1986).

[26] S. Albeverio, R. Høegh-Krohn, Quasi invariant measures, symmetric diffusion processes and quantum fields; Les méthodes mathématiques de la théorie quantique des champs, Colloques Internationaux du C.N.R.S., no. 248, Marseille, 23-27 juin 1975, C.N.R.S., (1976).

[27] S. Albeverio, R. Høegh-Krohn, Hunt processes and analytic potential theory on rigged Hilbert spaces, Ann. Inst. Henri Poincaré, Vol. XIII, no. 3, 269-291 (1977).

[28] S. Kusuoka, Dirichlet forms and diffusion processes on Banach spaces, J. Fac. Sci. Univ. Tokyo, Sec IA 29, 79-95, (1982).

[29] J. T. Cox, On one-dimensional diffusions with time parameter set $(-\infty,\infty)$. The Annals of Probability 5, No. 5, 807-813 (1977).

[30] M. Röckner, A Dirichlet problem for distributions and specifications for random fields, Memoirs of the Am. Math. Soc. 54, 324 (1985).

[31] J. Neveu, Processus aléatoires Gaussiens, Les presses de l'université de Montreal (1968).

[32] K. R. Parthasarathy, Probability measures on metric spaces, Academic Press, New York (1967).
[33] H. Föllmer, Phase transition and Martin boundary, in Séminaire de probabilités IX-Strasbourg, Lecture Notes in Math., Vol. 465, Springer, Berlin, Heidelberg, New York (1975).
[34] C. Preston, Random fields, Lecture Notes in Math., Vol. 534, Springer, Berlin, Heidelberg, New York.
[35] M. Röckner, Specifications and Martin boundaries for $\mathcal{P}(\Phi)_2$-random fields, Commun. Math. Phys. 106, 105-135 (1986).

FINITENESS OF THE FAMILY OF SIMPLY CONNECTED QUADRATURE DOMAINS

Makoto Sakai

Department of Mathematics
Tokyo Metropolitan University
Fukasawa, Setagaya-ku, Tokyo, 158, Japan

1. INTRODUCTION

A quadrature domain D of a functional L is a domain with finite area satisfying

$$(1.1) \qquad \int_D f(w)\,dudv = L(f) \qquad (w=u+iv)$$

for every f in the class $AL^1(D)$ of functions analytic and integrable in D. A simple example is a disk: The unit disk is a quadrature domain of $f \mapsto \pi f(0)$. In this case, the quadrature domain is determined uniquely. But, in general, it is not easy to determine all quadrature domains of a given functional.

In this paper we deal with functionals of the form

$$(1.2) \qquad L(f) = \Sigma_{j=1}^{n} b_j f^{(j-1)}(0).$$

We can construct a functional L of form (1.2) such that there exists an m real parameter family of (m+1)-ply connected quadrature domains of L. The construction is given by modifications of an (m+1)-ply connected quadrature domain. Another and more detailed construction was given by Gustafsson (1983), see Theorem 12 of his article. What can we say about simply connected domains? We shall show

THEOREM- The number of quadrature domains of a functional of form (1.2) which are simply connected and having no cusps on the boundary is finite.

2. FORMULATION OF THE PROBLEM AND A QUADRATURE DATA MAPPLING

If there is a quadrature domain D of a functional L of form (1.2), then $b_1 = L(1) = \int_D dudv > 0$. Hence we always assume $b_1 > 0$ and $b_n \neq 0$. We identify the functional L and the quadrature data $(b_1,\ldots,b_n)$ in the whole space $W = \mathbb{R}_+\times\mathbb{C}^{n-2}\times(\mathbb{C}\setminus\{0\})$, where $\mathbb{R}_+$ denotes the set of positive real numbers and $\mathbb{C}$ denotes the set of complex numbers. We denote by Q the set of functionals $L = (b_1,\ldots,b_n)$ in W which have at least one simply connected quadrature domain.

Let D be a simply connected domain containing 0. By the Riemann mapping theorem, there is a unique one-to-one conformal mapping p of the unit disk U onto D such that $p(0) = 0$ and $p'(0) > 0$. We identify the domain D and the mapping p. If $p(z) = \Sigma_{j=1}^n a_j z^j$ is a polynomial of degree n, we identify D and $(a_1,\ldots,a_n)$ in W. We denote by P the set of polynomials $p = (a_1,\ldots,a_n)$ in W which are univalent in U. First we show

PROPOSITION 2.1 - For $p = (a_1,\ldots,a_n)$ in P, the image $D = p(U)$ of U under p is a simply connected quadrature domain of a functional in Q. The functional $L = (b_1,\ldots,b_n)$ is determined uniquely by $(a_1,\ldots,a_n)$ and expressed as

$$(2.1) \qquad b_k = \pi\Sigma_{j=k}^n j\bar{a}_j\left(\Sigma\frac{a_1^{m_1}\ldots a_n^{m_n}}{m_1!\ldots m_n!}\right),$$

where the second summation is taken over all n-tuple $(m_1,\ldots,m_n)$ of nonnegative integers m_ℓ satisfying $\Sigma_{\ell=1}^n m_\ell = k$ and $\Sigma_{\ell=1}^n \ell m_\ell = j$.

PROOF - Let $z = x+iy$ and $w = u+iv$. For a natural number k, we have

$$\int_D kw^{k-1}dudv = \int_U kp(z)^{k-1}|p'(z)|^2 dxdy = \int_U (p(z)^k)'\overline{p'(z)}dxdy.$$

Set $p(z)^k = \Sigma_{j=k}^{kn} c_{kj} z^j$ for $k = 1,2,\ldots,n$. Noting $\int_U z^{\ell-1}\bar{z}^{j-1}dxdy = \pi/j$ if $\ell = j$ and $= 0$ if $\ell \neq j$, we see that the integral is equal to $\Sigma_{j=k}^n((jc_{kj})\times(j\bar{a}_j)(\pi/j)) = \pi\Sigma_{j=k}^n j\bar{a}_j c_{kj}$. Since

$$p(z)^k = (\Sigma_{j=1}^n a_j z^j)^k = \Sigma\frac{k!}{m_1!\ldots m_n!}(a_1 z)^{m_1}\ldots(a_n z^n)^{m_n},$$

where the summation is taken over all n-tuple $(m_1,\ldots,m_n)$ of nonnegative integers m_ℓ satisfying $\Sigma_{\ell=1}^n m_\ell = k$, $c_{kj} = k!\Sigma(a_1^{m_1}\ldots a_n^{m_n})/(m_1!\ldots m_n!)$ and $(m_1,\ldots,m_n)$ satisfy an additional condition $\Sigma_{\ell=1}^n \ell m_\ell = j$. If we introduce numbers b_k defined by (2.1), we get

$$\int_D kw^{k-1}dudv = k!b_k$$

for k = 1, 2,...,n and so (1.1) holds for $L = (b_1,\dots,b_n)$ and for every f of the form $f(w) = \Sigma_{k=1}^{n} c_k w^{k-1}$.

The proof will be complete if we show $\int_D f dudv = 0$ for f in $AL^1(D)$ having a zero of order not less than n at w = 0. Let $U_r = \{z \in \mathbb{C};\ |z| < r\}$. Then

$$\int_{p(U_r)} f(w)dudv = \int_{U_r} f(p(z))p'(z)\overline{p'(z)}dxdy = \frac{1}{2i}\int_{|z|=r} f(p(z))p'(z)\overline{p(z)}dz.$$

Let $\bar{p}(z) = \overline{p(\bar{z})}$. Since f(p(z)) has a zero of order not less than n at z = 0 and $\bar{z} = 1/z$ on $\{|z| = 1\}$, we get

$$\int_D fdudv = \lim_{r\to 1}\frac{1}{2i}\int_{|z|=r} f(p(z))p'(z)\bar{p}(1/z)dz = 0.$$

Q.E.D.

For $n \leqq 5$, (2.1) was obtained by Ullemar (1980). Conversely, we have

PROPOSITION 2.2- If D = p is a simply connected quadrature domain of a functional in Q, then p is a polynomial in P.

This result is due to Davis (1974) and Aharonov and Shapiro (1976). We omit the proof, because we shall show a more general version of the proposition in Proposition 4.2.

We denote by q the mapping of P into Q defined by (2.1) and call it a quadrature data mapping. By Propositions 2.1 and 2.2, we see that Q = q(P), namely, Q is exactly the image of P under the quadrature data mapping q. Thus the family of simply connected quadrature domains of a given $L = (b_1,\dots,b_n)$ is nothing but the inverse image $q^{-1}((b_1,\dots,b_n))$ of $(b_1,\dots,b_n)$. We embed W into $\mathbb{R}^{2n-1}$, extend q onto $\mathbb{R}^{2n-1}$ by (2.1) and regard the extension, we denote it again by q, as an algebraic mapping of $\mathbb{R}^{2n-1}$ into $\mathbb{R}^{2n-1}$. The family of simply connected quadrature domains of $L = (b_1,\dots,b_n)$ is expressed as $q^{-1}((b_1,\dots,b_n))\cap P$. For the proof of Theorem, we need an additional condition. Let S be the class of polynomials in W which are locally univalent (schlicht) on the closure $\bar{U}$ of U, namely, the class of polynomials p in W satisfying $p'(z) \neq 0$ on $\bar{U}$. S is an open subset of W and the image D = p(U) of U under p in $P\cap S$ has no cusps on the boundary.

We shall show, in Section 6, that $q^{-1}((b_1,\ldots,b_n))\cap S$ is finite for every $(b_1,\ldots,b_n)$ in W. To do so, we prepare a proposition which asserts finiteness of the solutions of simultaneous algebraic equations.

3. FINITENESS OF THE SOLUTIONS OF SIMULTANEOUS ALGEBRAIC EQUATIONS

Let $y = (y_1,\ldots,y_m)$ be an algebraic mapping of $\mathbb{R}^m$, namely, let $y_1,\ldots,y_m$ be polynomials of $x = (x_1,\ldots,x_m)$ with real coefficients. In this section, we shall consider the simultaneous algebraic equations $y_1(x) = \ldots = y_m(x) = 0$ and discuss finiteness of the number of the solutions which are in a given open subset of $\mathbb{R}^m$. Set

$$V_k = \{ x \in \mathbb{R}^m;\ y_1(x) = \ldots = y_k(x) = 0 \},$$

$k = 1,2,\ldots,m$. $V_1,\ldots,V_m$ are algebraic sets and satisfy $V_1 \supset V_2 \supset \ldots \supset V_m = y^{-1}(0)$.

PROPOSITION 3.1 — Let S be an open subset of $\mathbb{R}^m$. Assume that y satisfies the following condition:

(3.1) For every k, every x_0 on $V_k\cap S$ and every neighborhood N of x_0, there exists a point x in N such that $y_1(x) = \ldots = y_{k-1}(x) = 0$ but $y_k(x) \neq 0$.

Then $V_m\cap S = y^{-1}(0)\cap S$ is finite.

PROOF — We may assume that $V_m\cap S$ is not empty. We express V_k as a finite disjoint union of smooth manifolds, denote by d_k the maximum of the dimensions of the manifolds having nonempty intersections with S and denote by M_k one of such manifolds of dimension d_k. To prove the proposition, it is sufficient to show that $m > d_1$ and $d_{k-1} > d_k$, $k = 2,\ldots,m$. For then $d_m = 0$ and so $V_m\cap S$ will be a finite union of manifolds of dimension zero. Hence $V_m\cap S$ will be finite.

Take a point x_0 on $M_1\cap S$. Then, by (3.1), we can find a point x in a neighborhood of x_0 such that $y_1(x) \neq 0$. Hence $m > d_1$. For $k = 2,\ldots,m$, since $M_k\cap S$ is not empty, we can take a ball B having nonempty intersection with M_k such that $B \subset S$ and $V_k\cap B = M_k\cap B$. Assume that $d_{k-1} = d_k$. Then $V_{k-1}\cap B$ is a finite disjoint union of (not necessarily connected) smooth manifolds of dimensions not greater than d_k and contains $M_k\cap B$, and so we can find a point x_0 on $M_k \cap B$ and a neighborhood N of x_0 such that $N \subset B$ and $V_{k-1} \cap N = M_k \cap N = V_k \cap N$. This implies that $N \cap (V_{k-1} \setminus V_k)$ is empty and contradicts to (3.1). Hence $d_{k-1} > d_k$. Q.E.D.

4. SIMPLY CONNECTED QUADRATURE RIEMANN SURFACES

At the end of Section 2, we have extended the quadrature data mapping onto W. If $p = (a_1,\ldots,a_n)$ belongs to P, namely, if $p(z) = \Sigma_{j=1}^n a_j z^j$ is univalent in U, the image $D = p(U)$ corresponds to p and it is a simply connected quadrature domain of a functional in Q stated as in Proposition 2.1. What can we say about p in $W \setminus P$? In this case, p is not univalent in U and so we take a simply connected Riemann surface $\tilde{D}$ determined by the inverse function of $p|_U$, where $p|_U$ denotes the restriction of p to U. There are a one-to-one conformal mapping $\tilde{p}$ of U onto $\tilde{D}$ and a projection mapping $\pi_{\tilde{D}}$ of $\tilde{D}$ onto $D = \pi_{\tilde{D}}(\tilde{D}) = p(U)$ such that $\pi_{\tilde{D}} \circ \tilde{p} = p$. Thus a simply connected Riemann surface $(\tilde{D}, \pi_{\tilde{D}})$ with projection mapping $\pi_{\tilde{D}}$ corresponds to p.

Next we define a quadrature Riemann surface. Let (R, π_R) be a Riemann surface with projection mapping π_R. We denote by $AL^1(R,\pi_R)$ the class of analytic functions f in R with finite integral $\int_R |f(\tilde{w})| dudv$, where $\pi_R(\tilde{w}) = u+iv$. Let $\tilde{0}$ be a fixed point in R which is not a point of ramification of π_R. We interpret π_R as a local uniformizing parameter in a neighborhood of $\tilde{0}$ and let

$$L(f) = \sum_{j=1}^{n} b_j \left. \frac{d^{j-1} f}{d\pi_R^{j-1}} \right|_{\tilde{w}=\tilde{0}}. \tag{4.1}$$

We call (R,π_R) a quadrature Riemann surface of L if the constant function 1 belongs to $AL^1(R,\pi_R)$ and if

$$\int_R f(\tilde{w}) dudv = L(f) \qquad (\pi_R(\tilde{w}) = u+iv) \tag{4.2}$$

for every f in $AL^1(R,\pi_R)$.

PROPOSITION 4.1 Let $p = (a_1,\ldots,a_n)$ be in W. Let $\tilde{D}$ be a simply connected Riemann surface determined by the inverse function of $p|_U$, let $\tilde{p}$ be one of one-to-one conformal mappings of U onto $\tilde{D}$ and let $\pi_{\tilde{D}}$ be a projection mapping of $\tilde{D}$ onto $p(U)$ satisfying $\pi_{\tilde{D}} \circ \tilde{p} = p$. Set $\tilde{0} = \tilde{p}(0)$. Then $(\tilde{D}, \pi_{\tilde{D}})$ is a simply connected quadrature Riemann surface of a functional L of form (4.1). The functional $L = (b_1,\ldots,b_n)$ is determined uniquely and given by (2.1).

PROOF— Since $\int_{\tilde{D}} f(\tilde{w})\,dudv = \int_U kp(z)^{k-1}|p'(z)|^2 dxdy$ for $f(\tilde{w}) = k\pi_{\tilde{D}}(\tilde{w})^{k-1} = kw^{k-1}$, (4.2) holds for every f of the form $f(\tilde{w}) = \Sigma_{k=1}^{n} c_k w^{k-1}$. The rest of the proof is the same as that of Proposition 2.1. Q.E.D.

Conversely, by using arguments due to Davis (1974), we have

PROPOSITION 4.2 — Let (R,π_R) be a simply connected quadrature Riemann surface of a functional L of form (4.1) with $b_n \neq 0$, where $\tilde{0}$ denotes a fixed point in R which is not a point of ramification of π_R and satisfies $\pi_R(\tilde{0}) = 0$. Assume that π_R has just a finite number of points of ramification in R. Then there are a unique rational function r and a unique one-to-one conformal mapping $\tilde{r}$ of U onto R satisfying $r = \pi_R \circ \tilde{r}$, $\tilde{r}(0) = \tilde{0}$ and $r'(0) > 0$. If π_R has no points of ramification in R, then r is a polynomial in W.

PROOF— Let $\tilde{r}$ be a unique one-to-one conformal mapping of U onto R such that $\tilde{r}(0) = \tilde{0}$ and $(\pi_R \circ \tilde{r})'(0) > 0$. We shall show that $r = \pi_R \circ \tilde{r}$ is a rational function of z. If it is so and if π_R has no points of ramification in R, then $r'(z) \neq 0$ in U. Hence, by using computations as in Proposition 2.1, we see that r is a polynomial of degree n and the proof will be complete, see pp. 158-159 of Davis (1974).

To prove that r is a rational function, we discuss the Bergman reproducing kernel differential. Let δ_R be the divisor of $d\pi_R$ and let $\Gamma(R,\delta_R)$ be the class of square-integrable analytic differentials ω in R which are multiples of δ_R. $\Gamma(R,\delta_R)$ becomes a Hilbert space by an inner product $(\omega_1,\omega_2) = (i/2)\int_R \omega_1 \overline{\omega_2}$. Let $\kappa_R(\tilde{t},\tilde{w})$ be the Bergman reproducing kernel differential of $\Gamma(R,\delta_R)$, namely, $\omega(\tilde{w}) = (\omega(\tilde{t}),\kappa_R(\tilde{t},\tilde{w}))_{\tilde{t}}$ for every ω in $\Gamma(R,\delta_R)$. Since the constant function 1 belongs to $AL^1(R,\pi_R)$, $f = \omega/d\pi_R$ belongs to $AL^1(R,\pi_R)$ for every ω in $\Gamma(R,\delta_R)$. Hence, by (4.2), the representing differential of the functional $\omega \mapsto L(\omega/d\pi_R)$ defined in $\Gamma(R,\delta_R)$ is equal to

$$d\pi_R(\tilde{w}) = \{L_{\tilde{t}}(\kappa_R(\tilde{t},\tilde{w})/d\pi_R(\tilde{t}))\}^{-}.$$

Integrating the differential from $\tilde{w} = \tilde{0}$ to $\tilde{w} = \tilde{w}$ in R, we get

$$w = \pi_R(\tilde{w}) = \{L_{\tilde{t}}((\overline{\int_{\tilde{0}}^{\tilde{w}} \kappa_R(\tilde{w},\tilde{t})})/d\pi_R(\tilde{t}))\}^{-}.$$

Let $\kappa_{\tilde{U}}(z,s)$ be the Bergman reproducing kernel differential of $\Gamma(U,\delta_R \circ \tilde{r})$. Then $\int_{\tilde{0}}^{\tilde{w}} \kappa_R(\tilde{w},\tilde{t}) = \int_0^z \kappa_U(z,s)$, where $\tilde{w} = \tilde{r}(z)$ and $\tilde{t} = \tilde{r}(s)$. Since $\kappa_U(z,s)/\overline{ds}$ is the differential of a rational function of z, $w = r(z)$ is a rational function. Q.E.D.

For the sake of completeness, we prove the fact mentioned at the end of the proof of Proposition 4.2.

LEMMA 4.3 — Let $\kappa_U(z,s)$ be the Bergman reproducing kernel differential of $\Gamma(U,\delta_U)$ for a divisor δ_U in U. Then $\kappa_U(z,s)/\overline{ds}$ is the differential of a rational function for fixed s.

PROOF — Let $\Gamma(U)$ be the class of square-integrable analytic differentials in U and let Γ_1 be the orthogonal complement of $\Gamma(U,\delta_U)$ in $\Gamma(U)$. Let $\kappa_1(z,s)$ and $\kappa(z,s)$ be the Bergman reproducing kernel differential of Γ_1 and $\Gamma(U)$, respectively. Since $\kappa_U(z,s) = \kappa(z,s)-\kappa_1(z,s)$ and $\kappa(z,s) = (1/\pi)\times (1-z\bar{s})^{-2}dz\overline{ds}$, it is sufficient to show that Γ_1 consists of the differentials of rational functions.

For the sake of simplicity, we assume $\delta_U = z_1 \ldots z_m$, where $z_1,\ldots,z_m$ are nonzero distinct points in U. Since $\Gamma(U,\delta_U)$ consists of differentials of the form $\{\Pi_{j=1}^m (z-z_j)\}g(z)dz$, where g denote analytic functions in U, Γ_1 is of m-dimensional. Set $r_k = z^k/(\Pi_{j=1}^m (1-\bar{z}_j z))$, $k = 1,2,\ldots,m$. Then $\{dr_k\}$ is linearly independent and contained in $\Gamma(U)$. Since

$$(dr_k(z),\{\Pi_{j=1}^m (z-z_j)\}z^{\ell}dz) = \int_U r_k'(z)\{\Pi_{j=1}^m (\bar{z}-\bar{z}_j)\}\bar{z}^{\ell}dxdy,$$

by using the computations as in Proposition 2.1, we see that the above inner product is equal to

$$(i/2)\int_{|z|=1} r_k(z)\{\Pi_{j=1}^m (\bar{z}-\bar{z}_j)\}\bar{z}^{\ell}d\bar{z} = (-i/2)\int_{|z|=1} z^{k-\ell-m-2}dz = 0$$

for every k with $1 \leq k \leq m$ and every $\ell \geq 0$. Hence each dr_k, $k = 1,2,\ldots,m$, belongs to Γ_1 and so $\{dr_k\}$ spans Γ_1. Q.E.D.

From the proof of Lemma 4.3, we see that the reflection points of poles of r with respect to the unit circle are contained in the set of zeros of r'. Hence, if $r'(z) \not\equiv 0$ in U, then r is a polynomial. This again proves the last assertion of Proposition 4.2.

We give here an interesting example of one real parameter family of simply connected quadrature Riemann surfaces of $f \mapsto \pi f(\tilde{0})$. From the example we see that the rational function in Proposition 4.2 can not be replaced by a polynomial, in general. We also see that there are infinitely many simply connected quadrature Riemann surfaces of a functional of form (4.1).

EXAMPLE— Let $0 < t < 1$ and set $r_t(z) = t(2-t^2)^{1/2}z(1-t^{-1}(2-t^2)^{-1}z)\times(1-tz)^{-1}$. Let R_t be a simply connected Riemann surface determined by the inverse function of $r_t|_U$, let $\tilde{r}_t$ be a one-to-one conformal mapping of U onto R_t and let π_t be a projection mapping of R_t onto $r_t(U)$ satisfying $\pi_t \circ \tilde{r}_t = r_t$. Set $\tilde{0} = \tilde{r}_t(0)$. Then, since $r_t'(t) = 0$, by using the computations as in Proposition 2.1, we get $\int_{R_t} f(\tilde{w})dudv = \pi f(\tilde{0})$ for every f in $AL^1(R_t,\pi_t)$ and so (R_t,π_t) is a simply connected quadrature Riemann surface of $f \mapsto \pi f(\tilde{0})$ for every t.

5. MODIFICATIONS OF RIEMANN SURFACES

Let $(\tilde{D},\pi_{\tilde{D}})$ be a simply connected Riemann surface with projection mapping $\pi_{\tilde{D}}$ determined by the inverse function of $p|_U$ of a polynomial p and let $\tilde{p}$ be a one-to-one conformal mapping of U onto $\tilde{D}$ satisfying $\pi_{\tilde{D}} \circ \tilde{p} = p$. Let μ be a finite positive measure on $\tilde{D}$ with compact support. We call a Riemann surface (R,π_R) with projection mapping π_R a modification of $(\tilde{D},\pi_{\tilde{D}})$ with respect to μ if

(5.1) $\quad \tilde{D} \subset R$ and $\pi_{\tilde{D}} = \pi_R|_{\tilde{D}}$.

(5.2) $$\int_{\tilde{D}} f(\tilde{w})dudv + \int_{\tilde{D}} fd\mu = \int_R f(\tilde{w})dudv \qquad (\pi_R(\tilde{w})=u+iv)$$
for every f in $AL^1(R,\pi_R)$.

We assume that $p'(z) \neq 0$ in a neighborhood of the unit circle and assume that the support of μ is contained in a fixed compact subset of $\tilde{D}$.

PROPOSITION 5.1— For μ with sufficiently small total measure $\|\mu\|$, there exists a simply connected modification (R,π_R) of $(\tilde{D},\pi_{\tilde{D}})$ with respect to μ.

PROOF— We shall modify U with respect to $\mu\circ\tilde{p}$ under the weight function $|p'(z)|^2$ and obtain a simply connected domain Ω such that $U \subset \Omega$ and $\int_U f(z)|p'(z)|^2dxdy + \int_U fd(\mu\circ\tilde{p}) = \int_\Omega f(z)|p'(z)|^2dxdy$ for every f in $AL^1(\Omega)$.

We extend $\tilde{p}$ onto Ω and extend $\pi_{\tilde{D}}$, denote it by π_R, onto $R = \tilde{p}(\Omega)$. Then (5.1) and (5.2) follow immediately.

To show the existence of the above domain Ω, we note that one of such domains is obtained as the noncoincidence set for the solution of an obstacle problem with zero pressure and with obstacle $G_B\{\chi_B|p'|^2 m - (\chi_U|p'|^2 m + \mu\circ\tilde{p})\}$, where B denotes a sufficiently large disk, G_B denotes the Green operator on B, χ_B denotes the characteristic function of B and m denotes the two-dimensional Lebesgue measure, see Theorem 7.5 of Sakai (1984/85). Since $\mu\circ\tilde{p}$ can be replaced by gm for a smooth function g with compact support in U, see Lemma 6 of Sakai (1983), the domain is nothing but $\{z \in \mathbb{C};\ u(z) > 0\}$ for a solution u of the following problem:

$$\left.\begin{aligned} &-\Delta u + \chi_{B\setminus U}|p'|^2 - g \geqq 0 \\ &u \geqq 0 \\ &(-\Delta u + \chi_{B\setminus U}|p'|^2 - g)u = 0 \end{aligned}\right\} \quad \text{a.e. in } B.$$

We note that $|p'|^2 \geqq \lambda$ in a neighborhood of $\{|z| = 1\}$ for a constant $\lambda > 0$ and that $\|g\|_{L^\infty(B)}$ is small for μ with small $\|\mu\|$ because the support of $\mu\circ\tilde{p}$ is contained in a fixed compact subset of U. By the results due to Caffarelli, see Theorem 10.6 of Chap. 2 of Friedman (1982), we see that the domain has a smooth boundary for μ with small $\|\mu\|$. If it has more than one boundary component, then we can reduce one of them to a continuum having no interier points by increasing $\|\mu\|$ suitably. This contradicts to the above fact and so the domain is surrounded by a smooth Jordan curve for μ with sufficiently small $\|\mu\|$. Q.E.D.

6. PROOF OF THEOREM AND A CONJECTURE

First we introduce the fundamental quadrature domain D_k of order k. If $k = 1$, we set $D_1 = U$. If $k = 2,\dots,n$, D_k is the image of U under the one-to-one conformal mapping of U defined by $z \mapsto z + z^k/k$. D_1 is a quadrature domain of $L_1 = (\pi,0,\dots,0)$ and D_k is a quadrature domain of $L_k = (\pi(1+1/k),0,\dots,0,\pi/k!,0,\dots,0)$, where $\pi/k!$ is the kth component of L_k, for each $k = 2,\dots,n$. For arbitrary $c \neq 0$, $cD_1 = \{cw;\ w \in D_1\}$ is a quadrature domain of $(|c|^2\pi,0,\dots,0)$ and $cD_k = \{cw;\ w \in D_k\}$ is a quadrature domain of $(|c|^2\pi(1+1/k),0,\dots,0,c^{k}\bar{c}\pi/k!,0,\dots,0)$ for each $k = 2,\dots,n$.

To show Theorem, we take $(b_1,\dots,b_n)$ in W and fix it. Let $(x_1,x_2,\dots,x_{2n-1}) = (a_1,a_2,\dots,a_n) = a$ and set $(y_1,y_2,\dots,y_{2n-1}) = (q_n(a)-b_n,q_{n-1}(a)-b_{n-1},\dots,q_1(a)-b_1)$, namely, $x_1 = a_1$, $x_{2k-2}+ix_{2k-1} = a_k$

for $k = 2,\ldots,n$, $y_{2k-1}+iy_{2k} = q_{n-k+1}(a)-b_{n-k+1}$ for $k = 1,\ldots,n-1$, and $y_{2n-1} = q_1(a)-b_1$, where $q(a) = (q_1(a),\ldots,q_n(a))$. We regard y as an algebraic mapping of $\mathbb{R}^{2n-1}$ into $\mathbb{R}^{2n-1}$. To prove $q^{-1}((b_1,\ldots,b_n))\cap S = y^{-1}(0)\cap S$ is finite, we apply Proposition 3.1. It is sufficient to show that y satisfies Condition (3.1).

We consider the case $k = 2n-1$. Let $x_0 = a_0 = p_0$ be a polynomial in $V_{2n-1}\cap S$ and let $(\tilde{D},\pi_{\tilde{D}})$ be the corresponding simply connected quadrature Riemann surface of $L = q(a_0)$ mentioned in Proposition 4.1. We denote by ρ the inverse conformal mapping of $\pi_{\tilde{D}}$ defined in a neighborhood of 0 satisfying $\rho(0) = \tilde{0}$. Take $c \neq 0$ with small $|c|$ and let (R,π_R) be a simply connected modification of $(\tilde{D},\pi_{\tilde{D}})$ with respect to a measure $\chi_{\rho(cD_1)}dudv$; the existence of (R,π_R) is guaranteed by Proposition 5.1. Then (R,π_R) is a simply connected quadrature Riemann surface of $L = q(a_0)+(|c|^2\pi,0,\ldots,0)$ and so, by Proposition 4.2, R is a simply connected Riemann surface determined by the inverse function of $r|_U$ of a rational function r. Since $p_0'(z) \neq 0$ on $\overline{U}$, $r'(z) \neq 0$ in U for c with small $|c|$. Hence, by the last assertion of Proposition 4.2, r is a polynomial. Let x be the identified point with r. Then $y_k(x) = 0$, $k = 1,\ldots,2n-2$, but $y_{2n-1}(x) = |c|^2\pi \neq 0$. Thus y satisfies (3.1) for $k = 2n-1$.

For the case $k = 2j-1$ or $k = 2j$, $j = 1,\ldots,n-1$, we take x_0 in $V_k\cap S$ and apply the above arguments replacing cD_1 by cD_{n-j+1}. By choosing a suitable complex number c, we see that y satisfies (3.1) for $k = 2j-1$ or $2j$. This completes the proof of Theorem.

Finally we discuss the proof of Theorem. Suppose that p_0 does not belong to S, namely, p_0' vanishes at least at one point on $\overline{U}$. Then r' vanishes at some point in U and so r may not be a polynomial. Thus we can not apply our arguments to p_0 not in S, accordingly, to a quadrature domain having cusps on the boundary. Nevertheless it is very plausible that p_0 can be modified to a polynomial. The new modifications will be for domains in the w-plane and not for the domains in the z-plane as above. If we can handle modifications of domains having cusps on the boundary, Proposition 3.1 will be also applied to the case and the following natural conjecture will be true:

CONJECTURE — The number of simply connected quadrature domains of a functional of form (1.2) is finite.

REFERENCES

Aharonov, D., and Shapiro, H. S., 1976, Domains on which analytic functions satisfy quadrature identities, J. Anal. Math., 30:39.

Davis, P. J., 1974, "The Schwarz Function and its Applications," Carus Math. Monographs, No. 17, Math. Assoc. Amer., Washington, DC.

Friedman, A., 1982, "Variational Principles and Free-Boundary Problems," John Wiley & Sons, New York.

Gustafsson, B., 1983, Quadrature identities and the Schottky double, Acta Appl. Math., 1:209.

Sakai, M., 1983, Applications of variational inequalities to the existence theorem on quadrature domains, Trans. Amer. Math. Soc., 276:267.

Sakai, M., 1984/85, Solutions to the obstacle problem as Green potentials, J. Anal. Math., 44:97.

Ullemar, C., 1980, A uniqueness theorem for domains satisfying a quadrature identity for analytic functions, TRITA-MAT-1980-37, Dept. of Math., Royal Inst. of Tech., Stockholm.

ON AN INVERSE PROBLEM FOR THE NEWTONIAN POTENTIAL

Sandro Salsa

Department of Mathematics
Politecnico di Milano
via Bonardi 7, 20133 Milano (Italy)

INTRODUCTION

In a joint work with C. Maderna and C. Pagani,[1] we have considered the following classical inverse problem of potential theory:

to determine the figure of a Homogeneous material body G whose shape is unknown, from measurements of the Newtonian potential created by G, taken on the surface of a sphere containing G in its interior.

We assume that G belongs to a class of star-shaped domains parametrized by a smooth function

$$u: S^2 \longrightarrow \mathbb{R}, \qquad |u| < \text{constant} < 1$$

as follows. Let $\varphi_u : S^2 \longrightarrow \mathbb{R}^3$ be the differential embedding defined by

$$S^2 \ni \omega \longrightarrow \varphi_u(\omega) = (1 + u(\omega));$$

then we set $\Gamma_u = \varphi_u(S^2)$ (Γ_1 is the surface $|x| = 2$) and denote by G_u the bounded domain whose boundary is Γ_u.

Consider now the potential created by a mass of unitary density distributed over G_u:

$$V_u(x) = \int_{S^2} d\omega \int_0^{|\varphi_u(\omega)|} |x - t\omega|^{-1}\, t^2\, dt$$

where $d\omega$ is the surface element on S^2. We measure this potential on Γ_1 and we read it as a function on S^2. Let v be the difference between this measured potential and the potential created by G_0 on Γ_1 (i.e. $\pi/3$). Put $B(u)(\omega) = V_u(2\omega) - \pi/3$. Then we obtain the equation

$$B(u) = v \qquad \text{on } S^2$$

that we want to solve for u when v is given.

Questions of this kind arise in several fields of applied sciences, e.g. in geophisics or in geodesy. Many of them lead to unstable problems and many efforts are done in order to select classes of stable solutions [2,3], or to construct stable algorythms producing approximate solutions.

Existence results can be found in the survey of Prilepko[2] and in his recent paper in Diff. Eq. (1987). Our contribution is in this direction and it is contained in the theorem stated at the end of the next section.

Besides the result itself we think that the method used in constructing the solution presents some features of independent interest which could be useful (we hope) in other situations.

Our method is a variant of the Nash-Moser iteration technique, as it is presented in an abstract setting by L. Nirenberg[4]. In the paper of Nirenberg two families of Banach spaces X_s and Y_s are chosen depending on the parameter s (0< s< 1) in such a way that $X_{s'} \subset X_s$ and $Y_{s'} \subset Y_s$ when $s' > s$, with continuous injection. The problem consists in solving the equation F(u)=0 for u in some X_s, where F is a map acting continuously from a ball in $X_{s'}$ into Y_s for every s'>s. To this purpose one constructs in a suitable way (as we do too here) a sequence of approximated solutions $u_k \in X_{s_k}$, with $s_k \downarrow s$.

Since, in some sense, the parameter s measures the degree of regularity of u_k, we see that each step in the iteration causes a loss of regularity: certain additional hypotheses on F allow one to keep under control this loss and make the sequence u_k to converge to the desired solution.

In our situation we cannot choose as above two fixed scales of Banach spaces. More precisely: we choose as X-spaces certain classes of analytic functions; however, the definition of the Y-spaces (also classes of, analytic functions, of course) is part of the iteration procedure itself, in the sense that the approximate solution found at each step determines the Y-space for the next one.

The key point to make everything to work is then to estimate the Y-norm at a given step in terms of the Y-norm in the preceding one. In Nirenberg's case this estimate is assured by the continuous injection of $Y_{s'}$ into Y_s when s'>s.

In our case this difficulty is overcome by means of some estimates concerning analytic solutions of the Dirichlet and Cauchy problems for a second order elliptic operator (see lemmas 2 and 3).

The necessity of the above estimates forces the choice of the X-spaces: Sobolev or Hölder classes would not be sufficient to achieve the result.

Other versions of the Nash-Moser technique have already been used for solving problems in geodesy[5], in electrostatics[6], in potential theory[7]; this last paper deals with the same problem discussed here but the datum (the measured potential) is taken on the unknown surface Γ_u instead of Γ_1. In all the above situation one can work in scales of Sobolev or Hölder classes.

We conclude this introduction by remarking that some simplifications introduced in the position of the problem are by no way essential. For instance we could solve our equation in a neighborhood of a surface analytically diffeomorphic to S^2 or we could assume that G is not homogeneous, with density distributed according to a given law.

STATEMENT OF THE MAIN RESULT

We first introduce our relevant functional spaces.

Let $\{S_1\}$ be a finite open covering of S^2 and Ψ_j, $\Psi_j : \mathbb{R}^2 \supset D \longrightarrow S_j$, a system of analytic local charts defined in the unit disc $D=\{x \in \mathbb{R}^2, |x|<1\}$ admitting a bounded analytic extension to the polydisc $D \times D \subset \mathbb{C}^2$.

For s real, 0<s<1 and R>0 set

$$\mathbb{C}^2 \supset \mathcal{D}_s^R = \{z = x+i\xi, |x|<1, |\xi|<sR\}.$$

Now let f be a function defined on S^2. We say that $f \in X_s^R(S^2)$ if $f \circ \psi_j$ is analyitc and admits a bounded analytic extension to $\mathcal{D}_s^R$ for all maps ψ_j.
Instead of $X_s^R(S^2)$ we will simply write X_s, when no confusion arises.
So we have a one-parameter family of vector spaces that can be normed in different ways. We pose

$$\|f\|_s = \sup_j \sup_{z \in \mathcal{D}_s^R} |f \circ \psi_j(z)|$$

where the first supremum is taken with respect to a fixed atlas; to change the atlas gives an equivalent norm. Notice that these norms are increasing in strength with s. More generally, we shall use the norms

$$\|f\|_s^{(m)} = \sum_{|\alpha|=0}^{m} \|D_\omega^\alpha f\|_s \qquad (m>0, \text{integer}).$$

Here D_ω^α denotes a tangential derivative on S^2 of order $\alpha = (\alpha_1, \alpha_2)$. In local coordinates $D_\omega^\alpha f(z) = \partial_{z_1}^{\alpha_1} \partial_{z_2}^{\alpha_2} f \circ \psi_j(z)$ for some j.

Let now f and u be in X_s. We can solve uniquely the exterior Dirichlet problem

$$\Delta W = 0 \qquad \text{in } \mathbb{R}^3 \setminus \bar{G}_u$$
$$W = f \quad \text{on } \Gamma_u, W(x) \longrightarrow 0 \quad \text{as } |x| \longrightarrow \infty.$$

Call h the trace of W on Γ_1. The collection of all functions h defined in this way when f spans X_s forms a vector space hereby denoted $Y_{s,u}(S^2)$ or more simply $Y_{s,u}$. We define norms in $Y_{s,u}$ by putting

$$\|h\|_{s,u} = \|f\|_s.$$

We are now ready to state our main result:

THEOREM 1-Fix s, $0<s<1$. There exist two positive numbers R and ε, ε depending on s and R, such that for every $v \in Y_{1,0}$ with $\|v\|_{1,0} < \varepsilon$, there is a unique solution $u \in X_s^R(S^2)$ of the equation $B(u)=v$.

THE METHOD

To construct the solution we use the following Newton iterative scheme:

$$u_0 = 0, \quad u_{k+1} = u_k - L(u_k)[v - B(u_k)]$$

where L denotes the inverse of B', the first derivative of B.
Then we show that, for suitable R, the sequence u_k converges in X_s^R to a solution of $B(u)=v$ if v is small enough in $Y_{1,0}$.
As we have already mentioned in the introduction the main technical difficulty consists in controlling a $Y_{s,u}$ norm of an analytic function g, say, defined in G_u in terms of the $Y_{s,u}$ norm of its extension to G_u. In our case this can be achieved because g is harmonic in both the interior and exterior of G_u. Infact we need only to consider the case in which g is a single layer potential of the following type:

$$W_u(x) = \int_{S^2} \rho(\omega') |\varphi_u(\omega')|^2 |x - \varphi_u(\omega')|^{-1} d\omega'$$

with $\rho \in X_s$.

To realize our purpose the appropriate tools are the next two lemmas.

LEMMA 2-Let $u \in \dot{X}_{s''}$, with $\|u\|^{(2)}_{s'} < c$. Let W be harmonic in a neighborhood of Γ_u. Then there exist two positive constants c_0 and c_1, independent on s,s',s'', $0<s<s'<s''$, such that

$$\sup_{t<c_0(s'-s)} \|W\circ \Phi_u(\cdot,t)\|_s < c_1(s''-s')^{-1}\{\|W\circ\varphi_u\|_{s''} + \|\partial_{N_u} W\circ\varphi_u\|_{s''}\}$$

where $\Phi_u(\omega,t)=\omega(1-t)(1+u(\omega))$ and N_u is the unit exterior normal to Γ_u.

proof: consequence of well known results on the Cauchy problem[8].

LEMMA 3-Let $u \in X^R_s(S^2)$ such that $\|u\|^{(9)}_{s'} < \delta$, with δ, R sufficiently small. Then for a function W harmonic in G_u the estimate

$$\|D_1 W\circ\varphi_u\|_s < c(s'-s)^{-1} \|W\circ\varphi_u\|^{(7)}_{s'}$$

holds for every $s<s'$ and for every direction 1, with the constant c independent of s,s'.

The proof is a consequence of a general result concerning the solution of the following Dirichlet problem:

$$P(x,D)v = \sum_{|\alpha|=1,2} a_\alpha(x)D^\alpha v = 0 \qquad \text{in } \Sigma_n$$

$$v = f \qquad \text{on } S^{n-1}$$

where Σ_n is the unit ball in $\mathbb{R}^n$.
P(x,D) denotes a uniformly elliptic differential operator with coefficients analytic in Σ_n. We assume that these coefficients satisfy the additional hypothesis (H) described below.

Let $\{\psi_j\}$ be a set of admissible local charts, covering S^{n-1}. Write now the points $x\in\mathbb{R}^n$ in the form (x',x_n) with $x'\in\mathbb{R}^{n-1}$, $x_n\in\mathbb{R}$; introduce in correspondence to each ψ_j the transformation of variables T_j given by $T_j(y',y_n)=T_j(y)=(x',x_n-\psi_j(x'))$, where $y'\in D'$, the unit disc in $\mathbb{R}^{n-1}$ and $0<y_n<1/3$ (say). Clearly, the effect of the transformation T_j is to locally flatten S^{n-1}.
Set $\tilde{a}^j_\alpha(y) = a_\alpha(T_j^{-1}(x))$ and $n_0 = n + [n/2]+3$. Our hypothesis on the coefficients reads as follows:

(H): There exist two positive constants η, A_n, such that, for every $y=(y',y_n)$ with $|y'|<1$ and $0<y_n<\eta$, the following estimate holds

$$\sup_{1\leq|\gamma|\leq n_0} \left|D^\beta D^\gamma \tilde{a}^j_\alpha(y)\right| < A_n \frac{\beta!}{(sR)^{|\beta|}}$$

for every j and every multi-indexes α, β, $|\alpha|=1,2$.
Let now v be the solution of the Dirichlet problem and denote by v_1 the trace on S^{n-1} of the derivative of v along the direction 1.
The following theorem holds.

THEOREM 4-Assume that the hypothesis (H) on the coefficients a_α holds; assume furthermore that $f \in X_s^R(S^{n-1})$ with $\|f\|_s^{(n_0)}$ finite. Then if A_n and R are sufficiently small, for every direction l and every $s>s'>0$, we have

$$\|v_1\|_{s'} < c(s-s')^{-1}\|f\|_s^{(n_0)}$$

with c independent of s,s'.

proof (outline): performing the transformation of variables T_j it is possible to reduce our problem to a finite set of Dirichlet problems of the following type:

$$Q(x,D)v = \sum_{|\alpha|=1,2} A_\alpha(x)D^\alpha v = 0 \quad \text{in} \quad \Omega$$

$$v = F \qquad\qquad \text{on} \quad \partial\Omega$$

where Ω is a smooth domain in the half space $x_n>0$ whose closure contains the hemiball $E_{2L} = \{(x',x_n): x_n>0, |x'|<2L\}$ with L small positive number. The hypotheses on f and the a_α in the statement of the theorem correspond to the following inequalities for A_α and F:

$$\sup_{0\leq|\mu|\leq n_0} \left|D^\gamma D^\mu A_\alpha(x)\right| < A_n \frac{\gamma!}{(sR)^{|\gamma|}} \qquad \text{for every } \alpha, |\alpha|=1,2 \text{ and every } x\in\bar{\Omega};$$

F is analytic on $\partial\Omega$ and

$$\sup_{0\leq|\mu|\leq n_0} \left|D_{x'}^\gamma D_{x'}^\mu F(x')\right| < F_n \frac{\gamma!}{(sR)^{|\gamma|}}$$

for some constant F_n, for every (n-1) dimensional multiindex γ and every x' in $\sigma_{2L} = \{(x',x_n), x_n=0, |x'|<2L\}$.

The assertion of the theorem is then an easy consequence of lemma 5 below and the Sobolev immersion theorem.

LEMMA 5-Let $L/2<r<L$; if A_n and R are sufficiently small, depending on L-r, then there exists a constant c, dependent only on A_n such that

$$\left\{ \int_{E_r} \left|D_{x'}^\beta D^{\beta_0} v(x)\right|^2 dx \right\}^{1/2} < c\cdot\bar{F}_n \frac{\beta!}{(sR)^{|\beta|}}$$

for every multiindex β_0 with $|\beta_0| = [n/2]+2$, where $\bar{F}_n$ is a constant dependent on F_n and the norm of F in the Sobolev space $H^{n_0-\frac{1}{2}}(\partial\Omega)$.

To prove the lemma choose a test function $\chi=\chi(x)$ such that $\chi=1$ in E_r and $\chi=0$ outside $E_{r+\delta}$ where $L/2<r<r+\delta<L$.

For $w=\chi\cdot v$ we have:

$$Q(x,D)w = \sum_{0<|\beta|\leq 2} D^\beta\chi \cdot Q_\beta(x,D)v \quad \text{in } E_{r+\delta}, \; w=\chi v \text{ on } \sigma_{r+\delta}, \; w=0 \text{ on } \partial\Omega\setminus\sigma_{r+\delta}.$$

$Q_\beta(x,D)$ denotes a differential operator of order $2-|\beta|$.

Extending now w by zero in $\Omega \setminus E_{r+\delta}$ we can apply the Agmon-Douglis-Nirenberg estimates to obtain

$$\|w\|_{H^t(E_{r+\delta})} < c\left\{\|\chi\cdot F\|_{H^{t-1/2}(\sigma_{r+\delta})} + \|Q(x,D)w\|_{H^{t-2}(E_{r+\delta})}\right\}$$

for every $t \geq 2$.

Thanks to this estimates and the fact that $\|v\|_{H^1(\Omega)} < c\|F\|_{H^{1/2}(\partial\Omega)}$ it is clearly possible to fix c such that the estimate of the lemma holds for $|\beta| \leq 2$. The rest of the proof follows by induction on the length of β.

Using the above results one is able to prove the following crucial estimates for the operator B, for its first derivative B', given by the formula

$$B'(u)\rho(\omega) = \int_{S^2} |\varphi_u(\omega')|^2\, \rho(\omega')\, |2\omega - \varphi_u(\omega')|^{-1} d\omega',$$

and for the operator L, inverse of B'.

LEMMA 6 Let $o<s<s'<s'+r\leq a\leq 1$. There exist δ, R and c_2 such that if $\|u\|_a^{(9)} < \delta$, $\|\tilde{u}\|_a^{(9)} < \delta$ and $\|\tilde{u}-u\|_s < c_2 r$ then the following estimates hold:

i) $\|B(\tilde{u}) - B(u)\|_{s,u} < c(s'-s)^{-9}\|\tilde{u}-u\|_{s'+r}$

ii) $\|B'(u)\rho - B'(\tilde{u})\rho\|_{s,u} < c(s'-s)^{-10}\|\rho\|_{s'+r}$

iii) $\|L(u)h\|_s^{(9)} < c(s'-s)^{-18}\|h\|_{s,u}$.

proof (outline): i) follows easily from ii) so we give indications of how to prove ii) and iii).

ii): we have $\|B'(u)\rho - B'(\tilde{u})\rho\|_{s,u} = \|W_u \circ \varphi_u - W^*_{\tilde{u}} \circ \varphi_u\|_s < \|W_{\tilde{u}} \circ \varphi_{\tilde{u}} - W^*_{\tilde{u}} \circ \varphi_u\|_s +$

$$+ \|W_u \circ \varphi_u - W_{\tilde{u}} \circ \varphi_{\tilde{u}}\|_s$$

where $W^*_{\tilde{u}}$ denotes the analytic extension of $W_{\tilde{u}}$.

The first term in the above sum can be estimated in terms of the first t-derivative of the function $w = W_{\tilde{u}} \circ \Phi_{\tilde{u}}$, where Φ_u is defined as in lemma 2. More precisely we have

$$\|W_{\tilde{u}} \circ \varphi_{\tilde{u}} - W^*_{\tilde{u}} \circ \varphi_u\|_s < c\|u-\tilde{u}\|_s \cdot \sup_{|t| \leq c_o r} \|\partial_t w(\cdot,t)\|_s .$$

Using lemmas 2 and 3 we get

$$\sup_{|t| \leq c_o r} \|\partial_t w(\cdot,t)\|_s < c(s''-s)^{-8}\|W_{\tilde{u}} \circ \varphi_{\tilde{u}}\|_{s''+r} .$$

Rather straightforward computations show that $\|W_{\tilde{u}} \circ \varphi_{\tilde{u}}\|_{s''+r} < c(s'-s'')\|\rho\|_{s'+r}$ so that, choosing $s'-s''=(s'-s)/2$, the first term is done.

To estimate $W_u \circ \varphi_u - W_{\tilde{u}} \circ \varphi_{\tilde{u}}$ we write it as a potential of the following type:

$$\int_{S^2} b(\omega,\omega')\rho(\omega')|\omega-\omega'|^{-1}d\omega'$$

where b depends essentially on $|\omega-\omega'|$, $\varphi_u(\omega)-\varphi_u(\omega')$ and $\varphi_{\tilde{u}}(\omega)-\varphi_u(\omega')$. It is now just a matter of computation to prove the desired estimates.

iii):we write a fromula for L(u) by representing the solution of the equation $B'(u)\rho=h$ in terms of the potential W_u .Infact W_u is harmonic inside and outside G_u,continuous through Γ_u and its normal derivative suffers a jump across Γ_u of magnitude given by

$$\rho(\omega)=\{\partial_{N_u}W_u^+\circ\varphi_u(\omega)-\partial_{N_u}W_u^-\circ\varphi_u(\omega)\}\cdot|D\varphi_u(\omega)|\cdot|\varphi_u(\omega)|^{-2}$$

where W_u^+ and W_u^- are the restrictions of W_u to the regions inside and outside G_u,respectively,and $|D\varphi_u(\omega)|$ denotes the Jacobian of the map φ_u. The X_s norm of the normal derivatives are estimated by $\|W_u^+\circ\varphi_u\|_{s''}^{(7)}=\|W_u^-\circ\varphi_u\|_{s''}^{(7)}$ via lemma 3.Next, $\|W_u^+\circ\varphi_u\|_{s''}^{(7)}$ is estimated by the $X_{s'}$ norm of $W_u\circ\varphi_u$ times $(s'-s'')^{-7}$ which equals $\|h\|_{s',u}$ by definition. It is then easy to get the estimate iii).

The existence part of Theorem 1 goes now along lines similar to those in the paper of Nirenberg[4].The uniqueness part is a consequence of the results of Pagani[3].

REFERENCES

1. C. Maderna,C. D. Pagani,S. Salsa,Analytic solutions to the inverse problem of the Newtonian potential,to appear on Journal of Math. Anal. and Appl.

2. A. Prilepko, Uber Extistenz und Eindeutigheit von Lösungen inverser Probleme der Potentialtheorie, Math. Nachtr., 63 (1974).

3. C. D. Pagani,Stability of a surface determined from measures of potential,SIAM J.Math.Anal.17 (1)(1986).

4. L. Nirenberg,An abstract form of the nonlinear Cauchy-Kowalewski theorem,J.Diff.Geom. 6 (1972).

5. L. Hörmander,The boundary problem of physical geodesy,Arch.Rat.Mec.Anal. 62 (1976).

6. D. G. Shaeffer,The capacitor problem,Indiana Univ.Math.J. 24 (12)(1975).

7. C. Maderna,C. D. Pagani,S. Salsa,Existence results in an inverse problem of potential theory,Nonlinear Analysis T.M.A. 10 (1986).

8. J. F. Trèves,"Basic Linear Partial Differential Equations",Academic Press New York (1975).

ON THE INTEGRATION OF ELASTOSTATIC DISPLACEMENT EQUATION

Abdullah Shidfar

Department of Mathematics
Iran University of Science and Technology
Narmak, Tehran-16, Iran

1. INTRODUCTION

The Cauchy-Navier equation of motion for the linear elastic isotropic continuum, for the vanishing body forces, may be written in the form

$$\Box_2^2 \underline{u} + \frac{1}{1-2\nu}\nabla(\nabla . \underline{u}) = 0, \qquad \nu = \frac{\lambda}{2(\lambda+\mu)}, \tag{1.1}$$

where $\underline{u}(\underline{r},t)$ is the displacement vector, λ, μ are Lamé elastic constants, ν $(0 < \nu < \frac{1}{2})$ is Poisson's ratio,

$$\Box_2^2 = \Delta - \frac{1}{c_2^2}\frac{\partial^2}{\partial t^2}; \qquad c_2^2 = \frac{\mu}{\rho}, \tag{1.2}$$

and ρ is the mass density. There is a solution to equation (1.1) in the form[1]

$$\underline{u}(\underline{r},t) = \Box_1^2 \underline{\chi} - 2\kappa\nabla(\nabla . \underline{\chi}); \qquad \kappa = \frac{1}{4(1-\nu)}, \tag{1.3}$$

where

$$\Box_1^2 = \Delta - \frac{1}{c_1^2}\frac{\partial^2}{\partial t^2}; \qquad c_1^2 = \frac{\lambda+2\mu}{\rho}, \tag{1.4}$$

and $\underline{\chi}$ is a solution of the repeated wave equation

$$\Box_1^2 \Box_2^2 \underline{\chi} = 0. \tag{1.5}$$

In the limiting meadium, i.e., when c_1^2, c_2^2 approach infinity and so define rigid-continuum,[2] then equation (1.1) becomes

$$\Delta\underline{u} + \frac{1}{1-2\nu}\nabla(\nabla.\underline{u}) = 0, \tag{1.6}$$

with the general solution

$$\underline{u} = \Delta\underline{\chi} - 2\kappa\nabla(\nabla.\underline{\chi}), \tag{1.7}$$

where $\underline{\chi}$ is a biharmonic function, which is in agreement with Galerkin's solution[3].

Papkovich[4] and Neuber[5] gave a family of solutions to (1.6) in the form

$$\underline{u} = \underline{h} - \kappa\nabla(\underline{r}.\underline{h}+f); \qquad \kappa = \frac{1}{4(1-\nu)} \qquad (\tfrac{1}{4}<\kappa<\tfrac{1}{2}), \tag{1.8}$$

where f and $\underline{h}$ are harmonic and $\underline{r}$ is the position vector with respect to a given origin.

In this paper we are going to consider Papkovich-Neuber's solutions of equation (1.6) and we shall discuss the elimination of certain potentials in the representation (1.8). In section 2 we investigate some biharmonic function properties. These properties will become very useful for the remaining of this paper. In section 3 we consider the elimination of scalar potential f or z-component of $\underline{h}$ in the Papkovich-Neuber representation (1.8). In section 4 a new solution to (1.1) is given and the equivalence of this solution with the Papkovich-Neuber representation is discussed. Some final remarks are made in section 5.

2. ON THE REPRESENTATION OF A BIHARMONIC FUNCTION IN TERMS OF TWO HARMONIC FUNCTIONS

A function χ is said to be biharmonic throughout a bounded domain D with a closed surface ∂D if it has continuous derivatives up to the fourth order in D, and satisfies the biharmonic equation

$$\Delta^2\chi = \Delta(\Delta\chi) = 0, \tag{2.1}$$

in D. Hadamard[6] has shown that χ exists and is uniquely determined in D if χ and its normal derivative χ' have prescribed values on ∂D. In an unbounded exterior domain the condition

$$\chi = O(r), \qquad \text{as} \qquad r \to \infty, \tag{2.2}$$

must be added to ensure uniqueness. It was first demonstrated by Almansi[7] that a biharmonic function could always be represented in terms of the harmonic functions. Thus, if χ is a biharmonic function in D, then we may write

$$\chi = x\phi + \psi, \qquad y\phi + \psi, \qquad z\phi + \psi, \tag{2.3}$$

where ϕ , ψ are harmonic functions. This work was interesting and important, particularly from the point of view of possible applications to two-dimensional of elastic systems. These representations became largely ignored, principally because of the subsequent development of Muskhelishvili's[8] complex variable approach. However, there exists an alternative form for this representation

$$\chi = r^2\phi + \psi \ ; \qquad \Delta\phi = \Delta\psi = 0, \qquad r^2 = x^2 + y^2 + z^2, \tag{2.4}$$

which has considerable advantages. This r^2-form has been discussed by Bergman and Schiffer,[9] and proved generally by Jaswon and Shidfar[10] also by Jaswon, Maiti and Symm[11] to provide numerical solutions to elastic stress and deformation problems in two dimensions. The 3-dimensional representation has an important application to Papkovich-Neuber representation (1.8).

The functions ϕ and ψ in (2.3) and (2.4) may be multi-valued functions. For instance in two dimensions we may see directly that:

$$x\log r = y\theta + (x\log r - y\theta),$$
$$y\log r = -x\theta + (y\log r + x\theta),$$
$$2x\log r = r^2\left(\frac{x\log r + y\theta}{r^2}\right) + x\log r - y\theta,$$

where the bracketed terms are harmonic functions. These transformations are important in the theory of two-dimensional Volterra dislocations.

For a given biharmonic function χ, $\Delta\chi$ is a harmonic function, and $\Delta\chi$ may be written as the derivative of another harmonic function, i.e.,

$$\Delta\chi = 2\frac{\partial\phi}{\partial x}, \tag{2.5}$$

where ϕ is a harmonic function to be determined. In fact

$$\phi = \frac{1}{2}\int^{x} \Delta\chi \, d\xi + f, \tag{2.6}$$

where f is a function of y and z to be determined, so that

$$\Delta\phi = \frac{1}{2}\Delta\int^{x} \Delta\chi \, d\xi + \Delta f = 0. \tag{2.7}$$

Therefore f satisfies the Poisson equation

$$\Delta f = -\frac{1}{2}\Delta\int^{x} \Delta\chi \, d\xi. \tag{2.8}$$

A solution to this equation always exists. Thus, there exists a harmonic function ϕ such that (2.5) is sarisfied. By this result, we can immediately conclude that, any harmonic function can be represented as a derivative of another harmonic function .

The representation (2.3) is non-unique for a given χ, because for any harmonic function $h(x,y)$, the x-form representation (2.3) may be represented by

$$\chi = x(\phi + h) + \psi - xh, \tag{2.9}$$

where $\phi + h$ and $\psi - xh$ are harmonic functions.

Now we are going to consider the uniqueness of representation of r^2-form (2.4).

For a given biharmonic function χ, if ϕ, ψ are the harmonic solutions of equation (2.4), then ϕ is the solution of

$$4\phi + 4r\frac{\partial\phi}{\partial r} = \Delta\chi, \qquad \text{in 2-dimensions}, \tag{2.10}$$

$$6\phi + 4r\frac{\partial\phi}{\partial r} = \Delta\chi, \qquad \text{in 3-dimensions}, \tag{2.11}$$

which are differential equations for ϕ in terms of the harmonic function $\Delta\chi$. It is convenient to replace equations (2.10) and (2.11) by the more general equation

$$\kappa\phi + r\frac{d\phi}{dr} = f, \tag{2.12}$$

where f is a given harmonic function and κ is any real constant. We seek a harmonic function ϕ which satisfies the differential equation (2.12). This equation, which plays a key role in the related theory of Papkovich-Neuber representation, is called the Bergman-Schiffer equation[10].

In order to find the harmonic function ϕ we note that equation (2.12) has the general soulution

$$\phi = r^{-\kappa}\int^{r} \rho^{\kappa-1} f(\rho,\theta,\psi)\, d\rho + r^{-\kappa} g(\theta,\psi), \tag{2.13}$$

where g is an arbitrary function of the polar variables θ, ψ. However ϕ is not necessarily a harmonic function. The appropriate condition is $\Delta\phi = 0$, which yields the equation for g

$$\Delta\{r^{-\kappa} g(\theta,\psi)\} = -\Delta\{r^{-\kappa}\int^{r} \rho^{\kappa-1} f\, d\rho\}. \tag{2.14}$$

This is a Poisson's equation for g which always has a solution, and accordingly there always exists a harmonic function of the form (2.13) which satisfies equation (2.12).

The complementary solutions of equation (2.10) are

$$ar^{-1}\cos\theta + br^{-1}\sin\theta, \tag{2.15}$$

where a, b are constants. These complementary solutions do not exist in any interior domain including the origin and neither do they exist in any exterior domain excluding the origin, because by virtue of (2.2), ψ does not exist in exterior case. Therefore for such domains the equation (2.4) has a unique solution and the r^2-form representation for such domains is unique.

A similar statement may be used in three-dimensional case to show that the solution of equation (2.4) is unique for any exterior domain and any interior domain including the origin.

In the remainder of this section we are going to consider the equivalence of the representation x, y, z, r^2-forms.

The x-form representation may be transformed into y-or z-forms. Given x-form $\chi = x\phi+\psi$; $\Delta\phi = \Delta\psi = 0$, by virtue (2.5) we may write

$$\chi = x\frac{\partial h}{\partial y} + \psi = y\frac{\partial h}{\partial x} + (x\frac{\partial h}{\partial y} - y\frac{\partial h}{\partial x} + \psi), \tag{2.16}$$

where h, $\frac{\partial h}{\partial y}$, $\frac{\partial h}{\partial x}$, $x\frac{\partial h}{\partial y} - y\frac{\partial h}{\partial x} + \psi$ are harmonic functions.

Now if the representation $\chi = r^2\phi + \psi$; $\Delta\phi = \Delta\psi = 0$ is given, then it can be transformed into x-form. In fact, by virtue of (2.5) we may write

$$\chi = r^2\frac{\partial h}{\partial z} + \psi; \qquad \Delta h = 0, \tag{2.17}$$

or identically

$$\chi = x(x\frac{\partial h}{\partial z} - z\frac{\partial h}{\partial x}) + y(y\frac{\partial h}{\partial z} - z\frac{\partial h}{\partial z}) + z(x\frac{\partial h}{\partial x} + y\frac{\partial h}{\partial y} + z\frac{\partial h}{\partial z}) + \psi, \tag{2.18}$$

where

$$\Delta(x\frac{\partial h}{\partial z} - z\frac{\partial h}{\partial x}) = \Delta(y\frac{\partial h}{\partial z} - z\frac{\partial h}{\partial z}) = \Delta(x\frac{\partial h}{\partial x} + y\frac{\partial h}{\partial y} + z\frac{\partial h}{\partial z}) = 0. \tag{2.19}$$

y and z forms on the right-hand side of (2.18) can be transformed into x-form. Hence, r^2-form can be transformed into x-form.

3. ELIMINATION OF CERTAIN POTENTIALS IN (1.8)

It has been proved by Eubanks and Sternberg,[12] and K. Millar[13] that the harmonic function f may be eliminated from (1.8) for some special values of κ. Here we show that f may be eliminated from (1.8) generally. For this purpose, we attempt to construct a scalar harmonic H such that

$$\nabla H - \kappa\nabla(\underline{r}.\nabla H) = -\kappa\nabla f. \tag{3.1}$$

This relation yields the equation

$$-\frac{1}{\kappa}H + r\frac{dH}{dr} = f + \text{const}. \tag{3.2}$$

By formula (2.13), equation (3.2) has the family of solutions

$$H = r^{\frac{1}{\kappa}} \int^{r} \rho^{-\frac{1}{\kappa}-1} f \, d\rho + r^{\frac{1}{\kappa}} g(\theta,\psi), \tag{3.3}$$

where $g(\theta,\psi)$ can be found such that H becomes harmonic. Substituting from (3.1) into (1.8) yields

$$\underline{u} = \underline{h}' - \kappa\nabla(\underline{r}.\underline{h}'), \tag{3.4}$$

where $\underline{h}' = \underline{h} + \nabla H$.

An alternative study of equation (3.1) has been made by Jaswon and Symm[14], using the spherical harmonic expansions of f and H. However the spherical expansion method breaks down for $\kappa=1/n$, when n is an integer in any domain.

Eubanks and Sternberg[12] have shown that the z-component of $\underline{h}$, say h, may be eliminated from (1.8) if and only if there exists a harmonic function H such that

$$\frac{\partial H}{\partial z} = h(x,y,z), \tag{3.5}$$

By section 2 of this paper, there is always a harmonic solution to equation (3.5). Hence, the z-component of $\underline{h}$ can be eliminated from (1.8).

A solution to equation (1.6) has also been given by Bergman and Schiffer[9] and the generality of this solution has been discussed by Shidfar[15]. In the next section we are going to give a new solution to (1.6) which is equivalence to the Papkovich-Neuber representation. This solution shows that the scalar harmonic function f may be eliminated from (1.8), provided that the harmonic solution of the Bergman-Schiffer equation has no singularities in a given domain.

4. A NEW SOLUTION TO EQUATION (1.6)

In this section we shall show that we can obtain an infinite set of solutions to (1.6) of the form

$$\underline{u} = \nabla(x\phi) + \underline{\psi} \tag{4.1}$$

where $\underline{\psi}$ is a harmonic vector and ϕ is a harmonic function which may be expressed in terms of $\underline{\psi}$. In fact, from (4.1), we obtain

$$\nabla.\underline{u} = 2\frac{\partial\phi}{\partial x} + \nabla.\underline{\psi} \tag{4.2}$$

and

$$\Delta\underline{u} = 2\nabla(\frac{\partial\phi}{\partial x}). \tag{4.3}$$

Substituting from (4.2) and (4.3) into (1.6), we find

$$2\nabla\left(\frac{\partial\phi}{\partial x}\right) + \frac{1}{1-2\nu}\left(2\frac{\partial\phi}{\partial x}+\nabla.\underline{\psi}\right) = 0. \tag{4.4}$$

We integrate this equation and obtain

$$4\frac{1-\nu}{1-2\nu}\frac{\partial\phi}{\partial x} = -\frac{1}{1-2\nu}\nabla.\underline{\psi} + \text{const.} \tag{4.5}$$

This equation has a family of solutions

$$\phi = \frac{1}{4(1-\nu)}\int^{x}\nabla.\underline{\psi}\, d\xi + g(y,z) \tag{4.6}$$

where g(y,z) may be determined so that (4.6) becomes a harmonic function. Conversely, any given solution of (1.6) can be written in the form (4.1). In fact, from (4.3) and (1.6), we have

$$2\nabla\left(\frac{\partial\phi}{\partial x}\right) = -\frac{1}{1-2\nu}\nabla(\nabla.\underline{u}). \tag{4.7}$$

Integrating this equation, we find

$$2\frac{\partial\phi}{\partial x} = -\frac{1}{1-2\nu}\nabla.\underline{u} + \text{const.} \tag{4.8}$$

whence

$$\phi = -\frac{1}{2(1-2\nu)}\int^{x}\nabla.\underline{u}\, d\xi + g(y,z). \tag{4.9}$$

For instance, if $\underline{u}$ is a Papkovich-Neuber solution of (1.6), then

$$\phi = \frac{1}{4(1-\nu)}\int^{x}\nabla.\underline{u}\, d\xi + g(y,z).$$

We now show that the repesentation

$$\underline{u} = -\kappa\nabla(\underline{r}.\underline{h}) + \underline{h} \tag{4.10}$$

and

$$\underline{u} = \nabla(x\phi) + \underline{\psi} \tag{4.11a}$$

$$\phi = \frac{\kappa}{4}\int^{x}\nabla.\underline{\psi}\, d\xi + g(y,z) \tag{4.11b}$$

are equivalent. From this result we deduce that the scalar harmonic function f can be determined from the Papkovich-Neuber representation (1.8).

Let (4.10) is given as a solution to (1.6). Putting $\underline{\psi} = \underline{h}$ in (4.11a), we obtain

$$\nabla(x\phi) = -\kappa\nabla(\underline{r}.\underline{h}). \tag{4.12}$$

using ∇. on both sides of (4.12) yields

$$2\frac{\partial\phi}{\partial x} = -\frac{\kappa}{2}(\nabla.\underline{h}) \tag{4.13}$$

which has a solution in the form (4.11b). Conversely, (4.11a) and (4.11b) can be easily transformed into (4.10) and the equivalence is complete.

5. CONCLUSIONS

It has been shown that the representation of bihatmonic functions in terms of harmonic functions as r^2-form is unique for any exterior domain and any interior domain including the origin. The representations as x-, y-, z-forms are non-unique in any domains. All of these representations can be transformed into each other. Our discussion shows that the scalar harmonic potential or z-component of harmonic vector potential in the Papkovich-Neuber representation can be eliminated.

REFERENCES

1. M. Iacovache, O extindere a metodei lui Galerkin Pentru sistemul ecuatiilor elasticitatii, Bul. Stiint. Acad. Rep. Pop. Romane, Ser. A:1, 593 (1949).
2. S. A. Abdullah, Uniformly moving elastic singularities in the classical theory of elasticity, ph.D. Thesis, The City University, London (1984).
3. B. Galerkin, Contribution à la solution générale du problème de la théorie de l'élasticité dans le cas de trois dimensions, C. R. Acad. Sci. Paris 190:1047 (1930).
4. P. F. Papkovich, Solution générale des équations differentielles fondamentales d'elasticite exprimee par trois functions harmoniques, C. R. Acad. Sci. Paris, 195:513 (1932).
5. H. Neuber, Ein neuer Ansatz zür losung räumlicher probleme der elastizitätstheorie. J. Appl. Math. Mech. (ZAMM), 14:203 (1934).
6. J. Hadamard, Mémoire sur le problème d'analyse relatif à l'équilibre des plaques elastiques encastrees, Mem. Pres. div. sav., Acad. Sci. Int. France 33(4): (1908).
7. E. Almansi, Sull'integrazione dell'equazione differentiale $\Delta^{2n} = 0$, Annali di matematica pure et applicata, series III, 2:1 (1887).

8. N. I. Muskhelishvili, "Some basic problems of the mathematical theory of elasticity," Noordhoff, Groningen (1953b).
9. S. Bergman and M. Schiffer, "Kernel functions and elliptic differential equations in mathematical physics," Academic Press (1953).
10. M. A. Jaswon and A. Shidfar, On a differential equation of Bergman and Schiffer, J. Maths. & Phys. Sci. 14:225 (1980).
11. M. A. Jaswon, M. Maiti and G. T. Symm, Numerical biharmonic analysis and some applications, Int. J. Solids strctures, 3:309 (1967).
12. R. A. Eubanks and E. Sternberg, On the completeness of Boussinesq-Papkovick stress functions, J. Rat. Mech. Anal. 5:735 (1956).
13. R. F. Millar, On the compltenesss of the Papkovich potentials, Quart. Appl. Math. XLI:385 (1984).
14. M. A. Jaswon and G. T. Symm, "Integral equation methods in potential theory and elastostatics," Academic Press (1977).
15. A. Shidfar, On the Papkovich-Neuber representation of the elastostatic displacement fields, Betech 86, Proceedings of the 2nd Boundary Element Tech. Conference, M.I.T., U.S.A. (1986).

NON-LINEAR POTENTIAL THEORY IN LEBESGUE SPACES WITH MIXED NORM

Tord Sjödin

Department of Mathematics
University of Umeå
S-901 87 Umeå, Sweden

0. INTRODUCTION

The classical potential theory for the Riesz kernel in $\mathbb{R}^d$ has been generalized in many different ways. In this paper we are concerned with the L^p-potential theory which appeared around 1970. See N.G. Meyers [Me] and V.G. Maz'ya and V.P. Havin [MH] for an introduction to this theory. Applications and further developments of the theory are found in the books V.G. Maz'ya [Ma], V.G. Maz'ya, T.O. Shaposhnikova [MS], the lecture notes D.R. Adams [A] and the survey article L.I. Hedberg [H]. See also [S1] and [S2].

We are going to extend the L^p-theory to Lebesgue spaces $L^{p,q}(\mathbb{R}^d)$ with mixed norm. A more detailed exposition of our results as well as applications of the theory will appear in a forthcoming paper [S3].

The basic concepts of this paper are the two set functions $C_{k,p,q}$ and $c_{k,p,q}$, where $1<p,q<\infty$ and k is a kernel (see Section 1 for the exact definitions). We are going to study the capacitary distributions and capacitary potentials of $C_{k,p,q}$ and $c_{k,p,q}$ in great detail, and we prove a formula for the $C_{k,p,q}$-capacitary potential for a compact set with finite capacity. Such potentials are called non-linear $L^{p,q}$-potentials of measures (Section 2).

In Section 3 we prove that, under certain assumptions, non-linear $L^{p,q}$-potentials satisfy a maximum principle.

The general theory for non-linear $L^{p,q}$-potentials is applied to the Bessel kernel G_α^d in Section 4.

1. NOTATION AND DEFINITIONS

Let $\mathbb{R}^d=\mathbb{R}^m\times\mathbb{R}^n$ be the d-dimensional Euclidean space, where m and n are positive integers and $d=m+n$. We denote points in $\mathbb{R}^d$ by ξ and η and we write $\xi=(x,s)$ and $\eta=(y,t)$, where $x,y\in\mathbb{R}^m$ and $s,t\in\mathbb{R}^n$.

This notation is kept throughout the paper. In any Euclidean space the norm is written $|\cdot|$.

We fix numbers p and q, $1<p<\infty$ and $1<q<\infty$, and define the dual indices p' and q' by $1/p+1/p'=1$, $1/q+1/q'=1$.

Measurability of sets and functions refers to Lebesgue measure. All sets and functions are assumed to be measurable. The k-dimensional Lebesgue measure is denoted by $|E|_k$ and we drop k when $k=d$. We use the subscript or superscript + to denote non-negative elements. Various constants depending on $a,b,\ldots$ are denoted by $c(a,b,\ldots)$. Constants only depending on d are denoted by c.

For $1<p,q<\infty$ we denote by $L^{p,q}(\mathbb{R}^d)$ the space of all equivalence classes of extended real valued functions f in $\mathbb{R}^d$ for which

$$||f||_{p,q} = \left(\int\left(\int|f(x,s)|^p dx\right)^{q/p} ds\right)^{1/q}$$

is finite. See [BP] for properties of such spaces.

A capacity is a non-negative set function C defined on all subsets of $\mathbb{R}^d$ such that

(i) $C(\emptyset)=0$, $\emptyset$ the empty set,

(ii) $A_1 \subset A_2$ implies $C(A_1) \leq C(A_2)$.

We define a kernel to be a non-negative and lower semi-continuous function $k=k(\xi,\eta)$ defined on $\mathbb{R}^d \times \mathbb{R}^d$.

Our notation and terminology for measures and integrals is that in [Me]. We let M be the space of Radon measures and L_1 the Banach space of measures μ with finite total variation $||\mu||_1$. In M and L_1 we take the usual weak topologies. We define

$$k(\nu,\mu) = \int k(\xi,\eta)d\sigma^+(\xi,\eta) - \int k(\xi,\eta)d\sigma^-(\xi,\eta),$$

where μ and ν belong to M and $\sigma=\sigma^+-\sigma^-$ is the Jordan decomposition of the tensor product $\sigma=\nu_\xi \oplus \mu_\eta$ and one of the terms on the right hand side is finite. If f is a non-negative function then

$$k(\xi,f) = \int k(\xi,\eta)f(\eta)d\eta,$$

and if $\mu \in M^+$

$$k(\mu,\eta) = \int k(\xi,\eta)d\mu(\xi).$$

Now we are ready to define the set functions $C_{k,p,q}$ and $c_{k,p,q}$.

Definition 1.1 – Let $1<p,q<\infty$ and let k be a kernel. For any set A we define

$$C_{k,p,q}(A) = \inf ||f||^p_{p,q},$$

where infimum is over every $f \in L_+^{p,q}(\mathbb{R}^d)$ such that $k(\xi,f) \geq 1$ for all ξ in A.

A function $f \in L_+^{p,q}(\mathbb{R}^d)$ such that $k(\xi,f) \geq 1$ on A except in a set of $C_{k,p,q}$-capacity zero ($C_{k,p,q}$-q.e.) is called a test function for $C_{k,p,q}(A)$. It is easy to see that $C_{k,p,q}$ is a capacity in the sense defined above.

Let S be the σ-algebra in $\mathbb{R}^d$ of sets which are measurable for all Borel measures in $\mathbb{R}^d$.

Definition 1.2. Let $1<p,q<\infty$ have dual indices p' and q' respectively and let k be a kernel. For A belonging to S we define

$$c_{k,p,q}(A) = \sup \|\mu\|_1,$$

where supremum is over all non-negative Borel measures μ concentrated on A such that

$$\|k(\mu,\cdot)\|_{p',q'} \leq 1.$$

Any such measure μ is called a test measure for $c_{k,p,q}(A)$.

2. CAPACITIES AND POTENTIALS

In this section we are going to establish the basic properties of the set functions $C_{k,p,q}$ and $c_{k,p,q}$. We begin with the following proposition.

Proposition 2.1 – Let $1<p,q<\infty$ and let k be a kernel. Then

(i) For any set A,

$$C_{k,p,q}(A) = \inf C_{k,p,q}(G),$$

where infimum is over all open sets G containing A.

(ii) $$\lim_{i\to\infty} C_{k,p,q}(K_i) = C_{k,p,q}(K),$$

where $K_i \searrow K$ are compact sets

(iii) $$\lim C_{k,p,q}(A_i) = C_{k,p,q}(A),$$

where $A_i \nearrow A$ are arbitrary sets.

Proposition 2.1 is proved as in [Me]. A set A is called $C_{k,p,q}$-capacitable if

$$\sup_{K\subset A} C_{k,p,q}(K) = C_{k,p,q}(A).$$

Theorem 2.2 – All analytic sets are $C_{k,p,q}$-capacitable.

This follows from Proposition 2.1 and the general Choquet Theory [Ch,

Theorem 1]. Our next proposition follows at once from the definition of the $C_{k,p,q}$-capacity.

Proposition 2.3 – Let $f \in L^{p,q}(\mathbb{R}^d)$, then $k(\xi,f)$ is well defined and finite $C_{k,p,q}$-q.e. and

$$C_{k,p,q}(\{\xi;\ |k(\xi,f)| \geq \varepsilon\}) \leq \varepsilon^{-p} \cdot \|f\|^p_{p,q},$$

all $\varepsilon > 0$.

The triangle inequality in $L^{p,q}(\mathbb{R}^d)$ gives that

$$C_{k,p,q}(\overset{\infty}{\underset{1}{\cup}} A_i)^{1/p} \leq \sum_1^\infty C_{k,p,q}(A_i)^{1/p}$$

for any sets A_i, $i=1,2,\ldots$ If $p \leq q$ then $C_{k,p,q}$ is also subadditive, cf. [Me, Theorem 1]. In case $q<p$ the subadditivity of $C_{k,p,q}$ seems to be an open question, cf. [A, p.54] for the case of Besov capacities.

Now we turn to the question of the existence of capacitary distributions and capacitary potentials for $C_{k,p,q}$ and $c_{k,p,q}$.

Definition 2.4 – Let A be any set with $C_{k,p,q}(A) < \infty$. Any test function f for $C_{k,p,q}(A)$ such that

$$\|f\|^p_{p,q} = C_{k,p,q}(A)$$

is called a $C_{k,p,q}$-capacitary distribution of A and $k(\xi,f)$ is called a $C_{k,p,q}$-capacitary potential of A.

Definition 2.5 – Let A belong to S. Then any test measure μ for $c_{k,p,q}(A)$ such that $\|\mu\|_1 = c_{k,p,q}(A)$ is called a $c_{k,p,q}$-capacitary distribution for A and $k(\mu,\eta)$ is called a $c_{k,p,q}$-capacitary potential of A.

The existence of capacitary distributions is proved as in [Me, Theorems 9 and 13]. Any set A with finite $C_{k,p,q}$-capacity has a unique $C_{k,p,q}$-capacitary distribution. Every compact set K with $c_{k,p,q}(K) < \infty$ has a $c_{k,p,q}$-capacitary distribution.

The relation between the two set functions is proved using von Neumann's Minimax Theorem [Me, p.274] and a method due to Fuglede [F].

Theorem 2.6 – For any analytic set A

$$C_{k,p,q}(A)^{1/p} = c_{k,p,q}(A).$$

We are now ready to establish the basic formula for the capacitary distributions and capacitary potentials.

Theorem 2.7 – Let K be a compact set with $c_{k,p,q}(K)<\infty$ and let f and μ be $C_{k,p,q}$-capacitary and $c_{k,p,q}$-capacitary distributions of K respectively. Let $u(\xi)=k(\xi,f)$. Then

(i) $f(\eta) = c_{k,p,q}(K)\cdot k(\mu,\eta)^{p'-1}\cdot K(t)^{q'-p'}$,

for a.e. $\eta=(y,t)$ in $\mathbb{R}^d$, where

$$K(t) = \left(\int k(\mu,\eta)^{p'}dy\right)^{1/p'},$$

(ii) μ is concentrated on the set $\{\xi;\ u(\xi)=1\}\cap K$ and $u(\xi)\leq 1$ on the support of μ.

Theorem 2.7 can be proved by the method in [Me, Theorem 16].

For any non-negative measure μ in M we call

$$u(\xi) = \int k(\xi,\eta)\cdot k(\mu,\eta)^{p'-1}\cdot K(t)^{q'-p'}d\eta,$$

where $K(t) = \left(\int k(\mu,\eta)^{p'}dy\right)^{1/p'}$ and $\eta=(y,t)$, the non-linear $L^{p,q}$-potential of the measure μ. When $p=q$ we get the usual non-linear L^p-potentials cf. [Me]. Note the new extra factor $K(t)$ in $u(\xi)$ for $p\neq q$.

3. A MAXIMUM PRINCIPLE FOR NON-LINEAR $L^{p,q}$-POTENTIALS

In Section 2 we defined non-linear $L^{p,q}$-potentials of measures for very general kernels k. Here we define a class of kernels, called radial kernels, for which we can prove a maximum principle.

Definition 3.1 – Let $k:[0,\infty[\to[0,\infty]$ be non-increasing and lower semicontinuous. Then $k(\xi,\eta)=k(|\xi-\eta|)$, $\xi,\eta\in\mathbb{R}^d$, is called a radial kernel.

It is easy to see that a radial kernel also is a kernel as defined in Section 1. When k is a radial kernel we write

$$k*f(\xi) = \int k(|\xi-\eta|)f(\eta)d\eta$$

instead of $k(\xi,f)$ and analogously for $k*\mu$. For $\mu\in M^+$ we get the formula

$$u(\xi) = \int k(|\xi-\eta|)\cdot k*\mu(\eta)^{p'-1}\cdot K(t)^{q'-p'}d\eta, \tag{3.1}$$

where

$$K(t) = \left(\int k*\mu(\eta)^{p'}dy\right)^{1/p'}$$

and $\xi\in\mathbb{R}^d$, $\eta=(y,t)\in\mathbb{R}^d$, for the non-linear $L^{p,q}$-potential of the measure μ.

The classical maximum principle was extended to non-linear L^p-potentials in [AM, Theorem 2.3] and [MH, Part I, §3]. We prove its analogue for certain non-linear $L^{p,q}$-potentials of measures when $q \leq p$.

Theorem 3.2– *Let* $1<q\leq p<\infty$ *and let* k *be a radial kernel. Let* $\mu \in M^{+}$ *have its support contained in* $\mathbb{R}^m \times \{0\}$ *and define* $u(\xi)$ by (3.1). *Then*

$$u(\xi) \leq \max(Q, Q^{p'-1}) \cdot \sup_{\eta \in \operatorname{supp}\mu} u(\eta),$$

for all $\xi \in \mathbb{R}^d$, *where* Q *only depends on* m.

The proof is in two steps, where we in each step use the method in [AM, Theorem 2.3].

4. BESSEL CAPACITIES AND BESSEL POTENTIALS

A particularly interesting case is when the kernel k is the Bessel kernel G_α^d of order $\alpha>0$ in $\mathbb{R}^d$. General properties of the Bessel kernel are found in [St, p.130]. We will denote the corresponding capacities $C_{k,p,q}$ and $c_{k,p,q}$ by $B_{\alpha,p,q}^d$ and $b_{\alpha,p,q}^d$ respectively. See [Me, p.279] for the case $p=q$.

Our results obtained for general kernels in Sections 2 and 3 now carry over to the case of the Bessel kernel. In particular we have the potential representation of the $B_{\alpha,p,q}^d$-capacitary distribution for a compact set given in Theorem 2.7 and the maximum principle (Theorem 3.2).

Properties of the Bessel kernel also give the following proposition. We let $B(\xi,r)$ denote the closed ball with center in ξ and radius $r>0$.

Proposition 4.1– *Let* $1<p,q<\infty$ *and* $0 < \alpha \leq \frac{m}{p} + \frac{n}{q}$. *Then*

(i) $0 < B_{\alpha,p,q}^d(B(\xi,r)) < \infty$, *all* $0<r<\infty$,

(ii) *If* E *is measurable and*

$B_{\alpha,p,q}^d(E) = 0$, *then* $|E|=0$,

(iii) $B_{\alpha,p,q}^d(\{\xi\}) = 0$, *all* ξ *in* $\mathbb{R}^d$.

It is also easy to prove that if $0 < \alpha \leq \frac{n}{q}$ then $B_{\alpha,p,q}^d(E)=0$ for all sets $E \subset \mathbb{R}^m \times \{0\}$. These two results are sharp as concerns the values of α. If $\alpha > \frac{m}{p} + \frac{n}{q}$ then all non-empty sets have positive $B_{\alpha,p,q}^d$-capacity.

REFERENCES

[A] Adams, D.R., 1981, Lectures on L^p-potential theory, University of Umeå, Department of Mathematics, Report No. 2.

[BP] Benedek, A., Panzone, R., 1961, The spaces L^p with mixed norm, Duke Math. J. 28, 301-324.

[Ch] Choquet, G., 1959, Forme abstraite du théorème de capacitabilitié. Ann. Inst. Fourier (Grenoble) 9, 83-89.

[F] Fuglede, B., 1968, Applications du théorème minimax à l'étude de diverse capacités, C.R. Acad. Sci. Paris Sèr, A-B 266, A921-A923.

[H] Hedberg, L.I., 1986, Non-linear potential theory and Sobolev spaces, University of Linköping, Department of Mathematics, Report No 10.

[Ma] Maz'ya, V.G., 1985, Sobolev spaces, Springer Series in Soviet Math., Springer Verlag, Berlin-Heidelberg-New York-Tokyo.

[Me] Meyers, N.G., 1970, A theory of capacities for potentials of functions in Lebesgue classes, Math. Scand. 26, 255-292.

[MH] Maz'ya, V.G., Havin, V.P., 1972, Non-linear potential theory, Usp. Mat. Nauk 27, 67-138; Russian Math. Surveys 27 (1972), 71-148.

[MS] Maz'ya, V.G., Shaposhnikova, T.O., 1985, Theory of multipliers in spaces of differentiable functions, Monographs and Studies in Mathematics, Vol. 23, Pitman Publ. Co., New York.

[S1] Sjödin, T., 1975, Bessel potentials and extension of continuous functions on compact sets, Ark. Mat. 13, 2, 263-271.

[S2] Sjödin, T., 1978, Capacities of compact sets in linear subspaces of $\mathbb{R}^n$, Pac. J. of Math. 78, 1, 261-266.

[S3] Sjödin, T., 1987, Non-linear potential theory in Lebesgue spaces with mixed norm and extension of continuous functions on compact sets, Department of Mathematics, University of Umeå, Report No 5.

[St] Stein, E.M., 1970, Singular integrals and differnetiability properties of functions, Princeton University Press, Princeton, New Jersey.

SCHROEDINGER EQUATIONS WITH DISCONTINUOUS SOLUTIONS

Theo Sturm

Mathematisches Institut
Universität Erlangen-Nürnberg
Federal Republic of Germany

O. INTRODUCTION

In this paper we study the Schrödinger equation

$$(-\frac{1}{2}\Delta + \mu)h = 0$$

on a (possibly unbounded) domain U of R^d ($d \geq 1$). μ will be a signed measure with $\mu^+ \in D^{loc}$ and $\mu^- \in D_1$. By definition, a measure $\nu \geq 0$ is in the class D^{loc} if it is associated with an increasing continuous additive functional A_ν of standard Brownian motion $(P^x, X_t : x \in R^d, t \geq 0)$ such that for some $t > 0$ and every compact set $C \subset R^d$

$$\sup_{x \in R^d} E^x \left[A_\nu(t) \cdot 1_C(X_t)\right] < \infty.$$

We write $\nu \in D$ if this is also true for $C = R^d$, i.e. if

$$\sup_{x \in R^d} E^x \left[A_\nu(t)\right] < \infty.$$

Finally we say that ν is in the class D_1 if for some $t > 0$

$$\sup_{x \in R^d} E^x \left[\exp(A_\nu(t))\right] < \infty.$$

Obviously, $D \supset D_1 \supset K$, where K is the (generalized) Kato class (cf. [1], [4]) consisting of those measures $\nu \geq 0$ for which

$$\lim_{t \to 0} \sup_{x \in R^d} E^x \left[A_\nu(t)\right] = 0.$$

The present article is divided in three sections. In the first, we give a short description of the classes D and D_1. The second section is devoted to the investigation of the Schrödinger operator $-\frac{1}{2}\Delta + \mu$ on U. We describe the spectrum, formulate an (inhomogeneous) Dirichlet problem and state necessary and sufficient conditions for existence, uniqueness and positiveness of solutions. In particular, several results of K.L.Chung [2], N.Falkner [3] and W.Hansen/H.Hueber [4] for bounded domains U and $\mu \in K - K$ (i.e. $\mu = \mu^+ - \mu^-$ with $\mu^+, \mu^- \in K$) are extended to arbitrary domains U and $\mu \in D^{loc} - D_1$.

The only additional condition is that the Green kernel for U associated with $-\frac{1}{2}\Delta + \mu^+$ be a compact operator on the set $B_b(R^d)$ of bounded Borel-measurable functions on R^d. The discussion of this assumption is deferred to the last section.

This research was supported by the German Science Foundation (Research Group "Stochastic Analysis and its Applications").

I. THE CLASSES D AND D_1

Let A be an arbitrary continuous additive functional and τ a stopping time. We define kernels P_τ^A and G_τ^A by

$$P_\tau^A f(x) = E^x \left[f(X_\tau) \cdot \exp(-A(\tau)) \cdot 1_{\{\tau<\infty\}}\right],$$
$$G_\tau^A f(x) = E^x \left[\int_0^\tau f(X_t) \cdot \exp(-A(t))\, dt\right]$$

($x \in R^d, f \in B_+(R^d)$).
If $A = A_\mu := A_{\mu^+} - A_{\mu^-}$ and $\tau = \tau(V) := \inf\{t > 0 : X_t \notin V\}$ for some open set V we simplify notation and write P_V^μ and G_V^μ. We also give a meaning to $G_V^\mu \phi$ if ϕ is a signed measure in $D - D$:

$$G_V^\mu \phi(x) = G_V^\mu \phi^+(x) - G_V^\mu \phi^-(x)$$

(provided the difference is not $\infty - \infty$), where

$$G_V^\mu \phi^\pm(x) = E^x \left[\int_0^{\tau(V)} \exp(-A_\mu(t))\, dA_{\phi^\pm}(t)\right].$$

By means of the following propositions we try to show that the classes D and D_1 are quiet natural for our purpose. Let the continuous additive functional A be increasing.

Proposition 1- For $t > 0$ the following statements are equivalent:

a) $\sup\limits_{x\in R^d} E^x[A(t)] < \infty$;

b) $\inf\limits_{x\in R^d} E^x[\exp(-A(t))] > 0$.

Proposition 2- The following statements are equivalent:

a) $\|P_t^{-A}\| := \sup\limits_{x\in R^d} E^x[\exp(A(t))] < \infty$ for some $t > 0$;

b) $\|P_t^{-A}\|_{p,q} < \infty$ for every $t > 0$ and every $p, q \in [1, \infty]$ such that $p \le q$;

c) $\sup\limits_{x,y\in R^d} P_{B_r(y)}^{-A} 1(x) < \infty$ for some $r > 0$;

d) $\|G_\infty^{\alpha-A}\| < \infty$ for some $\alpha \ge 0$.

Here $\|.\|_{p,q}$ is the norm of an operator from $L^p(R^d)$ to $L^q(R^d)$ and $\|.\| = \|.\|_{\infty,\infty}$. In d) we use the convention that a constant $\alpha \in R$ defines an additive functional $t \to \alpha \cdot t$ which we also denote by α.

In order to give some analytic characterization of the classes D and D_1, let ν be an arbitrary nonnegative measure. For $r > 0$ we define the following kernels from $R^d \times B_+(R^d)$ to $B_+(R^d)$:

$$G_r\nu: \quad (x,f) \longmapsto \int_0^r P_t(f\cdot\nu)(x)\,dt,$$

$$\text{where } P_t(f\cdot\nu)(x) = \int (2\pi t)^{-d/2}\cdot\exp(-\frac{|x-y|^2}{2t})\cdot f(y)\,\nu(dy),$$

$$G^{1/r}\nu: \quad (x,f) \longmapsto \int_0^\infty \exp(-t/r)\cdot P_t(f\cdot\nu)(x)\,dt,$$

$$G_{B_r}\nu: \quad (x,f) \longmapsto \int_{B_r(x)} [N(|x-y|) - N(r)]\cdot f(y)\,\nu(dy),$$

$$\text{where} \quad N(r) = \begin{cases} \frac{1}{2}\pi^{-d/2}\cdot\Gamma(\frac{d}{2}-1)\cdot r^{2-d}, & d\geq 3,\\ -\frac{1}{\pi}\cdot\ln r, & d=2,\\ -r, & d=1.\end{cases}$$

Proposition 3

a) $\nu \in D \iff \lim_{r\to 0} \|G_r\nu\| < \infty$.

b) $\lim_{r\to 0}\|G_r\nu\| = \lim_{r\to 0}\|G^{1/r}\nu\| = \lim_{r\to 0}\|G_{B_r}\nu\|$.

c) In a) and b) every $\lim_{r\to 0}$ can be replaced by $\inf_{r>0}$.

Proposition 4 - Assume that $\nu \in D$.

a) $\nu \in D_1 \iff \lim_{\substack{r\to 0\\ n\to\infty}} \|(G_r\nu)^n\|^{1/n} < 1$.

b) $\lim_{\substack{r\to 0\\ n\to\infty}} \|(G_r\nu)^n\|^{1/n} = \lim_{\substack{r\to 0\\ n\to\infty}} \|(G^{1/r}\nu)^n\|^{1/n} = \lim_{\substack{r\to 0\\ n\to\infty}} \|(G_{B_r}\nu)^n\|^{1/n}$.

c) In a) and b) every $\lim_{\substack{r\to 0\\ n\to\infty}}$ can be replaced by $\inf_{\substack{r>0\\ n\in N}}$.

II. THE SCHROEDINGER EQUATION

In the sequel we fix a domain U and a signed measure $\mu \in D^{loc} - D_1$. Throughout this section we assume that $G_U^{\mu^+}$ is a compact operator on $B_b(R^d)$. This assumption will be discussed in the next section. At this moment we mention only that it is always fulfilled if U is bounded. Our first aim is to give a reasonable formulation of the Dirichlet problem for the Schrödinger equation $(-\frac{1}{2}\Delta + \mu)h = 0$ on U. Since in general this equation may have not enough continuous solutions, one has to consider the finely continuous Borel-measurable solutions. Let $f \in B_b(R^d)$ and $\phi \in D - D$. We say that a function h on U is a solution of the Dirichlet problem (U, μ, f, ϕ) iff $h \in B_b(R^d)$, h is finely continuous on U,

$$(-\frac{1}{2}\Delta + \mu)h = \phi \qquad \text{in } U, \qquad \text{and}$$

$$\lim_{t\to\tau(U)} h(X_t) = f(X_{\tau(U)}) \qquad P^x\text{-}a.s.\ (\text{for } x \in U).$$

By $H_o^\mu(U)$ we denote the set of solutions of the Dirichlet problem $(U, \mu, 0, 0)$. The set

$$\text{spec}_U(-\frac{1}{2}\Delta + \mu) := \{\alpha \in R : H_o^{\mu-\alpha}(U) \neq \{0\}\}$$

is called the spectrum of the Schrödinger operator $-\frac{1}{2}\Delta + \mu$ on U.

Remark: It should be mentioned that under the additional assumption $\mu, \phi \in K^{loc} -$

K^{loc}, every solution of the above Dirichlet problem (U, μ, f, ϕ) is continuous in U. Furthermore we have

$$\lim_{x \to z} h(x) = f(z)$$

for every regular boundary point $z \in \partial U$ in which f is continuous.

Theorem 1

a) $\alpha(U,\mu) := \inf \operatorname{spec}_U(-\frac{1}{2}\Delta + \mu) \in R$;
$\alpha(U,\mu) = -\lim_{t\to\infty} \frac{1}{t} \cdot \ln \|P_t^{U,\mu}\| = \inf_{t>0} \frac{1}{t} \cdot \ln \|P_t^{U,\mu}\|$,
where $\|P_t^{U,\mu}\| = \sup_{x\in R^d} E^x \left[\exp(-A_\mu(t)) \cdot 1_{\{t<\tau(U)\}}\right]$.

b) $G_U^{\mu-\alpha}$ is a positivity preserving, compact operator on $B_b(R^d)$ for every $\alpha < \alpha(U,\mu)$ (with spectrum $\operatorname{spec}(G_U^{\mu-\alpha})$);
$\operatorname{spec}_U(-\frac{1}{2}\Delta + \mu) = \left[\operatorname{spec}(G_U^{\mu-\alpha}) \setminus \{0\}\right]^{-1} + \alpha$.

c) $\operatorname{spec}_U(-\frac{1}{2}\Delta + \mu)$ is a countable set without accumulation points. For every eigenvalue $\alpha \in \operatorname{spec}_U(-\frac{1}{2}\Delta + \mu)$ the eigenspace $H_o^{\mu-\alpha}(U)$ has finite dimension. For $\alpha = \alpha(U,\mu)$ there exists some $h_o > 0$ such that $H_o^{\mu-\alpha}(U) = R \cdot h_o$.

Theorem 2 - Assume $0 \notin \operatorname{spec}_U(-\frac{1}{2}\Delta + \mu)$.

a) For every $f \in B_b(R^d)$ and $\phi \in D - D$, there exists a unique solution $h = h(U,\mu,f,\phi)$ of the Dirichlet problem (U,μ,f,ϕ).

b) $\alpha(U,\mu) > 0$ if and only if, in the above situation, $f \geq 0$ and $\phi \geq 0$ imply $h \geq 0$. In this case we have the representation

$$h(U,\mu,f,\phi) = P_U^\mu f + G_U^\mu \phi$$

c) In the general case we have for every $\alpha < \alpha(U,\mu)$

$$\begin{aligned} h(U,\mu,f,\phi) &= \left(I + \alpha \cdot G_U^{\mu-\alpha}\right)^{-1} h(U,\mu-\alpha,f,\phi) = \\ &= \left(I + \alpha \cdot G_U^{\mu-\alpha}\right)^{-1} \left(P_U^{\mu-\alpha} f + G_U^{\mu-\alpha}\phi\right). \end{aligned}$$

Remark: It is a immediate consequence of the above theorem that $\alpha(U,\mu) > 0$ implies $\|P_U^\mu\| < \infty$ and $\|G_U^\mu\| < \infty$ (moreover, $\|G_U^\mu \phi\| < \infty$ for every $\phi \in D$).

Our next aim is to obtain sufficient conditions for $\alpha(U,\mu) > 0$. For this purpose we define the set of nonnegative μ-superharmonic functions on U

$$\begin{aligned} S_+^\mu(U) = \{&\text{finely continuous } h \in B_+(U) \setminus \{\infty\} : \\ &h \geq P_B^\mu h \quad \text{for every ball } B \text{ such that } \bar{B} \subset U \text{ and } \|P_B^\mu\| < \infty\}. \end{aligned}$$

Theorem 3

a) $\alpha(U,\mu) < 0 \iff S_+^\mu(U) = \{0\}$;

b) $\alpha(U,\mu) = 0 \iff S_+^\mu(U) = R_+ \cdot h_o, \quad H_o^\mu(U) = R \cdot h_o$ with some $h_o > 0$;

c) $\alpha(U,\mu) > 0 \iff S_+^\mu(U) \not\subset H_o^\mu(U)$.

Corollary - Each of the following conditions implies that $\alpha(U,\mu) > 0$.

a) $0 \not\equiv P_U^\mu f \not\equiv \infty$ in U for some $f \in B_+(R^d)$.

b) $0 \not\equiv G_U^\mu \phi \not\equiv \infty$ in U for some $\phi \in D$.

c) $0 < s(x) \cdot E^{x,s}\left[\exp(-A_\mu(\tau(U))) \cdot 1_{\{\tau(U)<\infty\}}\right] < \infty$ for some $x \in U$ and $s \in S_+^0(U)$. Here $E^{x,s}$ and $\tau(U)$ denote the expectation and lifetime of Brownian motion starting in x, killed at the first exit of U, and conditioned by s.

III. COMPACTNESS

Let U be a domain and $\nu \in D^{loc}$. We give criteria for G_U^ν to be a compact operator on $B_b(R^d)$.

Theorem 4. The following statements are equivalent:

a) G_U^ν is compact;

b) $G_U^{\nu+\alpha}$ is compact for some $\alpha > 0$;

c) $\lim_{x\to\infty} G_U^\nu 1(x) = 0$.

Corollary. G_U^ν is compact under each of the following conditions:

a) $\nu \geq v \cdot \lambda^d$ with $\lim_{x\to\infty} v(x) = \infty$ (λ^d = Lebesgue measure on R^d);

b) G_U^0 is compact.

A domain U with compact Green kernel G_U^0 is called Green-compact ([5]). Due to Theorem 4, U is Green-compact iff $\lim_{x\to\infty} E^x[\tau(U)] = 0$. Obviously every bounded domain is Green-compact but, for example, every domain with finite Lebesgue measure is as well. The next theorem gives another characterization of Green- compactness.

Theorem 5. The following statements are equivalent:

a) $\lim_{x\to\infty} E^x[\tau(U)] = 0$;

b) $\lim_{x\to\infty} R_U^\delta(x) = 0$ for some $\delta > 0$,

where $R_U^\delta(x) := \inf\left\{r > 0 : \dfrac{\operatorname{cap}(B_r(x)\setminus U, B_{2r}(x))}{\operatorname{cap}(B_r(x), B_{2r}(x))} > \delta\right\}$.

REFERENCES

[1] M.Aizenman, B.Simon: Brownian motion and Harnack inequality for Schrödinger operators. Comm. Pure Appl. Math. **35** (1982), 209-273.

[2] K.L.Chung: Notes on the inhomogeneous Schrödinger equation. Seminar on Stochastic Processes **1984**, ed. by E.Cinlar, K.L.Chung, and R.K.Getoor. Birkhäuser, Boston (1986), 1-29.

[3] N.Falkner: Feynman-Kac functional and positive solutions of $\frac{1}{2}\Delta u + qu = 0$. Z. Wahrsch. Verw. Gebiete **65** (1983), 55-62.

[4] W.Hansen, H.Hueber: Eigenvalues in potential theory. Preprint, Bielefeld (1986).

[5] T.Sturm: On the Dirichlet-Poisson problem for Schrödinger operators. C.R. Math. Rep. Acad. Sci. Canada **IX** (1987), 149-154.

POLYNOMIAL HULLS AND ENVELOPES OF HOLOMORPHY

John Wermer

Department of Mathematics
Brown University
Providence, R.I. 02912, USA

This is a report on recent joint work with Herbert Alexander.

Let X be a compact set in $\mathbb{C}^2$. The <u>polynomial hull</u> of X denoted $\hat{X}$ is the set consisting of all points (λ_0, w_0) in $\mathbb{C}^2$ such that for every polynomial Q on $\mathbb{C}^2$

$|Q(\lambda_0, w_0)| \leq \max |Q|$ taken over the set X.

Let π denote the coordinate projection: $(\lambda, w) \to \lambda$ and if S is a subset of the complex plane, write

$$\pi^{-1}(S) = \{(\lambda, w) \in \hat{X} \mid \lambda \in S\}.$$

We fix a point (λ_0, w_0) in X such that $\pi(\lambda_0, w_0) \in \mathbb{C} \setminus \pi(X)$. Let Δ be a closed disk contained in $\mathbb{C} \setminus \pi(X)$ with center at λ_0. In view of Rossi's Local Maximum Modulus Principle (see e.g. [1], Chapter 9), it follows that $\pi^{-1}(\Delta)$ is the polynomial hull of $\pi^{-1}(\partial\Delta)$, where ∂ denotes "boundary". Also $\pi^{-1}(\Delta)$ is a neighborhood of (λ_0, w_0) in $\hat{X}$. Thus locally near (λ_0, w_0) $\hat{X}$ is the hull of $\pi^{-1}(\partial\Delta)$. The set $\pi^{-1}(\partial\Delta)$ lies over the circle $\partial\Delta$. We are thus led to the following

<u>Problem</u>: <u>Given a closed disk Δ in $\mathbb{C}$ and given a compact set Y in $\mathbb{C}^2$ which is over the circle $\partial\Delta$, i.e. such that $\pi(Y) \subseteq \partial\Delta$, describe the polynomial hull of Y.</u>

We shall assume without loss of generality that Δ is the closed unit disk: $|\lambda| \leq 1$. We put $\Gamma = \partial\Delta = \{\lambda \mid |\lambda| = 1\}$.

Example: We choose a continuous complex-valued function f on Γ and we put Y_f equal to the graph of f over Γ. If f is the boundary function of a function f analytic on $|\lambda| < 1$, then $\hat{Y}_f$ is the graph of f over $|\lambda| \leq 1$, while if f is not such a boundary function, then $\hat{Y}_f = Y_f$.

Let now Y be an arbitrary compact set in $\mathbb{C}^2$ lying over Γ. Without loss of generality, we assume that Y is contained in the set $\{(\lambda,w) \mid |w| < 1\}$.

We denote by Ω_0 the product domain: $|\lambda| < 1$, $|w| > 1$. If f is a function analytic and bounded on Ω_0, we define $f(\lambda,w)$ for a.a. λ on $|\lambda| = 1$ by $f(\lambda,w) = \lim f(t\lambda)$ as $t \to 1-$.

Definition: $\mathcal{F}(Y)$ is the space of all functions F analytic and bounded on Ω_0 such that for a.a. λ in Γ $F(\lambda,w)$ admits the representation

$$F(\lambda,w) = \int_{Y_\lambda} \frac{d\sigma_\lambda(\xi)}{w - \xi}, \quad |w| > 1, \tag{1}$$

where σ_λ is a probability measure on $Y_\lambda = \{w \mid (\lambda,w) \in Y\}$.

We note that for fixed λ in Γ $F(\lambda,\cdot)$ has an analytic extension, given by (1), to $\mathbb{C} \setminus Y_\lambda$.

Theorem 1: *Let Y be a compact set in $\mathbb{C}^2$ lying over Γ. Fix (λ_0,w_0) in $\mathbb{C}^2$ with $|\lambda_0| < 1$. Then (λ_0,w_0) lies in $\hat{Y}$ if and only if there exists a function F in $\mathcal{F}(Y)$ such that*

$$F(\lambda_0,w) = (w - w_0)^{-1}, \quad |w| > 1.$$

We shall now test Theorem 1 on the set $Y = Y_f$ of the example above, where f is a function continuous on Γ and analytic on $|\lambda| < 1$. We choose (λ_0,w_0) in Y_f with $|\lambda_0| < 1$. As we said above, then, $w_0 = f(\lambda_0)$. We now put

$$F(\lambda,w) = (w - f(\lambda))^{-1}, \quad (\lambda,w) \in \Omega_0.$$

One verifies at once then that $F \in \mathcal{F}(Y)$ and that we may take σ_λ to be the unit point mass at the point $f(\lambda)$ for λ in Γ and obtain the representation (1). So F satisfies the assertion of the Theorem.

We know of no necessary and sufficient *geometric* condition on the point (λ_0,w_0) stated in the space $\mathbb{C}^2$ in order that

(λ_0, w_0) belong to the polynomial hull $\hat{Y}$. However, if we are willing to go into $\mathbb{C}^n$ for $n \geqq 2$, then we can obtain such a geometric statement, in the following manner.

Let Y be a compact set in $\mathbb{C}^2$ lying over Γ. If $\xi \in \mathbb{C}$, we denote by V_ξ the vector $(\xi, \xi^2, \ldots, \xi^n)$ in $\mathbb{C}^n$. We write

$K(n,\lambda)$ = the closed convex hull in $\mathbb{C}^n$ of the set of all vectors V_ξ with $\xi \in Y_\lambda$.

<u>Theorem 2</u>: <u>Fix</u> (λ_0, w_0) <u>in</u> $\mathbb{C}^2$, $|\lambda_0| < 1$. <u>Then</u> $(\lambda_0, w_0) \in \hat{Y}$ <u>if and only if for each positive integer</u> n <u>there exists an analytic and bounded graph</u>: $W = \Phi(\lambda)$, $|\lambda| < 1$, <u>in</u> $\mathbb{C}^{n+1}$, <u>where</u> $W = (w_1, \ldots, w_n)$, $\Phi = (\Phi_1, \ldots, \Phi_n)$, <u>such that</u>

(2) <u>the graph passes through the point</u> $(\lambda_0, w_0, w_0^2, \ldots, w_0^n)$, <u>i.e.</u> $\Phi(\lambda_0) = (w_0, w_0^2, \ldots, w_0^n)$, <u>and</u>

(3) <u>for</u> a.a. λ <u>in</u> Γ, $\Phi(\lambda) \in K(n,\lambda)$.

$\mathcal{F}(Y)$ is a space of analytic functions on Ω_0. By an <u>envelope of holomorphy</u> of $\mathcal{F}(Y)$ we mean a domain Ω in $\mathbb{C}^2$ with $\Omega_0 \subset \Omega \subset \{|\lambda| < 1\}$ such that each F in $\mathcal{F}(Y)$ can be analytically continued as a single valued function from Ω_0 to Ω, and that Ω is maximal with this property.

<u>Theorem 3</u>: <u>The domain</u> $\Omega = \{(\lambda, w) \mid |\lambda| < 1\} \setminus \hat{Y}$ <u>is an envelope of holomorphy for</u> $\mathcal{F}(Y)$.

In view of Theorem 1 it is clear that no larger domain than Ω admits analytic continuation for all F in $\mathcal{F}(Y)$. What has to be shown is that each F in $\mathcal{F}(Y)$ can be given a single-valued analytic continuation from Ω_0 to Ω.

<u>Note</u>: It is of course possible that $\hat{Y} = Y$, i.e. that Y is <u>polynomially convex</u> in $\mathbb{C}^2$. We have the following: Y is <u>polynomially convex if and only if the space</u> $\mathcal{F}(Y)$ <u>is empty</u>.

Theorems 1 and 2 are proved by the authors in [2] and Theorem 3 is proved in [3]. Related results were proved earlier by the authors in [4], and more recent related results are in [5], [6].

The key fact that, if Y is a compact set in $\mathbb{C}^2$ lying over Γ, then the domain $\{|\lambda| < 1\} \setminus \hat{Y}$ is pseudoconvex was proved by Slodkowski, as part of his general theory of <u>analytic multivalued functions</u>. See [7] and [8] and [9].

REFERENCES

1. J. Wermer, "Banach Algebras and Several Complex Variables", Springer-Verlag (1976).
2. J. Wermer, Balayage and Polynomial Hulls, in: Proceedings of the Special Year in Complex Analysis, University of Maryland, 1985/86, Springer-Verlag, Lecture Notes in Math. 1276 (1987).
3. H. Alexander and J. Wermer, Envelopes of Holomorphy and Polynomial Hulls, to appear in Math. Ann.
4. H. Alexander and J. Wermer, Polynomial hulls with convex fibers, Math. Ann. 271 (1985).
5. F. Forstneric, Polynomially convex hulls with piecewise smooth boundaries, Math. Ann. 276 (1986).
6. F. Forstneric, Polynomial Hulls of Sets Fibered over the Circle (to appear).
7. Z. Slodkowski, Analytic Set-Valued Functions and Spectra, Math. Ann. 256 (1981).
8. Z. Slodkowski, Uniform Algebras and Analytic Multifunctions, Atti della Accad. Nazion. dei Lincei, 75 (1983).
9. Z. Slodkowski, Polynomial hulls with convex sections and interpolating spaces, Proc. AMS 96 (1986).

GREEN FUNCTION AND UNIFORM HARNACK INEQUALITY

Rainer Wittmann

Katholische Universität Eichstätt
Mathematisch-Geographische Fakultät

Abstract. We show that any harmonic space which possesses a Green function with certain properties satisfies the uniform inequality. Also a lower bound for the harmonic measure of balls is given.

1. INTRODUCTION

In the sequel $(X, \mathcal{H}^*)$ will be a $\mathcal{P}$-harmonic space in the sense of [**2**] and d will be a metric on X. For the sake of simplicity we assume that 1 is harmonic on X. We denote by $\mathcal{H}(U)$ (resp. $\mathcal{H}^*(U)$) the set of harmonic (resp. hyperharmonic) functions on U. We denote by $\mathcal{H}_+(U)$ and $\mathcal{H}^*_+(U)$ the subsets of nonnegative elements. We also define

$$B(x,r) := \{y \in X : d(x,y) < r\}, \quad \overline{B}(x,r) := \{y \in X : d(x,y) \leq r\}$$

We say that $(X, \mathcal{H}^*)$ satisfies the *uniform Harnack inequality* if there exists $C_H, R_H > 0$ such that for any $x \in X$ and any $0 < r \leq R_H$

$$u(y_1) \leq C_H u(y_2) \quad (u \in \mathcal{H}_+(B(x,2r)),\ y_1, y_2 \in \overline{B}(x,r))$$

provided that $\overline{B}(x,2r)$ is compact. The uniform Harnack inequality turned out to be crucial in proving non–tangential limit theorems, identity of Martin boundary and topological boundary, boundary Harnack principle etc. (cf. Wittmann [**10**],[**11**],[**12**],[**13**]) in the setting of harmonic spaces. Though the uniform Harnack inequality is well known for nearly all classes of elliptic second order differential equations, sometimes only information about the Green function is known. This was the case with the very important class of hypoelliptic operators of Hörmander type when we wrote our paper [**11**]. Meanwhile we learned that a much more general result has been proved by Franchi,Laconelli [**4**] using the Moser iteration process. Nevertheless it seems worthwile to establish a theorem which relates a certain (weak) estimate for the Green function with the uniform Harnack inequality. This is the principal aim of this paper. The uniform Harnack inequality for the Hörmander class is then an immediate consequence of Sanchez-Calle [**7**],Theorem I. In some sense our main result can be regarded as a quantitive variant of similar qualitative results in Boukricha [**1**] and Schirmeier [**8**]. Also our method of proof is inspired by these papers. By a comparison argument a lower bound for the harmonic measures of balls which was used in [**11**], [**12**] is given in sect. 3.

We use a very weak notion of Green function which was first used by Schirmeier [**8**] :

1.1. $g : X \times X \longrightarrow \overline{\mathbf{R}}_+$ is called a *weak Green function* on $(X, \mathcal{H}^*)$ if there exists a resolvent $(V_\alpha)_{\alpha>0}$ of submarkovian kernels on X and a σ–finite measure μ on X such that

(i) $\mathcal{H}_+^*(X)$ coincides with the cone of excessive functions of (V_α)

(ii) For any $y \in X$ the function $g_y := g(\cdot, y)$ is a potential with carrier $\{y\}$

(iii) $$\int g(x,y)f(y)\,\mu(dy) = Vf(x) := \sup_{\alpha>0} V_\alpha f(x) \quad (f \in M_+(X)^1)$$

If the subsequent condition 1.2 is satisfied then, by [8],Satz 2, any Green function g is already a Green function in the usual sense. We often use the following consequence of (ii) and the harmonicity of 1 :

$$\sup\{g(x,y) : d(x,y) \geq r\} \leq \sup\{g(x,y) : x \in \partial B(y,r)\} \quad (y \in X)$$
$$g(z,y) \geq \inf\{g(x,y) : x \in \partial B(y,r)\} \quad (y \in X,\ z \in B(x,r))$$

The following condition on a weak Green function g will be essential :
1.2. There exist $R_g > 0$, $\alpha_g, C_g > 1$ such that for any $0 < r \leq R_g$ and $y \in X$ we have

$$\begin{aligned}(1 - C_g^{-1})^{-1} \sup\{g(x,y) : x \in \partial B(y, \alpha_g r)\} &\leq \inf\{g(x,y) : x \in \partial B(y,r)\} \\ \leq \sup\{g(x,y) : x \in \partial B(y,r)\} &\leq C_g \inf\{g(x,y) : x \in \partial B(y, \alpha_g r)\}\end{aligned}$$

provided $\overline{B}(y, \alpha_g r)$ is compact. Compared with our knowledge about the Green functions of elliptic equations this condition is rather weak.

2. Uniform Harnack inequality

2.1. THEOREM. *Let g be a weak Green function such that condition 1.2 is satisfied. Then there exists $C_H, R_H > 0, \alpha_H > 1$ depending only on C_g, α_g, R_g such that for any $x \in X$ and $0 < r \leq R_H$ we have*

$$u(y_1) \leq C_H u(y_2) \quad (u \in \mathcal{H}_+(B(x, \alpha_H r)),\ y_1, y_2 \in \overline{B}(x,r))$$

provided $\overline{B}(x, \alpha_H r)$ is compact.

PROOF: Let

$$m := \inf\{n \in \mathbf{N} : (1 - C_g^{-1})^n \leq \tfrac{1}{3},\ \frac{\alpha_g^m + 1}{\alpha_g^m - 1} < \alpha_g\}$$
$$\alpha_H := 6\alpha_g^{2m}$$
$$R_H := \alpha_H^{-1} R_g$$

Let now $x \in X$ and $0 < r \leq R_H$, $h \in \mathcal{H}_+(B(x, \alpha_H r))$ be given such that $\overline{B}(x, \alpha_H r)$ is compact. For the sake of simplicity we denote $U := B(x, \alpha_H r)$.

For any $y_1, y_2 \in \overline{B}(x,r)$ and any $z \in X \setminus B(x, \alpha_g^m r)$ we have

$$(\alpha_g^m - 1)r \leq d(y_1, z), d(y_2, z) \leq (\alpha_g^m + 1)r \leq \alpha_g(\alpha_g^m - 1)r$$

Hence, by 1.2,

$$\begin{aligned}C_g^{-1} \sup\{g_z(w) : w \in \partial B(z, (\alpha_g^m - 1)r)\} &\leq \inf\{g_z(w) : w \in \partial B(z, \alpha_g(\alpha_g^m - 1)r)\} \\ \leq g_z(y_1), g_z(y_2) &\leq \sup\{g_z(w) : w \in \partial B(z, (\alpha_g^m - 1)r)\}\end{aligned}$$

Thus we obtain

(1) $$g_z(y_1) \leq C_g g_z(y_2) \quad (y_1, y_2 \in \overline{B}(x,r),\ z \in X \setminus B(x, \alpha_g^m r))$$

---

[1] $M_+(X)$ denotes the cone of positive measurable functions on X

Next we want to localize the above "Harnack inequality" for the Green function. To this end we recall that

$$g_U(y,z) := g(y,z) - R_{g_z}^{X\setminus U}(y)$$

is a weak Green function of the harmonic space induced on U (see Schirmeier [8] for details). For $y, z \in B(x, 2\alpha_g^m r)$, $w \in X \setminus U$ we have $d(y,z) \leq 4\alpha_g^m r$ and $6\alpha_g^{2m} r \leq d(x,w)$ whence

$$d(z,w) \geq (6\alpha_g^{2m} - 2\alpha_g^m)r \geq \alpha_g^m d(y,z)$$

Applying 1.2 m times we obtain

$$\sup\{g_z(w) : w \in X \setminus U\} \leq (1 - C_g^{-1})^m g_z(y) \leq \frac{1}{3} g_z(y) \quad (y, z \in B(x, 2\alpha_g^m r))$$

Since 1 is harmonic this implies

(2) $$g_U(y,z) \geq \frac{2}{3} g_U(y,z) \quad (y, z \in B(x, 2\alpha_g^m r))$$

Together with (1) we obtain

(3) $$g_U(y_1,z) \leq \frac{3}{2} C_g g_U(y_2,z) \quad (y_1, y_2 \in \overline{B}(x,r),\ z \in A := \overline{B}(x, 2\alpha_g^m r) \setminus \overline{B}(x, \alpha_g^m r))$$

Now we need the balayage operation relative to U :

$$^U R_s^A := \inf\{t \in \mathcal{H}_+^*(U) : I_A s \leq t\} \quad (s \in \mathcal{H}_+^*(U))$$

Let $u \in \mathcal{H}_+(U)$ be given. Since g_U is a weak Green function Theorem 6.6 in Hunt [6] (see Hansen [5], Wittmann [9],p. 69, Cornea,Wittmann [3] for analytic proofs) yields a sequence (f_n) of positive measurable functions such that

(4) $$\{f_n \neq 0\} \subset A$$

(5) $$^U R_u^A(y) = \sup_{n \in \mathbb{N}} \int g_U(y,z) f_n(z)\, \mu(dz)$$

Since $^U R_u^A = u$ on $\overline{B}(x,r)$ we can combine (3), (4) and (5) to obtain

$$u(y_1) \leq \frac{3}{2} C_g u(y_2) \quad (y_1, y_2 \in \overline{B}(x,r))$$

Setting $C_H := \frac{3}{2} C_g$ the assertion follows.

If we would have $\alpha_H \leq 2$ then the above Theorem would imply the uniform Harnack inequality. By slightly altering the notion of Harnack chain one could save the results of [**10**],[**11**],[**12**],[**13**] if only the above slightly weaker condition is satisfied. On the other hand the subsequent corollary shows that under a natural condition on d the full uniform Harnack inequality follows from the weaker statement.

2.2. COROLLARY. *Assume that there exists $C_d < \frac{2}{3}$, $R_d > 0$ such that for any $x, y \in X$ with $d(x,y) \leq R_d$ and $\overline{B}(x, 2d(x,y))$ compact there exists $z \in X$ with $d(x,z), d(y,z) \leq C_d d(x,y)$.*

If there exists a weak Green function satisfying 1.2 then $(X, \mathcal{H}^*)$ satisfies the uniform Harnack inequality with constants C_H, R_H depending only on $C_g, \alpha_g, R_g, C_d, R_d$.

PROOF: Let $x, y \in X$ with $d(x,y) \leq R_d$ and $\overline{B}(x, 2d(x,y))$ compact. We want to construct inductively $(y_{ni})_{0 \leq i \leq 2^n}$ such that

$$(1) \qquad y_{n,0} = x,\ y_{n,2^n} = y$$
$$(2) \qquad d(y_{n,i-1}, y_{n,i}) \le C_d^n d(x,y) \quad (1 \le i \le 2^n)$$
$$(3) \qquad d(x, y_{ni}) \le \sum_{j=1}^{n} C_d^j d(x,y) \quad (0 \le i < 2^n)$$

For $n = 0$ we simply choose $y_{0,0} := x$, $y_{0,1} := y$ and the above conditions are obviously satisfied. Assume now that $(y_{n-1,i})_{0 \le i \le 2^{n-1}}$ has already been constructed. Then for any $1 \le i \le 2^{n-1}$ there exists $y_{n,2i-1} \in X$ with

$$d(y_{n-1,i-1}, y_{n,2i-1}), d(y_{n-1,i}, y_{n,2i-1}) \le C_d d(y_{n-1,i-1}, y_{n-1,i})$$

Setting $y_{n,2i} := y_{n-1,i}$ for any $0 \le i \le 2^{n-1}$ we see that (1) and (2) are satisfied. Condition (3) follows from

$$\begin{aligned} d(x, y_{n,2i-1}) &\le d(y_{n,2i-1}, y_{n-1,i-1}) + d(y_{n-1,i-1}, x) \\ &\le C_d d(y_{n-1,i-1}, y_{n-1,i}) + \sum_{j=1}^{n-1} C_d^j d(x,y) \\ &\le \sum_{j=1}^{n} C_d^j d(x,y) \end{aligned}$$

By Theorem 2.1 there exists $\tilde{C}_H, R_H > 0, \alpha_H > 1$ depending only on C_g, α_g, R_g such that for any $x \in X$ and $0 < r \le R_H$ we have

$$(4) \qquad u(y_1) \le \tilde{C}_H u(y_2) \quad (u \in \mathcal{H}_+(B(x, \alpha_H r)),\ y_1, y_2 \in \overline{B}(x,r))$$

provided $\overline{B}(x, \alpha_H r)$ is compact. If $\alpha_H \le 2$ then the assertion is trivial. Thus we may assume that $\alpha_H > 2$. Let $K := \sum_{j=1}^{\infty} C_d^j < 2$ (since $C_d < \frac{2}{3}$) and

$$m := \inf\{n \in \mathbf{N} : \alpha_H C_d^n < 2 - K\}$$

Let now $x \in X$, $0 < r \le R_H$, $u \in \mathcal{H}_+(B(x,2r))$, $y \in \overline{B}(x,r)$ be given such that $\overline{B}(x,2r)$ is compact. As shown above there exists $(y_{mi})_{0 \le i \le 2^m}$ such that (1), (2) and (3) are satisfied. From (2) and (3) we get for any $1 \le i \le 2^m$

$$\begin{aligned} &y_{m,i-1} \in \overline{B}(y_{mi}, C_d^m r) \\ &B(y_{mi}, \alpha_H C_d^m r) \subset B(x, 2r) \end{aligned}$$

Hence, by (4),

$$\tilde{C}_H^{-1} u(y_{mi}) \le u(y_{m,i-1}) \le \tilde{C}_H u(y_{mi}) \quad (1 \le i \le 2^m)$$

Using (1) this implies

$$(5) \qquad \tilde{C}_H^{-m} u(x) \le u(y) \le \tilde{C}_H^{m} u(x)$$

Finally if $y_1, y_2 \in \overline{B}(x,r)$ then (5) yields

$$\tilde{C}_H^{-2m} u(y_2) \le u(y_1) \le \tilde{C}_H^{2m} u(y_2)$$

Choosing $C_H := \tilde{C}_H^{2m}$ the assertion follows.

REMARKS: (a) By the triangle inequality we have always $C_d \ge \frac{1}{2}$. One may choose $C_d = \frac{1}{2} + \varepsilon$ when d is defined by the minimal length (with repect to a Riemannian metric or other kinds of "local" metrics) of connecting arcs.

(b) In the above proofs we have only used that 1 is harmonic in a neighbourhood of $\overline{U}$ where U is a certain ball. Assume that 1 is no more harmonic and $W, W' \subset X$ are open sets with $W \subset \overline{W} \subset W'$. If there exists also a positive harmonic function h on W' such that

$$\inf_{x \in W'} h(x) > 0, \quad \sup_{x \in W'} h(x) < \infty$$

then the above statements remain true for the harmonic structure induced on W. Of course the constants then depend also on h and W, W'. To translate the above proofs one has to consider the sheaf

$$\tilde{\mathcal{H}}^*(U) := \{\frac{u}{h} : u \in \mathcal{H}^*(U)\} \quad (U \subset W \text{ open})$$

3. Lower bounds for the harmonic measure of balls

In the sequel we write μ_x^r instead of $\mu_x^{B(x,r)}$.

3.1. PROPOSITION. *Let $R_d > 0$ be such that the following two conditions are satisfied.*

(i) *For any $x \in X$, $0 < r \leq R_d$, $y \in B(x,r)$ there exists $z \in X$ with $d(x,z), d(y,z) \leq r/2$ provided $\overline{B}(x,r)$ is compact.*

(ii) *For any $x \in X$, $0 < r \leq R_d$, $y \in X \setminus \overline{B}(x,r)$ there exists $w \in X \setminus B(x,2r)$ with $d(y,w) \leq d(x,y)$ provided $\overline{B}(x,3r)$ is compact.*

*If there exists a weak Green function satisfying 1.2 then $(X, \mathcal{H}^*)$ satisfies the condition (HB) of [**11**],[**12**], i. e. there exists $R_H > 0$ and constants $C_\beta > 0$, $\beta > 0$ such that for any $x \in X$, $0 < r \leq R_H$, $y \in \partial B(x,r)$ we have*

$$\mu_x^r(B(y, \beta r)) \geq C_\beta \quad (\beta > 0)$$

provided $\overline{B}(x,4r)$ is compact.

We separate the rather technical "metric" details, which are trivial for the Euclidean metric, from the above proof :

3.2. LEMMA. *(a) If 3.1(i) is satisfied then for any $x \in X$, $0 < r \leq R_d$, $y \in B(x,r)$ such that $\overline{B}(x,2r)$ is compact, there exists $\omega : [0,1] \longrightarrow \overline{B}(x,r)$ with*

(i) $$d(\omega(s), \omega(t)) = |s-t| d(x,y)$$

(ii) $$\omega(0) = x, \quad \omega(1) = y$$

(b) If also 3.1(ii) is satisfied then for any $x \in X$, $0 < r \leq R_d$, $y \in B(x,r)$ there exists $\omega : [0,2] \longrightarrow \overline{B}(x,2r)$ such that (i), (ii) hold provided $\overline{B}(x,3r)$ is compact.

PROOF: For the proof of (a) assume that $x \in X$, $0 < r \leq R_d$, $y \in B(x,r)$ with $\overline{B}(x,2r)$ compact is given. We define $\omega(0) := x$ and $\omega(1) := y$. Assume ω has already been defined on $D_n := \{i2^{-n} : 0 \leq i \leq 2^n\}$ such that (i) holds for all $s,t \in D_n$. Then for any $0 \leq i < 2^n$ and sufficiently small $\varepsilon > 0$ the ball $\overline{B}(\omega(i2^{-n}), 2(2^{-n}+\varepsilon)d(x,y))$ is compact and $\omega((i+1)2^{-n}) \in B(\omega(i2^{-n}), (2^{-n}+\varepsilon)d(x,y))$. By our assumption there exists $z_\varepsilon \in X$ with $d(z_\varepsilon, \omega(i2^{-n})), d(z_\varepsilon, \omega((i+1)2^{-n})) \leq (2^{-n-1}+\frac{\varepsilon}{2})d(x,y) = \frac{1}{2}(d(\omega(i2^{-n}), \omega((i+1)2^{-n}))+\varepsilon)$. Let now $\omega((2i+1)2^{-n-1})$ be a limit point of z_ε. Then $d(\omega(i2^{-n}), \omega((2i+1)2^{-n})), d(\omega((i+1)2^{-n}), \omega((2i+1)2^{-n})) = 2^{-n-1}d(x,y)$. Thus we can define ω also on D_{n+1} such that (i) is satisfied. Consequently we can define ω on $\bigcup_{n=0}^\infty D_n$ and (i) is still satisfied. Assertion (a) then follows by continuous extension.

For the proof of (b) assume that $x \in X$, $0 < r \leq R_d$, $y \in \overline{B}(x,r)$ with $\overline{B}(x,3r)$ compact is given. Replacing r by $d(x,y)-\varepsilon$ $(0 < \varepsilon < r)$ the assumption of (b) yields $w_\varepsilon \in X \setminus \overline{B}(x, 2(d(x,y)-\varepsilon))$ with $d(y, w_\varepsilon) \leq d(x,y)$. Let now w be a limit point of $(w_\varepsilon)_{\varepsilon>0}$. Then we have $2d(x,y) = d(x,w) = 2d(y,w)$. By (a) there exist $\omega_1 : [0,1] \longrightarrow \overline{B}(x,r)$ and $\omega_2 : [0,1] \longrightarrow \overline{B}(y,r)$ with

$$d(\omega_i(s), \omega_i(t)) = |s-t| d(x,y) \quad (i = 1,2,\ s,t \in [0,1])$$
$$\omega_1(0) = x, \quad \omega_1(1) = y = \omega_2(0), \quad \omega_2(1) = w$$

Since $d(x,w) = d(x,y) + d(y,w)$ the function ω defined by

$$\omega(t) := \begin{cases} \omega_1(t) & \text{if } t \in [0,1] \\ \omega_2(2t-1) & \text{if } t \in [1,2] \end{cases}$$

satisfies the required properties. ∎

PROOF OF PROPOSITION 3.1: Let R_H be the minimum of $R_d/3$, R_g and the constant R_H of Corollary 2.2 (the assumptions of this corollary are fulfilled by (i)). It is sufficient to prove the assertion for $0 < \beta < 1$. To this end let $x \in X$, $0 < r \leq R_H$, $y \in \partial B(x,r)$ with $\overline{B}(x,4r)$ compact be given. By Lemma 3.2(b) there exists $\omega : [0,2] \longrightarrow \overline{B}(x,2r)$ with

(1) $$d(\omega(s), \omega(t)) = |s-t| d(x,y)$$

(2) $$\omega(0) = x, \quad \omega(1) = y$$

By (1), (2) and the triangle inequality we have

(3) $$d(\omega(t), \partial B(x,r)) = d(\omega(t), y) = |t-1| r \quad (t \in [0,2])$$

Setting $\tilde{y} := \omega(1 + \frac{\beta}{2\alpha_g^2})$, $z := \omega(1 + \frac{\beta}{2\alpha_g^2} - \frac{\beta}{2\alpha_g})$ we get from (3) that

(4) $$d(\tilde{y}, \overline{B}(x,r)) = d(\tilde{y}, y) = \frac{\beta}{2\alpha_g^2} r$$

(5) $$d(z, \tilde{y}) = \frac{\beta}{2\alpha_g} r$$

(6) $$d(z, X \setminus B(x,r)) = (\frac{\beta}{2\alpha_g} - \frac{\beta}{2\alpha_g^2}) r$$

From (4) we get

(7) $$d(\tilde{y}, \overline{B}(x,r) \setminus B(y, \beta r)) \geq \frac{\beta r}{2}$$

Let u be the restriction of the function

$$(\sup\{g_{\tilde{y}}(w) : w \in \partial B(\tilde{y}, \frac{\beta r}{2\alpha_g^2})\})^{-1} (g_{\tilde{y}} - \sup\{g_{\tilde{y}}(w) : d(\tilde{y}, w) \geq \frac{\beta r}{2}\})$$

to $B(x,r)$. By (4) and (7) we have

(8) $$\limsup_{w \to \overline{y}} u(w) \leq 1 \quad (\overline{y} \in \partial B(x,r))$$

(9) $$\limsup_{w \to \overline{y}} u(w) \leq 0 \quad (\overline{y} \in \partial B(x,r) \setminus B(y, \beta r))$$

From 1.2 and (5) we get

(10) $$u(z) \geq C_g^{-2}$$

Let now

$$h(w) := \mu_w^{B(x,r)}(B(y, \beta r))$$

By (8), (9), (10) and the maximum principle we have

$$h(z) \geq C_g^{-2} \tag{11}$$

Let

$$m := \inf\{n \in \mathbf{N} : \frac{1}{n} < \frac{1}{2}(\frac{\beta}{2\alpha_g} - \frac{\beta}{2\alpha_g^2})\}$$

$$z_i := \omega(\frac{i}{m}(1 + \frac{\beta}{2\alpha_g^2} - \frac{\beta}{2\alpha_g})) \quad (0 \leq i \leq m)$$

By (3) we have

$$d(z_{i-1}, z_i) < \frac{1}{2} d(z_i, X \setminus B(x,r))$$

$$z_0 = x, \quad z_n = z$$

In other words $(z_i)_{0 \leq i \leq m}$ is Harnack chain connecting x and z. Hence the uniform Harnack inequality of Corollary 2.2 implies

$$\mu_x^r(B(y, \beta r)) = h(x) \geq C_H^{-m} h(z) \geq C_H^{-m} C_g^{-2}$$

Setting $C_\beta := C_H^n C_g^2$ the assertion follows (note that m is independent of x, y, r). ∎

REMARK: If $\mathcal{H}^*$ is the sheaf generated by a hypoelliptic operator as in Sanchez–Calle [7] and the space $X = \mathbf{R}^n$ is endowed with the metric defined in [7],p. 150 then 1.2 is satisfied by [7],Theorem I and the assumptions on the metric in Proposition 3.1 follow from the definition in [7],p. 150. Hence in this case the uniform Harnack inequality as well as condition (HB) are satisfied.

REFERENCES

1. A. Boukricha, *Das Picard-Prinzip und verwandte Fragen bei Störung von harmonischen Räumen*, Math. Ann. **239** (1979), 247–270.
2. C. Constantinescu, A. Cornea, "Potential theory on harmonic spaces," Springer Verlag, Berlin – Heidelberg – New York, 1972.
3. A. Cornea, R. Wittmann, *An approximation theorem for cones of potentials*, Analysis **3** (1983), 317–335.
4. B. Franchi, E. Laconelli, *Hölder regularity theorem for a class of linear non-uniformly elliptic operators with measurable coefficients*, Ann. Sc. Norm. Super. Pisa Serie IV **10** (1983), 523–541.
5. W. Hansen, *Kernels on harmonic spaces*, (unpublished) (1979).
6. G. A. Hunt, *Markoff processes and potentials I*, Ill. J. Math. **1** (1957), 42–93.
7. A. Sanchez–Calle, *Fundamental solutions and geometry of the sum of squares of vector fields*, Invent. math. **78** (1984), 143–160.
8. U. Schirmeier, *Konvergenzeigenschaften in harmonischen Räumen*, Invent. math. **55** (1979), 71–95.
9. R. Wittmann, *Kacsche Potentialtheorie für Resolventen, Markoffsche Prozesse und harmonische Räume*, Abh. Bayer. Akad. Wiss. **161** (1982).
10. R. Wittmann, *Positive harmonic functions on nontangentially accessible domains*, Math. Z. **190** (1985), 419–438.
11. R. Wittmann, *A non-tangential limit theorem*, Osaka Math. J. **24** (1987), 61–76.
12. R. Wittmann, *Nichttangentiales Randverhalten von harmonischen Funktionen in irregulären Randpunkten*, Arch. Math. **48** (1987), 526–529.
13. R. Wittmann, *A uniform boundary Harnack inequality on non-tangentially accessible domains*, Preprint (April 1987).

Ostenstr. 26–28, D-8078 Eichstätt, Federal Republic of Germany

ON THE ASYMPTOTIC BEHAVIOR OF SOLUTIONS OF A SYSTEM OF INTEGRAL EQUATIONS OF MIXED BOUNDARY VALUE PROBLEM OF PLANE ELASTICITY IN A NEIGHBORHOOD OF CORNER POINTS OF THE CONTOUR

Stepan Zargaryan

Yerevan, USSR

In the papers [1-2] a method for investigation of asymptotics of solutions near singularities of the boundary of boundary integral equations, arising in problems of potential theory was proposed. This method is based on the fact that solutions of integral equations can be expressed in terms of solutions of some exterior and interior boundary value problems. In the author's papers [3-4] asymptotics of solutions of boundary integral equations near corner points of the contour in plane problems of elasticity of the first two boundary value problems for the Lame's system was obtained.

In the present lecture we study the mixed boundary value problem of elasticity in domains with piecewise smooth boundaries. We construct some new boundary value problem for expression of solutions of a system of singular equations arising in investigated problem and as a result, determine the asymptotics in a neighborhood of corner points of the contour. These points coincide with the points of changing boundary conditions.

1. Let $\Omega^{(i)}$ be a domain in R^2 with compact closure and with piecewise smooth boundary $\partial\Omega$, $\Omega^{(e)} = R^2 \setminus \bar{\Omega}^{(i)}$. We suppose that $\partial\Omega$ has a set M of finite number of corner points $P_1, P_2, \dots, P_{2m}$. We consider exterior mixed boundary value problem of two dimensional elasticity:

$$\begin{gathered} A\vec{u}^{(e)} \equiv G\Delta\vec{u}^{(e)} + (\tilde{\lambda} + G) \operatorname{grad} \operatorname{div} \vec{u}^{(e)} = 0 \quad \text{in } \Omega^{(e)}, \\ \vec{u}^{(e)}|_{\Gamma_1} = \vec{g}, \quad T(\partial_x, n)\, \vec{u}^{(e)}|_{\Gamma_2} = \vec{h} \end{gathered} \tag{1}$$

where $\vec{u}^{(e)}$ is the displacement vector, $T(\partial_x,n)$ is the matrix with elements $T_{ij} = \tilde{\lambda}\, n_i(x) \frac{\partial}{\partial x_j} + G n_j(x) \frac{\partial}{\partial x_i} + G n_j(x) \frac{\partial}{\partial n(x)}$ and $n(n_1,n_2)$ is outer normal to $\Omega^{(i)}$. Let Γ_1 denote the union of segments $\widehat{P_{2j-1}P_{2j}}$, $1 \leq j \leq m$, and let the displacement components be given on Γ_1 and the external stress components on the remaining part $\Gamma_2 = \partial\Omega \setminus \bar{\Gamma}_1$ of the boundary.

The solution of (1) can be expressed as the sum

$$\vec{u}^{(e)} = \int_{\Gamma_1} [T(\partial_y, n)\Gamma(y-x)]^* \overrightarrow{\varphi(y)}\, d_y s + \int_{\Gamma_2} \Gamma(y-x)\overrightarrow{\psi(y)}\, d_y s + \vec{C}, \quad (2)$$

where $*$ denotes the adjoint operator, Γ is the Kelvin-Somigliana tensor and $\vec{C}$ is an unknown constant vector. The matrix $[T(\partial_y, n)\Gamma(y-x)]^*$ may be writen in the form [5]

$$[T(\partial_y, n)\Gamma(y-x)]^* = \{a_1 R(x,y) + a_2 E\} \frac{\partial}{\partial n_y} \log \frac{1}{r} + a_2 S(x,y),$$

$$R(x,y) = \begin{Vmatrix} \left(\frac{\partial r}{\partial y_1}\right)^2, & \frac{\partial r}{\partial y_1}\frac{\partial r}{\partial y_2} \\ \frac{\partial r}{\partial y_1}\frac{\partial r}{\partial y_2}, & \left(\frac{\partial r}{\partial y_2}\right)^2 \end{Vmatrix}, \qquad S(x,y) = \frac{1}{r}\begin{Vmatrix} 0, & -\frac{dr}{d_y s} \\ \frac{dr}{d_y s}, & 0 \end{Vmatrix},$$

$$\frac{1}{r}\frac{dr}{d_y s} = \frac{\partial}{\partial y_1} \log \frac{1}{r} \cos(n_y, y_2) - \frac{\partial}{\partial y_1} \log \frac{1}{r} \cos(n_y, y_1), \quad (3)$$

$$r = \sqrt{(x_1-y_1)^2 + (x_2-y_2)^2}, \qquad a_1 = \frac{2(\tilde{\lambda}+G)}{\tilde{\lambda}+2G}, \qquad a_2 = \frac{G}{\tilde{\lambda}+2G},$$

E is the unit matrix.

The displacement vector (2) has a logarithmic growth for large $|x|$. If the vector valued function $\overrightarrow{\psi(y)}$ satisfies the orthogonality condition on Γ_2

$$\int_{\Gamma_2} \overrightarrow{\psi(y)}\, d_y s = 0, \quad (4)$$

then, displacement vector (2) is bounded at infinity.

Boundary conditions in (1) give the system of singular integral equations, on using (2),

$$\vec{\varphi} + \int_{\Gamma_1} [T(\partial_y, n)\Gamma(y-x)]^* \overrightarrow{\varphi(y)}\, d_y s + \int_{\Gamma_2} \Gamma(y-x)\overrightarrow{\psi(y)}\, d_y s + \vec{C} = \vec{g}, \qquad x \in \Gamma_1,$$

$$-\vec{\psi} + \int_{\Gamma_1} T(\partial_x, n)[T(\partial_y, n)\Gamma(y-x)]^* \overrightarrow{\varphi(y)}\, d_y s + \int_{\Gamma_2} T(\partial_x, n)\Gamma(y-x)\, \overrightarrow{\psi(y)}\, d_y s = \vec{h}, \quad x \in \Gamma_2,$$

$$\vec{C} = \vec{u}^{(e)}(\infty). \quad (5)$$

The kernels of the system of singular integral equations (5) are hypersingular in corner points of the contour.

In order to deduce asymptotic formulas of solutions of the system (5), we express the solution $(\vec{\varphi}, \vec{\psi}, \vec{C})$ in terms of the solutions of some auxiliary mixed boundary value problems.

Let $\vec{u}^{(i)}$ be a solution of the problem

$$A\vec{u}^{(i)} = 0 \quad \text{in} \quad \Omega^{(i)}, \quad \vec{u}^{(i)}\Big|_{\Gamma_1} = \vec{g}, \quad T(\partial_x, n)\vec{u}^{(i)}\Big|_{\Gamma_2} = \vec{h}. \qquad (6)$$

Applying the Green formula for all $x \in \partial\Omega \setminus M$, we obtain

$$\vec{u}^{(i)} = - \int_{\Gamma_1} \{\vec{g}[T(\partial_y, n)\, \Gamma(y-x)]^* - \Gamma(y-x)T(\partial_y, n)\vec{u}^{(i)}\} d_y s -$$
$$- \int_{\Gamma_2} \{\vec{u}^{(i)} [T(\partial_y, n)\, \Gamma(y-x)]^* - \Gamma(y-x)\, \vec{h}\} d_y s, \qquad (7)$$

$$\vec{u}^{(e)} = \int_{\Gamma_1} \{\vec{g}[T(\partial_y, n)\, \Gamma(y-x)]^* - \Gamma(y-x)T(\partial_y, n)\vec{u}^{(e)}\} d_y s +$$
$$+ \int_{\Gamma_2} \{\vec{u}^{(e)} [T(\partial_y, n)\, \Gamma(y-x)]^* - \Gamma(y-x)\, \vec{h}\} d_y s + 2\, \vec{u}^{(e)}(\infty). \qquad (8)$$

From (7) and (8) for all $x \in \Gamma_1$

$$2\, \overrightarrow{g(x)} = \int_{\Gamma_1} \Gamma(y-x)[T(\partial_y, n)\vec{u}^{(i)} - T(\partial_x, n)\vec{u}^{(e)}]\, d_y s -$$
$$- \int_{\Gamma_2} (\vec{u}^{(i)} - \vec{u}^{(e)})[T(\partial_y, n)\, \Gamma(y-x)]^* d_y s + 2\, \vec{u}^{(e)}(\infty). \qquad (9)$$

Let us define an integral representation of $T(\partial_x, n)\vec{u}^{(i)}$ and $T(\partial_x, n)\vec{u}^{(e)}$ for all $x \in \partial\Omega \setminus M$. From (7) and (8) it follows that

$$T(\partial_x, n)\vec{u}^{(i)}(x) = - \int_{\Gamma_1} \{\vec{g}\, T(\partial, n)[T(\partial_y, n)\, \Gamma(y-x)]^* -$$
$$- T(\partial_x, n)\, \Gamma(y-x)T(\partial_y, n)\vec{u}^{(i)}\} d_y s -$$
$$- \int_{\Gamma_2} \{\vec{u}^{(i)} T(\partial_x, n)[T(\partial_y, n)\, \Gamma(y-x)]^* - T(\partial_x, n)\, \Gamma(y-x)\, \vec{h}\} d_y s, \qquad (10)$$

$$T(\partial_x, n)\vec{u}^{(e)}(x) = \int_{\Gamma_1} \{\vec{g}\, T(\partial_x, n)[T(\partial_y, n)\, \Gamma(y-x)]^* -$$
$$- T(\partial_x, n)\, \Gamma(y-x)T(\partial_y, n)\vec{u}^{(e)}\} d_y s +$$
$$+ \int_{\Gamma_2} \{\vec{u}^{(e)} T(\partial_x, n)[T(\partial_y, n)\, \Gamma(y-x)]^* - T(\partial_x, n)\, \Gamma(y-x)\vec{h}\} d_y s, \qquad (11)$$

where n is the outer normal to $\Omega^{(i)}$.

Composition of stresses (10) and (11) at the points $x \in \Gamma_2$ reduces to

$$2\,\vec{h}(x) = \int_{\Gamma_1} T(\partial_x,n)\,\Gamma(y-x)\,[T(\partial_y,n)\vec{u}^{(i)} - T(\partial_y,n)\vec{u}^{(e)}]\,d_y s\; -$$

$$- \int_{\Gamma_2} (\vec{u}^{(i)} - \vec{u}^{(e)})\,T(\partial_x,n)[T(\partial_y,n)\,\Gamma(y-x)]^* d_y s. \qquad (12)$$

Now we introduce the solution of the problem

$$A\vec{v}^{(i)} = 0 \quad \text{in} \quad \Omega^{(i)}, \quad \vec{v}^{(i)}\Big|_{\Gamma_2} = \frac{1}{2}(\vec{u}^{(i)} - \vec{u}^{(e)})\Big|_{\Gamma_2},$$

$$T(\partial_x,n)\vec{v}^{(i)}\Big|_{\Gamma_1} = \frac{1}{2}\,(T(\partial_x,n)\vec{u}^{(i)} - T(\partial_x,n)\vec{u}^{(e)})\Big|_{\Gamma_1}, \qquad (13)$$

which will be used in the sequel. Taking into account the boundary conditions of the problem (13) we rewrite (9) and (12) in the form

$$\vec{g}(x) = \int_{\Gamma_1} \Gamma(y-x)\,T(\partial_x,n)\vec{v}^{(i)}\,d_y s\; -$$

$$- \int_{\Gamma_2} \vec{v}^{(i)}[T(\partial_y,n)\,\Gamma(y-x)]^* d_y s + \vec{u}^{(e)}(\infty), \quad x \in \Gamma_1,$$

$$\vec{h}(x) = \int_{\Gamma_1} T(\partial_x,n)\,\Gamma(y-x)\,T(\partial_y,n)\vec{v}^{(i)}\,d_y s\; -$$

$$- \int_{\Gamma_2} \vec{v}^{(i)}\,T(\partial_x,n)[T(\partial_y,n)\,\Gamma(y-x)]^* d_y s, \quad x \in \Gamma_2.$$

From these expressions and from identities

$$\vec{v}^{(i)}(x) = \int_{\partial\Omega}\{\vec{v}^{(i)}[T(\partial_y,n)\,\Gamma(y-x)]^* - \Gamma(y-x)\,T(\partial_y,n)\vec{v}^{(i)}\}\,d_y s,$$

$$x \in \Gamma_1,$$

$$T(\partial_x,n)\vec{v}^{(i)}(x) = \int_{\partial\Omega}\{\vec{v}^{(i)}\,T(\partial_x,n)[T(\partial_y,n)\,\Gamma(y-x)]^* -$$

$$- T(\partial_x,n)\,\Gamma(y-x)\,T(\partial_y,n)\vec{v}^{(i)}\}\,d_y s, \quad x \in \Gamma_2,$$

we obtain

$$\vec{g}(x) = \vec{v}(x) + \int_{\Gamma_1} \vec{v}^{(i)}[T(\partial_y,n)\,\Gamma(y-x)]^* d_y s\; -$$

$$- \int_{\Gamma_2} \Gamma(y-x)\,T(\partial_y,n)\vec{v}^{(i)}\,d_y s + \vec{u}^{(e)}(\infty), \quad x \in \Gamma_1,$$

$$\vec{h}(x) = T(\partial_x,n)\vec{v}^{(i)}(x) + \int_{\Gamma_1} \vec{v}^{(i)}\,T(\partial_x,n)[T(\partial_y,n)\,\Gamma(y-x)]^* d_y s\; -$$

$$- \int_{\Gamma_2} T(\partial_x,n)\,\Gamma(y-x)\,T(\partial_y,n)\vec{v}^{(i)}\,d_y s, \quad x \in \Gamma_2. \qquad (14)$$

An examination of equalities (14) and the integral equations (5) shows that equalities

$$\vec{\varphi} = \vec{v}^{(i)}\big|_{\Gamma_1}, \quad \vec{\psi} = -T(\partial_x, n)\vec{v}^{(i)}\big|_{\Gamma_2}, \quad \vec{C} = \vec{u}^{(e)}(\infty) \qquad (15)$$

determine the solution of the system (5) Therefore solvability of the system of singular integral equations (5) follows from solvability of the problem (13).

If resultant force of the stresses applied on the boundary $\partial\Omega$ is equal to zero then orthogonality conditions (4) are satisfied.

Indeed it is obvious that

$$\int_{\Gamma_2} \vec{\psi}\, d_y s = -\int_{\Gamma_2} T(\partial_y, n)\vec{v}^{(i)}\, d_y s = \int_{\Gamma_1} T(\partial_y, n)\vec{v}^{(i)}\, d_y s ,$$

because the resultant force of the stresses arising in problem (13) is equal to zero. Since $T(\partial_x, n)\vec{u}^{(e)} = T(\partial_x, n)\vec{u}^{(i)} = \vec{h}$ on Γ_2 we now see that

$$\int_{\Gamma_2} T(\partial_y, n)\vec{v}^{(i)}\, d_y s = \frac{1}{2}\int_{\Gamma_1} [T(\partial_y, n)\vec{u}^{(i)} - T(\partial_y, n)\vec{u}^{(e)}]\, d_y s =$$

$$= \frac{1}{2}\int_{\partial\Omega} T(\partial_y, n)\vec{u}^{(i)} d_y s - \frac{1}{2}\int_{\partial\Omega} T(\partial_y, n)\vec{u}^{(e)}\, d_y s = 0.$$

Here the first summand is equal to zero as a result of solvability of the problem (6). The second summand is also equal to zero owing to assumption about being zero the resultant force of the stresses applied on the boundary $\partial\Omega$.

2. The solvability in some Hölder weight spaces of the system of boundary integral equations (5) in domains with piecewise smooth boundaries has been proved in [6].

The asymptotics of solutions of a system (5) in a neighborhood of corner points of the contour depends on the root of the equation

$$4\lambda^2 \sin^2\alpha + 4\varkappa \sin^2 \lambda\alpha - (\varkappa + 1)^2 = 0 \qquad (16)$$

with the least positive real part; real part of λ decreases as α increases. Here $\alpha \in (0, 2\pi)$ be the angle between the tangent to $\partial\Omega$ at the corner point P_j from the side of $\Omega^{(e)}$, $\varkappa = 3-4\nu$, where ν is the Poisson's ratio.

We define a norm in the space $C^{1,\gamma}_{\beta}(\Omega^{(i)})$ of functions with $0 < \gamma < 1$ and $\beta \in R^1$ by the equality

$$\|u;\, C^{1,\gamma}_{\beta}(\Omega^{(i)})\| = \sup_{x,y \in \Omega^{(i)}} \frac{|d^{\beta}(x)\, \nabla u(x) - d^{\beta}(y)\, \nabla u(y)|}{|x-y|^{\gamma}} +$$

$$+ \sup_{x \in \Omega^{(i)}} |u(x)|$$

where $d(x)$ is the distance from the point x to M, $M = \bigcup_{j=1}^{2m} P_j$.

Replacing here $\Omega^{(i)}$ by $\Omega^{(e)}$ and $\sup|u(x)|$ by $\sup(1+x)|u(x)|$. we obtain the definition of the space $C_{\beta}^{1,\gamma}(\Omega^{(e)})$. By $C_{\beta}^{1,\gamma}(\partial\Omega)$ we denote the space of traces on $\partial\Omega \setminus M$ of functions in $C_{\beta}^{1,\gamma}(\Omega^{(i)})$ or $C_{\beta}^{1,\gamma}(\Omega^{(e)})$.

Let us introduce another space $C_{\beta}^{0,\gamma}(\partial\Omega)$ of functions on $\partial\Omega \setminus M$ with the norm

$$\|u;\ C_{\beta}^{0,\gamma}(\partial\Omega)\| = \sup_{x,y\in\partial\Omega\setminus M} \frac{|d^{\beta}(x)u(x)-d^{\beta}(y)u(y)|}{|x-y|^{\gamma}} +$$

$$+ \sup_{x\in\partial\Omega\setminus M} |d^{\beta-\gamma}(x)u(x)|.$$

It is proved in [6] that if γ and β satisfy

$$0 < 1 - \beta + \gamma < \min\{1, \operatorname{Re}\lambda(p_j)\} ,$$

then the integral operator of the system (5) realizes continuous mapping

$$\overset{\circ}{C}{}_{\beta}^{1,\gamma}(\Gamma_1) \times C_{\beta}^{0,\gamma}(\Gamma_2) \ni (\vec{\varphi}, \vec{\psi}) \to (\vec{g},\vec{h}) \in C_{\beta}^{1,\gamma}(\Gamma_1) \times C_{\beta}^{0,\gamma}(\Gamma_2).$$

We denote by $\overset{\circ}{C}{}_{\beta}^{1,\gamma}(\Gamma_1)$ the subspace of the space $C_{\beta}^{1,\gamma}(\Gamma_1)$ with functions which vanish on M.

Restricting the discussion for simplicity to a smooth functions $\vec{g}$ and $\vec{h}$, we shall assume with no loss of generality that $\Omega^{(e)}$ coincides with sector $\{x_1 + ix_2 = \rho e^{i\theta};\ 0 < \rho < \delta,\ 0 < \theta < \alpha\}$

Suppose that $0 < \alpha < \pi$. According to [7] for $x \in \Omega^{(e)}$ and $\rho \to 0$

$$\vec{u}^{(e)}(x) = \vec{g}(0) + \vec{O}(\rho^{1+\varepsilon}),\ \varepsilon \geqq 0,\ \text{if}\ 0 < \alpha \leqq \alpha_0,$$

$$\vec{u}^{(e)}(x) = \vec{g}(0) + \vec{O}(\rho^{\lambda^e}),\ 0 < \operatorname{Re}\lambda^e < 1,\ \text{if}\ \alpha_0 < \alpha < \pi. \tag{17}$$

Similarly, for $x \in \Omega^{(i)}$ and $\rho \to 0$

$$\begin{aligned}\vec{u}^{(i)}(x) - \vec{g}(0) \sim\ & \vec{e}_{\rho}\operatorname{Re}\{A\rho^{\lambda^i}[A_1\sin(\lambda^i+1)\theta + A_2\cos(\lambda^i+1)\theta - \\ & - A_3\sin(\lambda^i-1)\theta + A_4\cos(\lambda^i-1)\theta]\} + \\ & + \vec{e}_{\theta}\operatorname{Re}\{A\rho^{\lambda^i}[A_1\cos(\lambda^i+1)\theta - A_2\sin(\lambda^i+1)\theta + \\ & + \nu_1 A_3\cos(\lambda^i-1)\theta + \nu_1 A_4\sin(\lambda^i-1)\theta]\},\ 0 < \theta < 2\pi - \alpha ,\end{aligned}$$

$$A_1 = \nu_1 P,\ A_2 = -R,\ A_3 = -P,\ A_4 = R, \tag{18}$$

where λ^e and λ^i are the complex roots of the equation

$4\lambda^2\sin^2\omega + 4\varkappa\sin^2\lambda\omega - (\varkappa+1)^2 = 0$ with the least positive real parts if $\omega = \alpha$ and $\omega = 2\pi - \alpha$, respectively [8]. It is known that $0 < \operatorname{Re}\lambda^i < \operatorname{Re}\lambda^e < 1$, and the corresponding value of α derived from (16) if $\lambda = 1$ is $\alpha_0 = \arcsin\sqrt{1-\nu}$. Here

$$P = 2\lambda^i \cos(\lambda^i+1)(2\pi-\alpha) - (\lambda^i+1)(1-\nu_1)\cos(\lambda^i-1)(2\pi-\alpha),$$

$$R = 2\lambda^i\sin(\lambda^i+1)(2\pi-\alpha) + (\lambda^i+1)(1-\nu_1)\sin(\lambda^i-1)(2\pi-\alpha),$$

$$\nu_1 = (\varkappa+\lambda^i)/(\varkappa-\lambda^i).$$

Substituting (17) and (18) in the (13) we obtain for $x \in \Omega^{(i)}$ and $\rho \to 0$

$$\begin{aligned}
\vec{v}^{(i)} \sim{}& \vec{e}_\rho \operatorname{Re}\{B\rho^{\lambda^i}[-\tfrac{1}{2}(\lambda^i-1)(1-\nu_1)\, S\sin(\lambda^i+1)\theta - \\
&- \tfrac{1}{2}(\lambda^i+1)(1-\nu_1)\, Q\cos(\lambda^i+1)\theta + S\sin(\lambda^i-1)\theta + \\
&+ Q\cos(\lambda^i-1)\theta]\} + \vec{e}_\theta \operatorname{Re}\{B\rho^{\lambda^i}[-\tfrac{1}{2}(\lambda^i-1)(1-\nu_1) S\cos(\lambda^i+1)\theta + \\
&+ \tfrac{1}{2}(\lambda^i+1)(1-\nu_1)\, Q\sin(\lambda^i+1)\theta - \nu_1 S\cos(\lambda^i-1)\theta + \\
&+ \nu_1 Q\sin(\lambda^i-1)\theta]\} + \tfrac{1}{2}\vec{e}_\rho \operatorname{Re}\{A\rho^{\lambda^i}[\nu_1 P\sin(\lambda^i+1)\theta - \\
&- R\cos(\lambda^i+1)\theta + P\sin(\lambda^i-1)\theta + R\cos(\lambda^i-1)\theta]\} + \\
&+ \tfrac{1}{2}\vec{e}_\theta \operatorname{Re}\{A\rho^{\lambda^i}[\nu_1 P\cos(\lambda^i+1)\theta + R\sin(\lambda^i+1)\theta - \\
&- \nu_1 P\cos(\lambda^i-1)\theta + \nu_1 R\sin(\lambda^i-1)\theta]\}
\end{aligned} \tag{19}$$

and

$$\begin{aligned}
T(\partial_x, n)\vec{v}^{(i)} \sim{}& \vec{e}_\rho G\operatorname{Re}\{(1-\nu_1)\, B\rho^{\lambda^i-1}[-(\lambda^i-1)\, S\cos(\lambda^i+1)\theta + \\
&+ (\lambda^i+1)\, Q\sin(\lambda^i+1)\theta + (\lambda^i-1)\, S\cos(\lambda^i-1)\theta - \\
&- (\lambda^i-1)\, Q\sin(\lambda^i-1)\theta]\} + \vec{e}_\theta G\operatorname{Re}\{(1-\nu_1)B\rho^{\lambda^i-1} \times \\
&\times[(\lambda^i-1)\, S\sin(\lambda^i+1)\theta + (\lambda^i+1)\, Q\cos(\lambda^i+1)\theta - \\
&- (\lambda^i+1)\, S\sin(\lambda^i-1)\theta - (\lambda^i+1)\, Q\cos(\lambda^i-1)\theta]\} + \\
&+ \tfrac{1}{2}\vec{e}_\rho G\operatorname{Re}\{A\rho^{\lambda^i-1}[2\lambda^i\nu_1 P\cos(\lambda^i+1)\theta + 2\lambda^i R\sin(\lambda^i+1)\theta + \\
&+ (\lambda^i-1)(1-\nu_1)\, P\cos(\lambda^i-1)\theta - (\lambda^i-1)(1-\nu_1)\, R\sin(\lambda^i-1)\theta]\} \\
&+ \tfrac{1}{2}\vec{e}_\theta G\operatorname{Re}\{A\rho^{\lambda^i-1}[-2\lambda^i\nu_1\, P\sin(\lambda^i+1)\theta + 2\lambda^i R\cos(\lambda^i+1)\theta - \\
&- (\lambda^i+1)(1-\nu_1)\, P\sin(\lambda^i-1)\theta - (\lambda^i+1)(1-\nu_1) R\cos(\lambda^i-1)\theta]\},
\end{aligned} \tag{20}$$

where

$$Q = \frac{(\lambda^i-1)(1-\nu_1)}{2\lambda^i} \cos(\lambda^i+1)(2\pi-\alpha) + \nu_1 \cos(\lambda^i-1)(2\pi-\alpha) ,$$

$$S = \frac{(\lambda^i+1)(1-\nu_1)}{2} \sin(\lambda^i+1)(2\pi-\alpha) + \nu_1 \sin(\lambda^i-1)(2\pi-\alpha) .$$

From (19) and (20), according to (15), we obtain for $x \in \Gamma_1$ and $\rho \to 0$

$$\vec{\varphi} - \vec{\varphi}(0) \sim \vec{e}_\rho \mathrm{Re}\{B\rho^{\lambda^i} Q[- \frac{(\lambda^i+1)(1-\nu_1)}{2\lambda^i} + 1]\} +$$

$$+ \vec{e}_\theta \mathrm{Re}\{B\rho^{\lambda^i} S[- \frac{(\lambda^i-1)(1-\nu_1)}{2\lambda^i} - \nu_1]\} . \qquad (21)$$

Similarly we find that for $x \in \Gamma_2$ and $\rho \to 0$

$$\vec{\psi} \sim -\vec{e}_\rho G \,\mathrm{Re}\{B\rho^{\lambda^i-1}(1-\nu_1)[-(\lambda^i-1)S \cos(\lambda^i+1)(2\pi-\alpha) +$$

$$+ (\lambda^i+1)Q \sin(\lambda^i+1)(2\pi-\alpha) + (\lambda^i-1)S \cos(\lambda^i-1)(2\pi-\alpha) -$$

$$- (\lambda^i-1) Q \sin(\lambda^i-1)(2\pi-\alpha)]\} -$$

$$- \vec{e}_\theta G \,\mathrm{Re}\{B\rho^{\lambda^i-1}(1-\nu_1)[(\lambda^i-1)S \sin(\lambda^i+1)(2\pi-\alpha) +$$

$$+ (\lambda^i+1)Q \cos(\lambda^i+1)(2\pi-\alpha) - (\lambda^i+1)S \sin(\lambda^i-1)(2\pi-\alpha) -$$

$$- (\lambda^i+1)Q \cos(\lambda^i-1)(2\pi-\alpha)]\} . \qquad (22)$$

Now we suppose that $\pi < \alpha < 2\pi$. In this case for $x \in \Omega^{(e)}$ and $\rho \to 0$ we have

$$\vec{u}^{(e)}(x) = \vec{g}(0) + \vec{e}_\rho \mathrm{Re}\{C\rho^{\lambda^e}[C_1 \sin(\lambda^e+1)\theta + C_2 \cos(\lambda^e+1)\theta -$$

$$- C_3 \sin(\lambda^e-1)\theta + C_4 \cos(\lambda^e-1)\theta]\} +$$

$$+ \vec{e}_\theta \mathrm{Re}\{C\rho^{\lambda^e}[C_1 \cos(\lambda^e+1)\theta - C_2 \sin(\lambda^e+1)\theta +$$

$$+ \nu_1 C_3 \cos(\lambda^e-1)\theta + \nu_1 C_4 \sin(\lambda^e-1)\theta]\} + \vec{O}(\rho),$$

$$2\pi - \alpha \leqq \theta \leqq 2\pi \qquad (23)$$

where coefficients C_j, $j = 1,2,3$ can be expressed through $C_4 = C$ from the solution of the system of equations

$$C_1\sin(\lambda^e+1)2\pi + C_2\cos(\lambda^e+1)2\pi - C_3\sin(\lambda^e-1)2\pi +$$

$$+ C\cos(\lambda^e-1)2\pi = 0,$$

$$C_1\cos(\lambda^e+1)2\pi - C_2\sin(\lambda^e+1)2\pi + \nu_1 C_3\cos(\lambda^e-1)2\pi +$$

$$+ \nu_1 C\sin(\lambda^e-1)2\pi = 0,$$

$$2\lambda^e C_1\cos(\lambda^e+1)(2\pi-\alpha) - 2\lambda^e C_2\sin(\lambda^e+1)(2\pi-\alpha) -$$

$$- (\lambda^e-1)(1-\nu_1) C_3\cos(\lambda^e-1)(2\pi-\alpha) -$$

$$- (\lambda^e-1)(1-\nu_1) C\sin(\lambda^e-1)(2\pi-\alpha) = 0,$$

$$2\lambda^e C_1\sin(\lambda^e+1)(2\pi-\alpha) + 2\lambda^e C_2\cos(\lambda^e+1)(2\pi-\alpha) -$$

$$- (\lambda^e+1)(1-\nu_1) C_3\sin(\lambda^e-1)(2\pi-\alpha) -$$

$$- (\lambda^e+1)(1-\nu_1) C\cos(\lambda^e-1)(2\pi-\alpha) = 0. \qquad (24)$$

This system was defined from boundary value problem (1) with homogeneous boundary conditions. From (24) we have

$$C_1 = m_1 C,\ C_2 = m_2 C,\ C_3 = m_3 C, \qquad (25)$$

because the rank of the matrix of the system (24) is equal to three. In (24) λ^e is the complex root of equation (16) with the least positive real part.

Similarly, for $x \in \Omega^{(i)}$ and $\rho \to 0$

$$\vec{u}^{(i)} = \vec{g}(0) + \vec{O}(\rho), \quad \text{if} \quad 2\pi - \alpha_0 \leq \alpha < 2\pi,$$
$$\vec{u}^{(i)} = \vec{g}(0) + \vec{O}(\rho^{\lambda^i}), \quad \text{if} \quad \pi < \alpha < 2\pi - \alpha_0, \qquad (26)$$

where λ^i is the complex root of the equation $4\sin^2\lambda(2\pi-\alpha) + 4\lambda^2\sin^2(2\pi-\alpha) - (\varkappa+1)^2 = 0$ with the least positive real part. It is obvious that $0 < \operatorname{Re}\lambda^e < \operatorname{Re}\lambda^i < 1$.

We shall represent the asymptotic of the solution of boundary value problem (13) in the form

$$\vec{v}^{(i)} \sim \vec{e}_\rho \operatorname{Re}\{\rho^\lambda[D_1\sin(\lambda+1)\theta + D_2\cos(\lambda+1)\theta - D_3\sin(\lambda-1)\theta +$$

$$+ D_4\cos(\lambda-1)\theta]\} + \vec{e}_\theta \operatorname{Re}\{\rho^\lambda[D_1\cos(\lambda+1)\theta - D_2\sin(\lambda+1)\theta +$$

$$+ \nu_1 D_3\cos(\lambda-1)\theta + \nu_1 D_4\sin(\lambda-1)\theta]\},\ 0 \leq \theta \leq 2\pi - \alpha. \qquad (27)$$

From (23) and (26), taking into account (24) and (25), for right hand sides of the boundary conditions of the problem (13) we get

$$\vec{v}^{(i)}\Big|_{\Gamma_2(\theta=2\pi-\alpha)} \sim -\frac{1}{2}\vec{e}_\rho \mathrm{Re}\{C\rho^{\lambda^e}[m_1\sin(\lambda^e+1)(2\pi-\alpha) +$$

$$+ m_2\cos(\lambda^e+1)(2\pi-\alpha) - m_3\sin(\lambda^e-1)(2\pi-\alpha) +$$

$$+ \cos(\lambda^e-1)(2\pi-\alpha)]\} - \frac{1}{2}\vec{e}_\theta \mathrm{Re}\{C\rho^{\lambda^e}[m_1\cos(\lambda^e+1)(2\pi-\alpha) -$$

$$- m_2\sin(\lambda^e+1)(2\pi-\alpha) + \nu_1 m_3\cos(\lambda^e-1)(2\pi-\alpha) +$$

$$+ \nu_1 \sin(\lambda^e-1)(2\pi-\alpha)]\},$$

$$T(\partial_x,n)\vec{v}^{(i)}\Big|_{\Gamma_1(\theta=0)} \sim \frac{1}{2}\vec{e}_\rho G\ \mathrm{Re}\{C\rho^{\lambda^e-1}[2\lambda\, m_1\cos(\lambda^e+1)2\pi -$$

$$- 2\lambda^e m_2\sin(\lambda^e+1)2\pi - (\lambda^e-1)(1-\nu_1)m_3\cos(\lambda^e-1)2\pi -$$

$$- (\lambda^e-1)(1-\nu_1)\sin(\lambda^e-1)2\pi]\} +$$

$$+ \frac{1}{2}\vec{e}_\theta G\ \mathrm{Re}\{C\rho^{\lambda^e-1}[-2\lambda^e\, m_1\sin(\lambda^e+1)2\pi - 2\lambda^e m_2\cos(\lambda^e+1)2\pi +$$

$$+ (\lambda^e+1)(1-\nu_1)m_3\sin(\lambda^e-1)2\pi - (\lambda^e+1)(1-\nu_1)\cos(\lambda^e-1)2\pi]\}.$$

Thus, the complex coefficients D_j, $j=1,2,3,4$, in the asymptotic (27) must be determined from the system of equations

$$D_1\sin(\lambda^e+1)(2\pi-\alpha) + D_2\cos(\lambda^e+1)(2\pi-\alpha) - D_3\sin(\lambda^e-1)(2\pi-\alpha) +$$

$$+ D_4\cos(\lambda^e-1)(2\pi-\alpha) = -\frac{1}{2}C[m_1\sin(\lambda^e+1)(2\pi-\alpha) +$$

$$+ m_2\cos(\lambda^e+1)(2\pi-\alpha) - m_3\sin(\lambda^e-1)(2\pi-\alpha) +$$

$$+ \cos(\lambda^e-1)(2\pi-\alpha)],$$

$$D_1\cos(\lambda^e+1)(2\pi-\alpha) - D_2\sin(\lambda^e+1)(2\pi-\alpha) +$$

$$+ \nu_1 D_3\cos(\lambda^e-1)(2\pi-\alpha) + \nu_1 D_4\sin(\lambda^e-1)(2\pi-\alpha) =$$

$$= -\frac{1}{2}C[m_1\cos(\lambda^e+1)(2\pi-\alpha) - m_2\sin(\lambda^e+1)(2\pi-\alpha) +$$

$$+ \nu_1 m_3\cos(\lambda^e-1)(2\pi-\alpha) + \nu_2\sin(\lambda^e-1)(2\pi-\alpha)],$$

$$D_1 + \frac{\lambda^e-1}{\varkappa-\lambda^e}D_3 = \frac{1}{2}C[m_1\cos(\lambda^e+1)2\pi - m_2\sin(\lambda^e+1)2\pi +$$

$$+ \frac{\lambda^e-1}{\varkappa-\lambda^e}m_3\cos(\lambda^e-1)2\pi + \frac{\lambda^e-1}{\varkappa-\lambda^e}\sin(\lambda^e-1)2\pi],$$

$$- D_2 + \frac{\lambda^e+1}{\varkappa-\lambda^e} D_4 = \frac{1}{2} C[-m_1\sin(\lambda^e+1)2\pi - m_2\cos(\lambda^e+1)2\pi -$$

$$- \frac{\lambda^e+1}{\varkappa-\lambda^e} m_3\sin(\lambda^e-1)2\pi + \frac{\lambda^e+1}{\varkappa-\lambda^e} \cos(\lambda^e-1)2\pi] \; . \qquad (28)$$

The determinant of this system is

$$4\varkappa\sin^2\lambda^e(2\pi-\alpha) + 4\lambda^2\sin^2(2\pi-\alpha) - (\varkappa+1)^2 \neq 0,$$

and in asymptotic (27) we have $\lambda = \lambda^e$. Let $D_k = I_k C$, $k = 1, 2, 3, 4$. From (27), taking into account (28), according to (15) we obtain for $\rho \to 0$

$$\vec{\varphi} - \vec{\varphi}(0) \sim \vec{e}_\rho \mathrm{Re}\{C\rho^{\lambda^e}(I_2+I_4)\} + \vec{e}_\theta \mathrm{Re}\{C\rho^{\lambda^e}(I_1+\nu_2 I_3) \quad , \qquad (29)$$

$$\vec{\psi} \sim \vec{e}_\rho G\ \mathrm{Re}\{C\rho^{\lambda^e-1}[2\lambda^e I_1\cos(\lambda^e+1)(2\pi-\alpha) -$$

$$- 2\lambda^e I_2\sin(\lambda^e+1)(2\pi-\alpha) -$$

$$- (\lambda^e-1)(1-\nu_1)I_3\cos(\lambda^e-1)(2\pi-\alpha) -$$

$$- (\lambda^e-1)(1-\nu_1)I_4\sin(\lambda^e-1)(2\pi-\alpha)]\} +$$

$$+ \vec{e}_\theta G\ \mathrm{Re}\{C\rho^{\lambda^e-1}[-2\lambda^e I_1\sin(\lambda^e+1)(2\pi-\alpha) -$$

$$- 2\lambda^e I_2\cos(\lambda^e+1)(2\pi-\alpha) +$$

$$+ (\lambda^e+1)(1-\nu_2)I_3\sin(\lambda^e-1)(2\pi-\alpha) -$$

$$- (\lambda^e+1)(1-\nu_2)I_4\cos(\lambda^e-1)(2\pi-\alpha)]\}. \qquad (30)$$

The coefficient B in the asymptotics (21) and (22) and also the coefficient C in the asymptotics (29) and (30) can be obtained by the method, applied to the solutions of integral equations for Zaremba problem in [1].

REFERENCES

1. S.S. ZARGARYAN and V.G. MAZ´YA, Singularities of the solution of the system of integral equations of potential theory, arising in Zaremba problem, Vestn. Leningrad Univ. Math. n 1 (1983). English transl. v. 16 (1984).
2. S.S. ZARGARYAN and V. MAZ´YA, On the asymptotic of solutions of integral equations of potential theory in a neighborhood of corner points of the contour, Prikl. Mat. Mekh. v. 48, n 1 (1984), English transl. in Y. Appl. Math. Mech. v. 48 (1984).
3. S.S. ZARGARYAN, On the asymptotic of solutions of the system of singular integral equations, arising in Lame equations, in a neighborhood of corner points of the contour, Dokl.Akad. Nauk Arm. SSR, v.77, n 1(1983).

4. S.S. ZARGARYAN, Singularities of the solution of the system of singular integral equations of plane problem of elasticity with given external stress components on the boundary, Dokl. Akad. Nauk Arm. SSR, v. 77,n 4 (1983).

5. N.S. KAKHNIASHVILI, Investigation of plane problems of elasticity by method of potential theory , Trudy Tbil.Univ. Mekh.Mat., v.50 (1953).

6. V.G. MAZ´YA, The potential theory for the Lame system in domains with piecewise smooth boundaries, Proc. of the Conference in 1982 on Partial Diff. Equations. Memoriam of I.N. VEKUA in Tbilisi, Tbilisi (1986).

7. V.A. KONDRAT´EV, Boundary value problems for elliptic equations in domains with conical or corner points, Trudy Moskov. Mat . Obshch. v. 16 (1967). English transl. in Trans.Moscow Math. Soc. v.16 (1967).

8. M.L. WILLIAMS, Stress singularities resulting from various boundary conditions in angular corners of plate in extension, J. Appl.Mech.,v.19, n 4 (1952).

INDEX